Mechanisms of Molecular Carcinogenesis – Volume 1

Johannes Haybaeck
Editor

Mechanisms of Molecular Carcinogenesis – Volume 1

Springer

Editor
Johannes Haybaeck
Department of Pathology
Medical Faculty
Otto von Guericke University Magdeburg
Magdeburg
Germany

Institute of Pathology
Medical University of Graz
Graz
Austria

ISBN 978-3-319-85208-9 ISBN 978-3-319-53659-0 (eBook)
DOI 10.1007/978-3-319-53659-0

Printed on acid-free paper

This Springer imprint is published by Springer Nature
The registered company is Springer International Publishing AG
The registered company address is: Gewerbestrasse 11, 6330 Cham, Switzerland

Preface

This book, entitled *Mechanisms of Molecular Carcinogenesis*, addresses the latest developments in the assessment of molecular carcinogenesis. Mechanistic insights gained by various in vitro and in vivo model systems need to be validated, so that the patients can effectively be assisted on their way into the clinics. Although the clinical relevance of model systems is sometimes not obvious, drug development is increasingly based on the knowledge of how these systems work, and targeted drugs are a first step toward individualized medicine.

Patient-derived model systems that faithfully recapitulate human cancer are critical for the identification and validation of innovative drug targets and particular drugs. Furthermore, understanding these model systems means to simultaneously gain a basic understanding of cancer. Traditional approaches most often fail in the late stages of drug development (i.e., clinical phase II/III) because the currently available preclinical models have substantial limitations and therefore predict tumor plasticity and heterogeneity in the human patient only inappropriately. Scientists seek to overcome these limitations in cancer research by utilizing a panel of clinically well-characterized tumor tissues for the generation of different patient-derived 3D cell culture models (PD3D). These contain either tumor cells alone or are combined with cancer-associated fibroblasts, as well as xenograft mouse models (PDX). The in-depth comparison of various models to define the stability of gene expression and their response to chemotherapy pose a critical challenge in applied cancer research. Therefore, next-generation sequencing is used to characterize well-described mutations and translocations of particular tumor entities in the original patient tissue, whereas their transcriptome is often analyzed using RNAseq.

Patient-derived tissues, as well as models, are often implemented into high-content analyses and screening platforms for high-throughput drug discovery, taking into account the influence of tumor stroma on drug treatment efficacy. Model systems, such as transgenic or knockout animals, PD3D models, or cell lines, are assumed to allow for high precision profiling of both mRNA expression patterns and protein levels of novel targets and furthermore enable the scientist to dissect and discover signaling pathways.

The emerging field of computational pathology, which is of high clinical relevance, provides the basis for the application of novel tools, allowing for the construction of computer-based models and the simulation of biological processes.

In-depth knowledge of the relevant molecular mechanisms of carcinogenesis is becoming increasingly important for targeted molecular therapy within the framework of personalized medicine and patient care. Therefore, the purpose of this book is to provide the reader with up-to-date insights into molecular and cellular mechanisms of cancer onset and progression, spread of cancer cells, and metastasis. It aims to fill the gap between basic cancer research and daily clinical practice, where the prescription and advancement of routinely applied treatment strategies can only be accomplished by individuals who have a deeper knowledge of the mode of action of the respective medications. Notably, the more advanced the tools for fighting cancer, the greater the need for a mechanistic understanding of medical approaches. Therefore, this book deals with molecular diagnostics and targeted and genetic therapies, as well as their usability. It draws a parallel to modern technology platforms and gives an outlook for the future.

This book aims at bridging the gap between basic and applied cancer research and the clinics, thereby trying to transfer knowledge from bench to bedside. A mechanistic understanding of carcinogenic events might be fundamental to the future of cancer research and treatment. Prognostic and predictive tumor biomarkers are extremely important and are therefore highlighted in various chapters of the book.

Only few medical areas have undergone such dramatic changes as did molecular pathology over the last few years. Thus, it is reasonable to look at this fascinating and very fast growing scientific field from different angles. Excellent books on molecular technologies, diagnostic approaches, and therapeutic algorithms have been written, but a book addressing all these areas and simultaneously shedding light on the molecular mechanisms related to tumorigenesis is still lacking.

This book not only provides a summary of the basic knowledge but also, which is more important, gives an overview of recent advances made in basic cancer research. We hope that it will serve as a comprehensive and concise source that provides the reader with knowledge of new developments and insights into carcinogenesis.

We wish that this book stimulates our readers and that they will be fascinated by this exciting and scientifically, as well as clinically relevant, emerging topic.

Magdeburg, Germany Johannes Haybaeck

Contents

The Endocannabinoid System in Carcinogenesis

1

Rudolf Schicho

Contents

Abstract

Cannabis sativa has been used in traditional medicine for thousands of years. The receptors responding to its ingredients and the mechanisms behind its effects, however, have only recently been revealed. The description of the cannabinoid receptors in the brain initiated a huge strive in unraveling the molecular actions of Δ^9-tetrahydrocannabinol (THC), the plant's main ingredient, leading to the characterization of the so-called endocannabinoid system. The discovery of the endocannabinoid system rekindled the ambition to focus on *Cannabis* as a medicinal plant and to exploit it for new pharmacological treatments [1]. Anectodal reports suggest that the plant's ingredients may have an impact on a variety of inflammatory

R. Schicho, Ph.D.
Institute of Experimental and Clinical Pharmacology, Medical University of Graz, Universitätsplatz 4, 8010 Graz, Austria
e-mail: rudolf.schicho@medunigraz.at

J. Haybaeck (ed.), *Mechanisms of Molecular Carcinogenesis – Volume 1*,
DOI 10.1007/978-3-319-53659-0_1

diseases, and even cancer. In the past years, basic researchers have therefore increasingly focused on investigating antitumorigenic effects of cannabinoids/endocannabinoids in cancer cells and mouse models of cancer.

The present chapter shortly describes the role of the endocannabinoid system in cancer and provides an insight into potential mechanisms behind the antitumorigenic effects of cannabinoids/endocannabinoids.

1.1 The Endocannabinoid System and Its Components

The physiological role of the endocannabinoid system is not yet clarified, but it is assumed to take part in the homeostasis of various organ systems, such as the central nervous system and systems in the periphery [2, 3]. It consists of the cannabinoid (CB) receptors 1 and 2 (CB1 and CB2), their endogenous ligands (the "endocannabinoids"), as well as their degrading and synthesizing enzymes. The effects of endo- and exogenous cannabinoids (the latter can be of synthetic or herbal origin) are predominantly brought about by the activation of CB receptors, although non-CB receptor-mediated effects of cannabinoids have been frequently observed. Endocannabinoids are short-lived bioactive lipids and chemical derivatives of arachidonic acid. Next to anandamide and 2-AG, which are the best characterized endocannabinoids (Fig. 1.1), other endocannabinoids of little known function have been described, such as virodhamine (*O*-arachidonoyl ethanolamine) and noladin ether (2-arachidonyl glyceryl ether).

Anandamide is formed on demand through rapid metabolization of membrane phospholipids by *N*-acyl phosphatidylethanolamine phospholipase D (NAPE-PLD), while 2-AG is synthesized by diacylglycerol lipase (DAGL). Endocannabinoids have originally been described as retrograde messengers in the nervous system, but they are also released from cancer cells, such as Caco-2 [4], and from macrophages and platelets [5, 6]. After activating CB receptors, 2-AG and anandamide are rapidly degraded by specific enzymes, i.e., by monoacylglycerol lipase (MAGL) and fatty acid amide hydrolase 1 (FAAH-1), respectively. A recently found FAAH-2 is more effective in degrading other fatty acid amides [7]. Enzymes like cyclooxygenase-2 (Cox-2) and

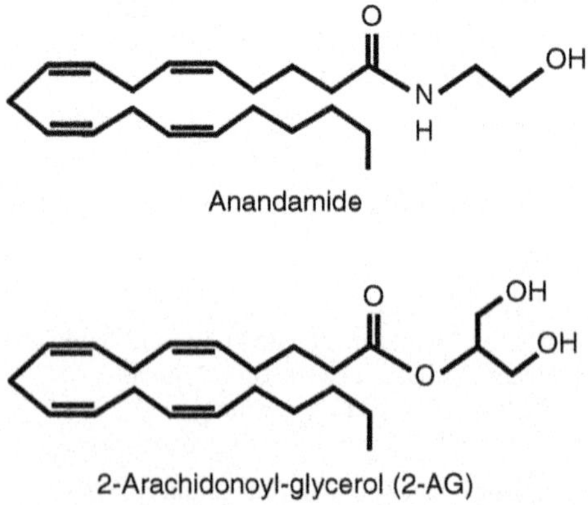

Fig. 1.1 Chemical structures of endocannabinoids

lipoxygenase (Lox) may also degrade endocannabinoids, giving rise to prostaglandin ethanolamides, glyceryl prostaglandins, and derivatives of hydroxyeicosatetraenoic (HETE) and hydroperoxyeicosatetraenoic acid (HPETE) [8, 9]. The pathophysiological importance of these mediators in cancer remains to be elucidated.

1.2 The Endocannabinoid System and Cancer

CB receptors have been investigated in a variety of cancer cell lines. Evidence now exists that endo- and exogenous cannabinoids produce anticarcinogenic effects via multiple mechanisms and pathways. Cannabinoids/endocannabinoids interfere with tumor cell proliferation through apoptosis, autophagy, and cell cycle arrest. They are also able to inhibit tumor cell migration, invasion, and angiogenesis, thus interfering with metastasis (Fig. 1.2). CB receptors are expressed in many aggressive tumor cell lines, such as of brain, breast, thyroid, prostate, pancreas, lung, skin, gastrointestinal, and hematopoietic cancer. Anti-tumorigenic effects of synthetic or natural CB receptor agonists or endocannabinoids have been demonstrated in these tumor cells in vitro [10]. In clinical studies, however, CB receptors, endocannabinoids, and endocannabinoid-degrading enzymes were often observed to correlate with tumor aggressiveness, suggesting high activity of the endocannabinoid system in aggressive tumors [11]. CB receptors are also present in endothelial cells, and antiangiogenic/antimetastatic actions by cannabinoids/endocannabinoids have been reported [12]. Cannabinoids/endocannabinoids, therefore, inhibit hallmark mechanisms of carcinogenesis not only by interfering with cancer cells but also with cells of the tumor microenvironment [13].

The role of the degrading enzymes, MAGL and FAAH, in cancer is still unclear. In prostate cancer, epithelial FAAH immunoreactivity has been shown to positively correlate with disease severity at diagnosis in cases with midrange CB1 immunoreactivity scores, but not in those with high CB1 immunoreactivity scores [14]. Expression of MAGL is elevated in colon cancer tissue [15], suggesting a high turnover of 2-AG in this type of cancer. Ovarian and melanoma cancer cells with MAGL deficiency exhibit lower levels of fatty acids [16]. It is believed that low

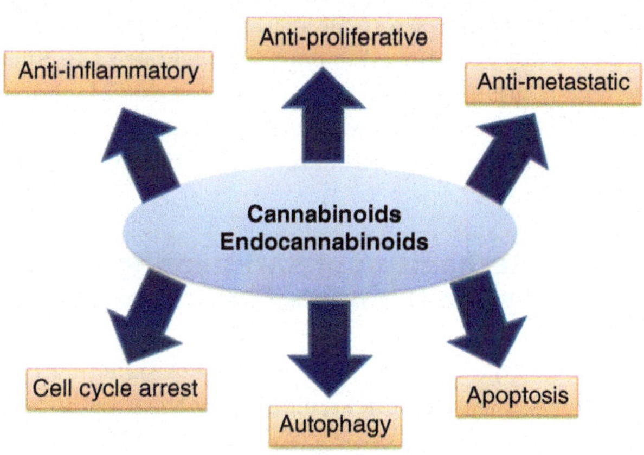

Fig. 1.2 Anticarcinogenic effects of cannabinoids/endocannabinoids

Table 1.1 Receptors of the endocannabinoid system in various types of cancer

Type of cancer	CB receptor regulation	Correlation with disease severity	References
Prostate carcinoma	CB1 increased TRPV1 increased CB1 increased	No Yes (increase with malignancy) Yes (poorer prognosis)	[21] [14]
Pancreatic carcinoma	CB1/2 increased	ND	[22]
Hepatocellular carcinoma	CB1/2 increased TRPV1 increased	Yes (improved prognosis) Yes (improved prognosis)	[23] [24]
Glioma	CB1 decreased CB2 increased	ND ND	[25]
Colon cancer	CB1 decreased	ND	[26]
Colon cancer	CB1 increased	In microsatellite stable stage II tumors	[27]
Breast cancer	CB2 increased	Correlation with Erb2 expression	[28]

levels of fatty acids may produce lower amounts of oncogenic signaling lipids that would normally contribute to pathogenicity.

1.3 Receptors of the Endocannabinoid System Involved in Carcinogenesis

The cannabinoid receptors CB1 and CB2 are seven-transmembrane, G protein-coupled receptors (GPCRs) and are the main receptors responsible for the actions of the endocannabinoids. Other receptors, such as the G protein-coupled receptors 55 and 18 (GPR55 and GPR18) [17, 18], the peroxisome proliferator-activated receptors (PPARs) [19], and the transient receptor potential vanilloid receptor 1 (TRPV1) [20], may also be activated by certain cannabinoids/endocannabinoids. In tumors of breast, pancreatic, prostate, and liver cancer, CB1 and CB2 are upregulated; however, a clear correlation between severity of tumors and levels of CB receptor expression has not been established yet (Table 1.1). While a positive correlation between CB2 expression and tumor aggressiveness was noted in breast cancer [28], no such correlation exists in pancreatic cancer [22]. High levels of CB1 receptors in prostate cancer have been linked to poor survival rate [14]. Similar to CB receptors, TRPV1 receptors are upregulated in human prostate [21] and oral squamous cell carcinoma [29]. TRPV1 correlates with tumor grades in prostate carcinoma [21], but no association with malignancy was observed in oral squamous cell carcinoma [29].

1.4 Mechanisms of Anti-tumorigenic Effects by Cannabinoids/Endocannabinoids

The majority of studies suggest that the signaling pathways induced by endocannabinoids generate anticarcinogenic effects, but pro-carcinogenic effects by endocannabinoids have been described as well [11]. The in vitro and in vivo mechanisms

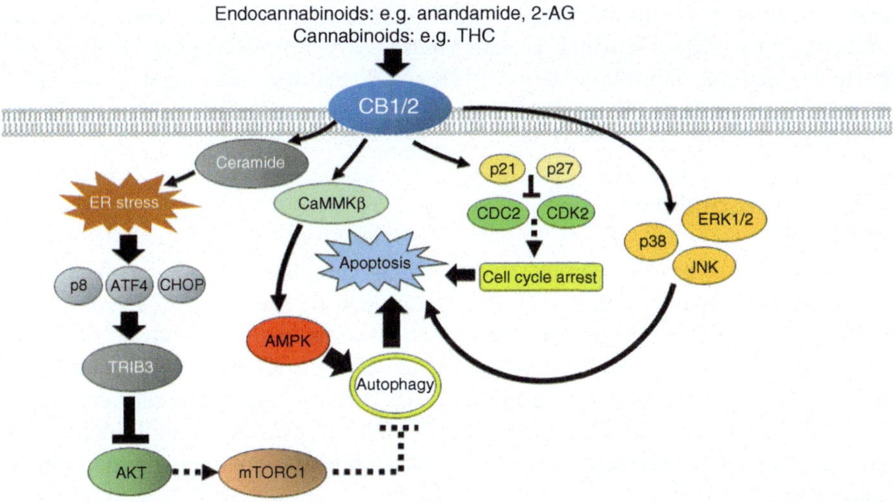

Fig. 1.3 Anticarcinogenic signaling pathways induced by cannabinoids/endocannabinoids. Stimulation of CB1 and/or CB2 receptors by cannabinoids/endocannabinoids induces autophagy and apoptosis in glioma cells. This involves de novo production of ceramide, which causes endoplasmic reticulum stress and the upregulation of transcription factor p8 and other downstream molecules, such as CHOP and ATF4, and the upregulation of TRIB3, which blocks the prosurvival AKT-mTORC1 pathway. In breast cancer cells, the induction of p38, JNK, and ERK1/2 may also lead to apoptosis. CB receptor-dependent cell cycle arrest was observed in breast, gastric, and prostate cancer cells, which involved p21 and p27 and the regulation of CDC2 and CDK2. In hepatocellular carcinoma cells, activation of AMPK via CaMMKβ represents another pathway that leads to autophagy and apoptosis. *2-AG* 2-arachidonoyl glycerol, *AKT* protein kinase B, *AMPK* AMP-activated kinase, *ATF4* activating transcription factor 4, *CaMMKβ* Ca^{2+}/calmodulin-dependent kinase kinase-β, *CB* cannabinoid, *CDC2* cyclin-dependent kinase 1, *CDK2* cyclin-dependent kinase 2, *CHOP* C/EBP-homologous protein, *ER* endoplasmic reticulum, *ERK* extracellular-signal-regulated kinase, *JNK* c-Jun N-terminal kinase, *mTORC1* mammalian target of rapamycin complex 1, *p38* mitogen-activated protein kinase p38, *p8* transcription factor p8, *p21* and *p27* cyclin-dependent kinase inhibitory proteins, *THC* tetrahydrocannabinol, *TRIB3* pseudo-kinase tribbles homolog 3

through which cannabinoids/endocannabinoids inhibit tumor cell growth have been characterized as antiproliferative, proapoptotic, and antimetastatic (Figs. 1.2 and 1.3). These effects may occur at different levels, e.g., at the level of CB receptor expression or endocannabinoid production/degradation, MAGL/FAAH activity, or at the level of receptor interaction, such as with GPR55.

1.4.1 Apoptosis

Apoptosis can be induced in glioma cells by THC via CB receptors [30]. Part of this mechanism is the de novo production of ceramide, a sphingolipid with proapoptotic properties. Ceramide production causes endoplasmic reticulum stress and the upregulation of transcription factor p8 and other downstream molecules, such as pseudo-kinase tribbles homolog 3 (TRB3), C/EBP-homologous protein (CHOP), and activating transcription factor 4 (ATF4) [30, 31]. In breast cancer cells, the induction of mitogen-activated protein kinase (MAPK) pathways (p38, JNK, and ERK1/2) and the inhibition of the PI3K–AKT pathway may lead to apoptosis through inhibition of antiapoptotic or stimulation of proapoptotic proteins [11].

1.4.2 Autophagy

Cannabinoids have been demonstrated to trigger autophagy in cancer cells through endoplasmic reticulum stress [32]. Autophagy can be seen here as a process that participates in the CB receptor-induced pathways that lead to apoptosis. In glioma cells, for instance, the THC-induced inhibition of the AKT-mTORC1 pathway has been implicated in triggering autophagy [30], while in hepatocellular carcinoma cells, activation of AMP-activated kinase (AMPK) via Ca^{2+}/calmodulin-dependent kinase kinase-β (CaMMKβ) may have been responsible for inducing autophagy/apoptosis [33].

1.4.3 Cell Proliferation and Cell Cycle Arrest

Cell proliferation is decreased in various cancer cell lines (colon, breast, gastric, and prostate) after stimulation of CB receptors and inhibition of AKT phosphorylation. CB receptor-dependent cell cycle arrest was observed in breast, gastric, and prostate cancer cells and involved the regulation of CDC2 (also known as cyclin-dependent kinase 1), cyclin-dependent kinase 2 (CDK2), and the cyclin-dependent kinase inhibitors p21 and p27 [32].

1.4.4 Metastasis

Angiogenesis and cancer cell migration are key mechanisms of metastasis, and there is evidence that the endocannabinoid system plays a role therein. Cannabinoids/endocannabinoids decrease adhesion and migration of various cancer cell lines through inhibition of vascular endothelial growth factor (VEGF) [12] and through the downregulation of VEGF receptors 1 and 2 (VEGFR1/R2) [32, 34]. Cannabinoids/endocannabinoids have also been shown to modulate the effects of matrix metalloproteinase-2 (MMP2) and tissue inhibitor of matrix metalloproteinases 1 (TIMP1) [12, 32]. In addition, cannabinoids are able to inhibit proliferation and migration of endothelial cells [12]. Following incubation with anandamide, breast and prostate cancer cells displayed reduced phosphorylation of the tyrosine kinases Fak and Src, which may lead to decreased cell migration [35]. Decreased migration through inhibition of EGF-induced phosphorylation of ERK1/2, JNK1/2, and AKT was observed in human non-small cell lung cancer cells after incubation with THC [36]. Another mechanism that cannabinoids utilize to inhibit migration involves the expression of adhesion molecules. Thus, cannabinoid agonist WIN

55,212-2 inhibits IL-1-induced vascular cell adhesion molecule 1 (VCAM-1) and intercellular adhesion molecule 1 (ICAM-1) expression in astrocytoma and glioblastoma cell lines [37].

1.5 Anti-tumorigenic Effects of Cannabinoids/ Endocannabinoids: Non-CB Receptor-Mediated

Evidence that cannabinoids/endocannabinoids exert anti-tumorigenic effects independently of CB receptors comes from a study in N1E-115 neuroblastoma cells. The study showed that in the presence of anandamide, blockade of FAAH led to a reduction in cell viability through toxic effects but not through mechanisms of apoptosis or necrosis [38]. The effect was not mediated by CB, TRPV1, or GPR55 receptors, but involved lipid rafts instead. In another study with various tumor cell lines in which MAGL was genetically downregulated, the amount of free fatty acids, as well as the migration/invasion of the tumor cells, was reduced by CB receptor-independent mechanisms [16]. As already mentioned, anandamide and 2-AG can be degraded by Cox-2. In this context, it is interesting to note that in cancer cells expressing high levels of Cox-2, such as in colon carcinoma cells, treatment with anandamide resulted in CB receptor-independent cell death [39]. High prostaglandin levels of the E series were observed in these cells, suggesting the conversion of anandamide into prostaglandin products. In tumors with high Cox-2 expression, therefore, endocannabinoids could be metabolized to prostaglandins which then promote cell death via non-CB receptor mechanisms.

1.6 Potential Pro-tumorigenic Effects of the Endocannabinoid System

Although a large amount of data favors an anti-tumorigenic role of cannabinoids/endocannabinoids in carcinogenesis, some reports suggest that the endocannabinoid system could also be involved in tumor growth. This notion is supported by studies showing elevated levels of endocannabinoids in various types of cancer, such as in glioblastoma and meningioma [40], pituitary adenoma [41], colon adenomas and carcinomas [4], and prostate and bladder tumors [42]. Increased expression of CB receptors that correlate with disease severity has been demonstrated in malignant vs. nonmalignant tissue (Table 1.1). In addition, FAAH has been shown to associate with the severity of prostate cancer [43], suggesting that the endocannabinoid system is overactive during tumorigenesis or even drive tumor growth. In an experimental model of hepatocellular carcinoma, CB1 and anandamide were found upregulated together with other tumor-promoting genes, such as the forkhead box protein M1 (FOXM1) and the growth factor receptor-bound protein 2 (GRB2) interactome, suggesting that CB1 is involved in tumor-promoting effects in this type of cancer [44].

1.7 Synopsis

The theory holds that the endocannabinoid system is protective and maintains homeostasis during excitotoxic and inflammatory conditions. Indeed, a large body of preclinical data indicate that cannabinoids/endocannabinoids possess anti-tumorigenic properties which could be exploited for pharmacological treatment. Pro-tumorigenic behavior by the endocannabinoid system, however, has been observed in some studies, suggesting that its role in cancer is not yet clearly defined. It should be considered that high concentrations of exogenous cannabinoids are often used in studies to demonstrate anti-tumorigenic effects, which may not always reflect the physiological actions of endocannabinoids. Also, anti-tumorigenic effects independent of CB receptors have been reported. It, therefore, seems that the role of the endocannabinoid system during cancer development is multifaceted and complex, and depends on many factors, such as cell type, lipid raft formation, receptor crosstalk, and an intricate interaction of signaling pathways, to name a few. Nevertheless, pharmacological interference with components of the endocannabinoid system certainly represents a new and interesting approach to control the growth and spread of tumors.

References

1. Pacher P, Bátkai S, Kunos G. The endocannabinoid system as an emerging target of pharmacotherapy. Pharmacol Rev. 2006;58(3):389–462.
2. Maccarrone M, Bab I, Bíró T, Cabral GA, Dey SK, Di Marzo V, Konje JC, Kunos G, Mechoulam R, Pacher P, Sharkey KA, Zimmer A. Endocannabinoid signaling at the periphery: 50 years after THC. Trends Pharmacol Sci. 2015;36(5):277–96.
3. Schicho R, Storr M. Targeting the endocannabinoid system for gastrointestinal diseases: future therapeutic strategies. Expert Rev Clin Pharmacol. 2010;3(2):193–207.
4. Ligresti A, Bisogno T, Matias I, De Petrocellis L, Cascio MG, Cosenza V, D'Argenio G, Scaglione G, Bifulco M, Sorrentini I, Di Marzo V. Possible endocannabinoid control of colorectal cancer growth. Gastroenterology. 2003;125(3):677–87.
5. Di Marzo V, De Petrocellis L, Sepe N, Buono A. Biosynthesis of anandamide and related acylethanolamides in mouse J774 macrophages and N18 neuroblastoma cells. Biochem J. 1996;316(3):977–84.
6. Varga K, Wagner JA, Bridgen DT, Kunos G. Platelet- and macrophage-derived endogenous cannabinoids are involved in endotoxin-induced hypotension. FASEB J. 1998; 12(11):1035–44.
7. Wei BQ, Mikkelsen TS, McKinney MK, Lander ES, Cravatt BF. A second fatty acid amide hydrolase with variable distribution among placental mammals. J Biol Chem. 2006;281(48):36569–78.
8. Alhouayek M, Muccioli GG. COX-2-derived endocannabinoid metabolites as novel inflammatory mediators. Trends Pharmacol Sci. 2014;35(6):284–92.
9. Moody JS, Kozak KR, Ji C, Marnett LJ. Selective oxygenation of the endocannabinoid 2-arachidonylglycerol by leukocyte-type 12-lipoxygenase. Biochemistry. 2001;40(4):861–6.
10. Flygare J, Sander B. The endocannabinoid system in cancer-potential therapeutic target? Semin Cancer Biol. 2008;18(3):176–89.
11. Pisanti S, Picardi P, D'Alessandro A, Laezza C, Bifulco M. The endocannabinoid signaling system in cancer. Trends Pharmacol Sci. 2013;34(5):273–82.

12. Blázquez C, Casanova ML, Planas A, Gómez Del Pulgar T, Villanueva C, Fernández-Aceñero MJ, Aragonés J, Huffman JW, Jorcano JL, Guzmán M. Inhibition of tumor angiogenesis by cannabinoids. FASEB J. 2003;17(3):529–31.
13. Hanahan D, Weinberg RA. Hallmarks of cancer: the next generation. Cell. 2011;144(5):646–74.
14. Chung SC, Hammarsten P, Josefsson A, Stattin P, Granfors T, Egevad L, Mancini G, Lutz B, Bergh A, Fowler CJ. A high cannabinoid CB(1) receptor immunoreactivity is associated with disease severity and outcome in prostate cancer. Eur J Cancer. 2009;45(1):174–82.
15. Ye L, Zhang B, Seviour EG, Tao KX, Liu XH, Ling Y, Chen JY, Wang GB. Monoacylglycerol lipase (MAGL) knockdown inhibits tumor cells growth in colorectal cancer. Cancer Lett. 2011;307(1):6–17.
16. Nomura DK, Long JZ, Niessen S, Hoover HS, Ng SW, Cravatt BF. Monoacylglycerol lipase regulates a fatty acid network that promotes cancer pathogenesis. Cell. 2010;140(1):49–61.
17. Console-Bram L, Brailoiu E, Brailoiu GC, Sharir H, Abood ME. Activation of GPR18 by cannabinoid compounds: a tale of biased agonism. Br J Pharmacol. 2014;171(16):3908–17.
18. Ross RA. The enigmatic pharmacology of GPR55. Trends Pharmacol Sci. 2009;30(3):156–63.
19. O'Sullivan SE, Kendall DA. Cannabinoid activation of peroxisome proliferator-activated receptors: potential for modulation of inflammatory disease. Immunobiology. 2010;215(8):611–6.
20. De Petrocellis L, Di Marzo V. Lipids as regulators of the activity of transient receptor potential type V1 (TRPV1) channels. Life Sci. 2005;77(14):1651–66.
21. Czifra G, Varga A, Nyeste K, Marincsák R, Tóth BI, Kovács I, Kovács L, Bíró T. Increased expressions of cannabinoid receptor-1 and transient receptor potential vanilloid-1 in human prostate carcinoma. J Cancer Res Clin Oncol. 2009;135(4):507–14.
22. Carracedo A, Lorente M, Egia A, Blázquez C, García S, Giroux V, Malicet C, Villuendas R, Gironella M, González-Feria L, Piris MA, Iovanna JL, Guzmán M, Velasco G. The stress-regulated protein p8 mediates cannabinoid-induced apoptosis of tumor cells. Cancer Cell. 2006;9(4):301–12.
23. Xu X, Liu Y, Huang S, Liu G, Xie C, Zhou J, Fan W, Li Q, Wang Q, Zhong D, Miao X. Overexpression of cannabinoid receptors CB1 and CB2 correlates with improved prognosis of patients with hepatocellular carcinoma. Cancer Genet Cytogenet. 2006;171(1):31–8.
24. Miao X, Liu G, Xu X, Xie C, Sun F, Yang Y, Zhang T, Hua S, Fan W, Li Q, Huang S, Wang Q, Liu G, Zhong D. High expression of vanilloid receptor-1 is associated with better prognosis of patients with hepatocellular carcinoma. Cancer Genet Cytogenet. 2008;186(1):25–32.
25. De Jesús ML, Hostalot C, Garibi JM, Sallés J, Meana JJ, Callado LF. Opposite changes in cannabinoid CB1 and CB2 receptor expression in human gliomas. Neurochem Int. 2010;56(6–7):829–33.
26. Wang D, Wang H, Ning W, Backlund MG, Dey SK, DuBois RN. Loss of cannabinoid receptor 1 accelerates intestinal tumor growth. Cancer Res. 2008;68(15):6468–76.
27. Gustafsson SB, Palmqvist R, Henriksson ML, Dahlin AM, Edin S, Jacobsson SO, Öberg Å, Fowler CJ. High tumour cannabinoid CB1 receptor immunoreactivity negatively impacts disease-specific survival in stage II microsatellite stable colorectal cancer. PLoS One. 2011;6(8):e23003.
28. Caffarel MM, Andradas C, Mira E, Pérez-Gómez E, Cerutti C, Moreno-Bueno G, Flores JM, García-Real I, Palacios J, Mañes S, Guzmán M, Sánchez C. Cannabinoids reduce ErbB2-driven breast cancer progression through Akt inhibition. Mol Cancer. 2010;9:196.
29. Marincsák R, Tóth BI, Czifra G, Márton I, Rédl P, Tar I, Tóth L, Kovács L, Bíró T. Increased expression of TRPV1 in squamous cell carcinoma of the human tongue. Oral Dis. 2009;15(5):328–35.
30. Salazar M, Carracedo A, Salanueva IJ, Hernández-Tiedra S, Lorente M, Egia A, Vázquez P, Blázquez C, Torres S, García S, Nowak J, Fimia GM, Piacentini M, Cecconi F, Pandolfi PP, González-Feria L, Iovanna JL, Guzmán M, Boya P, Velasco G. Cannabinoid action induces autophagy-mediated cell death through stimulation of ER stress in human glioma cells. J Clin Invest. 2009;119(5):1359–72.

31. Carracedo A, Gironella M, Lorente M, Garcia S, Guzmán M, Velasco G, Iovanna JL. Cannabinoids induce apoptosis of pancreatic tumor cells via endoplasmic reticulum stress-related genes. Cancer Res. 2006;66(13):6748–55.
32. Velasco G, Sánchez C, Guzmán M. Towards the use of cannabinoids as antitumour agents. Nat Rev Cancer. 2012;12(6):436–44.
33. Vara D, Salazar M, Olea-Herrero N, Guzmán M, Velasco G, Díaz-Laviada I. Anti-tumoral action of cannabinoids on hepatocellular carcinoma: role of AMPK-dependent activation of autophagy. Cell Death Differ. 2011;18(7):1099–111.
34. Portella G, Laezza C, Laccetti P, De Petrocellis L, Di Marzo V, Bifulco M. Inhibitory effects of cannabinoid CB1 receptor stimulation on tumor growth and metastatic spreading: actions on signals involved in angiogenesis and metastasis. FASEB J. 2003;17(12):1771–3.
35. Grimaldi C, Pisanti S, Laezza C, Malfitano AM, Santoro A, Vitale M, Caruso MG, Notarnicola M, Iacuzzo I, Portella G, Di Marzo V, Bifulco M. Anandamide inhibits adhesion and migration of breast cancer cells. Exp Cell Res. 2006;312(4):363–73.
36. Preet A, Ganju RK, Groopman JE. Delta9-Tetrahydrocannabinol inhibits epithelial growth factor-induced lung cancer cell migration in vitro as well as its growth and metastasis in vivo. Oncogene. 2008;27(3):339–46.
37. Curran NM, Griffin BD, O'Toole D, Brady KJ, Fitzgerald SN, Moynagh PN. The synthetic cannabinoid R(+)WIN 55,212-2 inhibits the interleukin-1 signaling pathway in human astrocytes in a cannabinoid receptor-independent manner. J Biol Chem. 2005;280(43):35797–806.
38. Hamtiaux L, Hansoulle L, Dauguet N, Muccioli GG, Gallez B, Lambert DM. Increasing antiproliferative properties of endocannabinoids in N1E-115 neuroblastoma cells through inhibition of their metabolism. PLoS One. 2011;6(10):e26823.
39. Patsos HA, Hicks DJ, Dobson RR, Greenhough A, Woodman N, Lane JD, Williams AC, Paraskeva C. The endogenous cannabinoid, anandamide, induces cell death in colorectal carcinoma cells: a possible role for cyclooxygenase 2. Gut. 2005;54(12):1741–50.
40. Petersen G, Moesgaard B, Schmid PC, Schmid HH, Broholm H, Kosteljanetz M, Hansen HS. Endocannabinoid metabolism in human glioblastomas and meningiomas compared to human non-tumour brain tissue. J Neurochem. 2005;93(2):299–309.
41. Pagotto U, Marsicano G, Fezza F, Theodoropoulou M, Grübler Y, Stalla J, Arzberger T, Milone A, Losa M, Di Marzo V, Lutz B, Stalla GK. Normal human pituitary gland and pituitary adenomas express cannabinoid receptor type 1 and synthesize endogenous cannabinoids: first evidence for a direct role of cannabinoids on hormone modulation at the human pituitary level. J Clin Endocrinol Metab. 2001;86(6):2687–96.
42. Schmid PC, Wold LE, Krebsbach RJ, Berdyshev EV, Schmid HH. Anandamide and other N-acylethanolamines in human tumors. Lipids. 2002;37(9):907–12.
43. Thors L, Bergh A, Persson E, Hammarsten P, Stattin P, Egevad L, Granfors T, Fowler CJ. Fatty acid amide hydrolase in prostate cancer: association with disease severity and outcome, CB1 receptor expression and regulation by IL-4. PLoS One. 2010;5(8):e12275.
44. Mukhopadhyay B, Schuebel K, Mukhopadhyay P, Cinar R, Godlewski G, Xiong K, Mackie K, Lizak M, Yuan Q, Goldman D, Kunos G. Cannabinoid receptor 1 promotes hepatocellular carcinoma initiation and progression through multiple mechanisms. Hepatology. 2015;61(5):1615–26.

MACC1, a Novel Player in Solid Cancer Carcinogenesis

2

Giridhar Mudduluru, Katharina Ilm, Mathias Dahlmann, and Ulrike Stein

Contents

G. Mudduluru • U. Stein (✉)
Experimental and Clinical Research Center, Charité University Medicine Berlin and Max-Delbrück-Center for Molecular Medicine in the Helmholtz Association, Robert-Rössle-Str. 10, 13125 Berlin, Germany

German Cancer Consortium (DKTK), Heidelberg, Germany
e-mail: ustein@mdc-berlin.de

K. Ilm • M. Dahlmann
Experimental and Clinical Research Center, Charité University Medicine Berlin and Max-Delbrück-Center for Molecular Medicine in the Helmholtz Association, Robert-Rössle-Str. 10, 13125 Berlin, Germany

J. Haybaeck (ed.), *Mechanisms of Molecular Carcinogenesis – Volume 1*,
DOI 10.1007/978-3-319-53659-0_2

Abstract

Cancer develops due to uncontrolled proliferation of cells initiated by genetic instability, mutations, and environmental stress. Cells acquire these fundamental abnormalities in a multistep process due to changes in complex multilayer molecular network signaling axes equipping them with increased capacity of proliferation, survival, extracellular matrix (ECM) degradation, migration, invasion, and metastasis. Recently, metastasis-associated in colon cancer 1 (MACC1) was identified with differential display RT-PCR by analyzing the normal mucosa, primary, and metastasis specimens of colon cancer. We discovered that MACC1 is an important transcriptional regulator of hepatocyte growth factor (HGF) receptor c-Met and showed that MACC1 plays an important role in tumorigenesis, migration, invasion, and distant metastasis. In this book chapter, we discuss in-depth the structure and function of MACC1 in different aspects of carcinogenesis like gene regulation, signaling, cell proliferation, apoptosis, migration, invasion, metastasis, angiogenesis, epithelial mesenchymal transition (EMT), its role in cell metabolism, and also the impact of MACC1 as predictive and prognostic marker. In addition, we describe MACC1 as druggable target molecule by different approaches to reduce tumorigenesis and metastasis.

2.1 MACC1: Regulation, Structure, Signaling

2.1.1 Transcriptional and Posttranscriptional Regulation of MACC1

Formation of distant metastases is the limiting factor for the therapy of colorectal cancer (CRC) patients and drops the 5-year survival rate from 90% to 10%. Therefore, the identification of biomarkers that allow for an individual diagnosis and prognosis of cancer patients becomes more and more important. By differential display RT-PCR of human colon cancer tissues, metastases, and normal colon mucosa, we identified a previously not described differentially expressed complementary DNA fragment. We cloned the full length DNA and identified a novel gene which was named metastasis-associated in colon cancer 1 MACC1 (GenBank gene ID 346389, European Molecular Biology Laboratory (EMBL) data bank accession code AJ313524) [1]. Tumors with UICC stages I, II, and III due to TNM classification (not distantly metastasized at the time point of surgery), which developed metachronous metastasis, showed a significantly higher MACC1 expression at the time point of diagnosis compared to non-metastasizing tumors. The 5-year survival rate for patients with low MACC1 mRNA expression in their primary tumors was 80% and only 15% for patients with high MACC1 expression. It was also shown that MACC1 is an important regulator of hepatocyte growth factor (HGF)/c-Met signaling, which is crucial for cell motility, invasiveness, and metastasis.

Apart from its inducing role in the metastatic process by transcriptional activation of the oncogene c-Met, evidence suggests MACC1 as an important signaling molecule contributing to CRC tumorigenesis. Especially the identification of the basal MACC1 promoter and transcription factors regulating the expression allowed further insights. Based on promoter luciferase constructs, we examined the promoter sequence and identified several binding sites for transcription factors [2]. The minimal essential promoter region within −426 to −18 harbors binding sites, e.g., activator protein 1 (AP1), specificity protein 1 (SP1), and CCAAT/enhancer-binding protein (c/EBP). The physical binding of these transcription factors to the MACC1 core promoter was validated using chromatin immunoprecipitation (ChIP) and electrophoretic mobility shift assay (EMSA). Knockdown of SP1, c-Jun, or c/EBP decreased the MACC1 expression in CRC cell lines. In addition to the regulation of the endogenous MACC1 expression by SP1 and c-Jun, we also demonstrated MACC1-specific reduction of migration by knockdown of these transcription factors. To demonstrate the biological dependency of MACC1 expression on these transcription factors, we performed qRT-PCR of 60 CRC tumor patients. We identified a significant positive correlation of these transcription factors with MACC1 expression. AP1 expression is known to be altered in several cancer entities including CRC. It controls cellular processes including differentiation, proliferation, and apoptosis. In tumors the AP1 expression levels are often increased and associated with a transformed phenotype. Besides AP1, also SP1 is known to be highly expressed in several cancer entities and involved in tumor development, proliferation, and metastasis. Further, in clinical studies, SP1 was shown to be prognostic for patient survival. The regulation of the MACC1 expression by these transcription factors involved in differentiation and tumorigenesis imparts an additional dimension of MACC1-associated phenotype.

In the meantime further transcriptional regulators of the MACC1 expression were identified in different cancer entities. Sex-determining region Y (SRY)-related HMG box factor 17 (SOX17) was identified as negative regulator of the MACC1 expression in esophageal cancer [3]. Promoter luciferase constructs were used to demonstrate the reduction of MACC1 transcription based on SOX17 expression. In addition, binding of SOX17 to the SRY-binding motif in the MACC1 promoter was verified by quantitative ChIP-PCR. SOX17 is a transcription factor which is important for esophagus tissue development and tumor suppressor that transcriptionally downregulates nibrin (NBN), nuclear factor of activated T-cells 5 (NFAT5), MACC1, metastasis-associated lung adenocarcinoma transcript 1 (MALAT1), casein kinase 1 alpha 1 (CSNK1A1), fibronectin 1 (FN1), and plasminogen activator inhibitor 1 (SERPINE1) mRNA-binding protein 1 (SERBP1) gene expression [3]. These genes are partly involved in cancer cell growth, proliferation, cell adhesion, migration, and invasion of esophageal cancer.

Altered MACC1 expression levels were previously also shown in ovarian cancer [4]. Recently, positive correlation of serine/threonine-protein kinase pim-3 (Pim-3) and MACC1 expression was shown in human ovarian cancer cell lines [5]. Overexpression of Pim-3 leads to increased MACC1 mRNA and protein expression. Pim-3 is a proto-oncogene with serine/threonine kinase activity that can prevent apoptosis and promote cell survival and protein translation [6]. To date, the regulatory mechanism of MACC1 expression by Pim-3 has not been described in detail.

MACC1 expression levels were also shown to be increased in nasopharyngeal carcinomas compared to normal tissue [7]. In addition, they showed that the Epstein-Barr virus-encoded latent membrane protein-1 (LMP1) regulates MACC1 expression. LMP1 was shown to mimic the TNFR family member CD40 and has a putative role in the pathogenesis of nasopharyngeal carcinomas [8].

The correlations and regulation of MACC1 expression with tumor suppressor genes and proto-oncogenes like SOX17, Pim-3, and LMP1 underline the role of MACC1 as an important signaling molecule during carcinogenesis of different cancer entities.

Regulatory expression networks are not limited to the 5′UTR and transcriptional activity. The 3′UTR of genes harbors binding sites for microRNAs (miRNAs). These small noncoding RNAs regulate the gene expression at the posttranscriptional level. Through binding to the corresponding regions in the 3′UTR of their target genes, the mRNA translation can be inhibited. MicroRNAs are involved in the regulation of tumorigenesis, differentiation, proliferation, and survival. Altered expression patterns of miRNAs are shown in several cancer entities. There are miRNAs described which regulate MACC1 expression and MACC1 induced migratory abilities in vitro and metastasis in vivo. The tumor suppressor miR-143 was the first described miRNA posttranscriptionally regulating the MACC1 expression and MACC1-dependent migration and invasion [9]. A negative correlation of miR-143 and MACC1 expression was identified in CRC patient samples and cell lines. Restoration of miR-143 expression in CRC cells reduced the MACC1 protein expression and migratory abilities. In cell lines with high miR-143, the use of miRNA inhibitors increased the MACC1 protein expression. The binding of the miRNA to the MACC1 3′UTR was verified by luciferase constructs harboring the endogenous or mutated binding sites for the miRNA. miR-143 acts as a tumor suppressor and targets genes involved in differentiation, proliferation, apoptosis, chemoresistance, and metastasis. The most prominent target is the proto-oncogene Kirsten rat sarcoma viral oncogene homolog (KRAS), which is a crucial player in the traditional CRC model following the adenoma-carcinoma sequence [10]. Additionally described targets are extracellular-regulated protein kinase 5 (ERK5) [11], DNA methyltransferase 3A (DNMT3A) [12], E twenty-six (ETS)-like transcription factor 1 (ELK1) [13], neuroblastoma RAS viral (V-Ras) oncogene homolog (NRAS) [14] and B-cell CLL/lymphoma 2 (Bcl-2) [15].

In the meantime, the relationship between MACC1 and miR-574-5p expression was investigated in CRC cell lines and in a liver metastatic mice model [16]. CRC cell lines with high MACC1 expression and low miR-574-5p had significantly increased lesions of liver metastasis. The endogenous MACC1 expression inversely correlated with miR-574-5p expression. Endogenous MACC1 protein expression levels were negatively regulated due to transfection with miRNA mimics or inhibitors in CRC cell lines. Previously described targets are the carbonyl reductase 1 (CBR1) [17] and Quaking 6/7 (Qki6/7). The negatively regulation of Qki6/7 leads to increased β-catenin/Wnt signaling and development of CRC [18].

Recently, the tumor suppressor miRNAs, miR-200a, miR-218 and miR-338-3p, were identified as posttranscriptional regulators of the MACC1 expression [19–21]. The miR-200a belongs to the miR-200 family, which is known to regulate EMT-associated proteins [22] by targeting transcription factors like zinc finger E-box-binding protein 1 and 2 (ZEB1 and ZEB2) [23]. miR-338-3p affects the behavior of the cancer cells by upregulating the epithelial marker E-cadherin and downregulating the mesenchymal markers via targeting the ZEB2 [20].

These findings underline the important role of MACC1 during cancer initiation, EMT, and tumor progression.

2.1.2 Structure and Domain Architecture of MACC1

The MACC1 protein consists of 852 amino acids. It was analyzed with online predictive tools [ELM, http://elm.eu.org [24]; PSIPRED, http://bioinf.cs.ucl.ac.uk/psipred [25]; PredictProtein, https://www.predictprotein.org [26]] and its structure was subsequently published [1, 27, 28]. Its schematic structure is represented in Fig. 2.1a.

The N-terminus of MACC1 is predicted to be highly disordered, which has high affinity for protein-protein interactions (Fig. 2.1b). The putative clathrin box LIDME (aa23-27) closely matches the canonical sequence pattern L[LI][DEN][LF][DE], which is known to interact with the N-terminal part of the clathrin heavy chain [29]. Three adjacent short motifs containing the tripeptides NPF or DPF are predicted to interact with proteins containing epsin15 homology (EH) domains [30, 31].

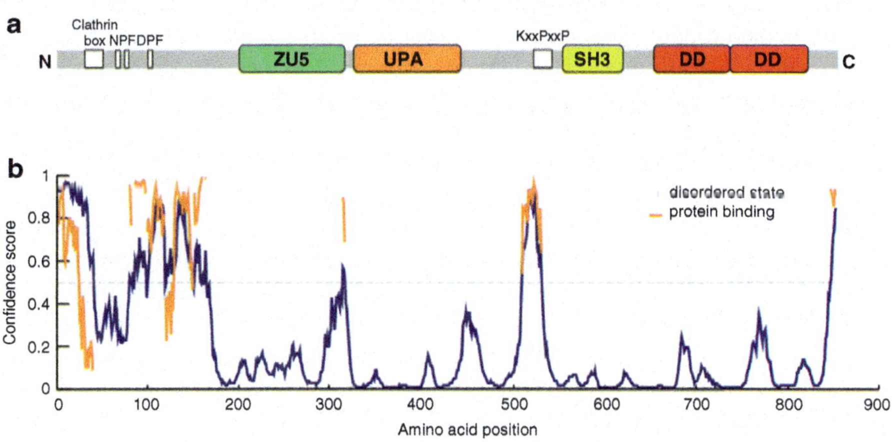

Fig. 2.1 (**a**) Schematic representation of predicted MACC1 protein domains. *Colored boxes*: predicted similarity to known structural domains. *White boxes*: predicted similarity to short protein interaction motifs; *ZU5* ZO-1/Unc5-like domain (type II), *UPA* domain present in Unc5, PIDD, and ankyrins, *SH3* variant Src homology-3 domain, *DD* death domain, *NPF/DPF* AP2-binding motif, *KxxPxxP* class I SH3 domain-interacting motif; (**b**) prediction of disordered regions and putative protein-binding areas of MACC1 by DISOPRED/PSIPRED

Kirchhausen et al. reported a function of the clathrin box in endocytosis [32], whereas the members of the EH domain-containing protein family determine the fate of the vesicle after internalization [33–35]. Both predicted interaction motifs strongly suggest a role of MACC1 in clathrin-mediated endocytosis (CME) of still unknown cargo proteins or in directed trafficking of vesicles.

The first two predicted structural domains of MACC1, C-terminal of the unstructured N-terminus, are highly similar to the ZU5 domain and to the UPA domain, respectively. The first was initially found in the proteins zona occludens (ZO) 1 and uncoordinated (Unc)-5. The latter was described to be part of the UPA domain, which is also present in Unc5, p53-induced death domain-containing (PIDD) protein, and ankyrins, respectively.

A very recent report suggests a common ancestor of the family of ZU5-like domains early in evolution, since highly similar domains have been found in bacteria and archaea [36]. A high number of these domains are capable of auto-proteolysis when a conserved histidine and serine (H/S) are present at the putative active site of proteolysis. Intriguingly, caspase recruitment domain-containing protein (Card) 8 and NACHT and leucine rich repeat (LRR) and pyrin (PYD) domains-containing protein (NLRP) 1 show active auto-proteolytic activity and share a unique putatively active site H/S pattern, which is also found in the predicted ZU5-like domain of MACC1 [36, 37]. It needs to be clarified whether or not this feature contributes to any cellular function of MACC1, since no auto-proteolytic activity of MACC1 has been reported so far.

ZU5 domains, which lost the auto-proteolytic activity during evolution, acquired important protein-protein interactions, especially in proteins involved in maintaining the cytoskeleton, cellular signaling, motility, and apoptosis [38–40]. As it has recently been shown for ankyrins, the two ZU5 domains occurring in tandem do not overlap in their function [41, 42]. Although structurally similar, the different surface charge of each respective ZU5 domain determines distinct interaction partners. While the ZU5A domain interacts with spectrin, the ZU5B domain binds to the adjacent UPA and death domains (DD). This intramolecular binding of a ZU5-UPA-DD module was earlier reported for Unc5b, regulating the signaling activity of the netrin receptor upon ligand binding [43]. MACC1 shares the general domain composition of Unc5b: a ZU5-UPA module and two C-terminal DDs, also predicted by sequence similarity. Thus, it may form intramolecular domain interactions as well, which contribute to protein stability or regulate cellular functions of MACC1.

A major structural difference of MACC1 to ZU5-UPA-DD-containing proteins is the presence of a canonical proline-rich interaction motif (PRM; KxxPxxP) for classical sarcoma proto-oncogene (Src)-homology 3 (SH3) domains, closely followed by a predicted variant SH3 domain [1, 27, 28]. Due to the high variance in the protein sequence of both PRMs and SH3 domains and the high selectivity of their interactions, a prediction of interaction partners of MACC1 via these motifs is hardly fruitful [44]. A first indication of the functional relevance of SH3-PRM interactions of MACC1 was the albeit weak binding of its SH3 domain to the PRM of tumor necrosis factor ligand superfamily member 6/Fas ligand (TNFL6/FasL) and a disintegrin and metalloprotease (ADAM) 10 [45, 46]. When both regions of MACC1 were altered, either by mutating the PRM (KxxAxxA) or by deleting the SH3 domain (ΔSH3), the ability to induce cell motility in vitro and tumor growth and metastasis formation in vivo was significantly reduced [1, 47]. This finding highlights the importance of the protein-protein interactions of this region for the functionality of the MACC1 protein.

The in silico analysis of the MACC1 protein sequence predicts interesting protein domains and protein-binding motifs indicating a variety of cellular interaction partners and predestining MACC1 as a signal transducer. It is yet to be shown how these features contribute to different cellular functions of MACC1 and whether they affect cancer initiation, progression, and metastasis.

2.1.3 MACC1 as a Signaling Molecule

We have shown for the first time that MACC1 transcriptionally regulates c-Met [1]. It is well documented in the literature that HGF is a ligand of the c-Met receptor tyrosine kinase [48], which is activated by serine proteases. c-Met is a disulfide-linked heterodimer, which arises from cleavage of a precursor to an extracellular alpha chain and a transmembrane beta chain. Intracellular regions of the beta chain contain the juxtamembrane catalytic region and multifunctional docking site. After HGF binding to c-Met, it gets activated by different transphosphorylations of tyrosine residues, first in the catalytic region and then Tyr1349 and Tyr1356, which are multifunctional docking sites [49]. These activated tyrosine residues create docking sites for a variety of adaptor proteins and kinases like growth factor receptor-bound protein 2 (GRB2), Grb2-associated adaptor protein (GAB1), Src, phosphatidylinositol-3-kinase (PI3K), Son of Sevenless (SOS), and signal transducer and activator of transcription 3 (STAT3), which leads to activation of downstream signaling pathways that include the mitogen-activated protein kinase (MAPK), PI3K/V-Akt murine thymoma viral oncogene homolog (AKT), and STAT pathways [48]. The tight control of the HGF/c-Met signaling axis is observed during embryonic development. However, this control is lost in multiple levels, for example, at the transcriptional level, inadequate degradation of proteins, and cross-talk in the downstream signaling axis in different cancer types, like carcinomas, musculoskeletal sarcomas, soft tissue sarcomas, hematopoietic malignancies, and other neoplasms. Deregulation of HGF/c-Met signaling results in tumorigenesis and metastasis. In-depth details of HGF/c-Met associations can be found in a searchable fully referenced table with this web link (http://www.vai.org/Met/Index.aspx).

MACC1 is one of the key carcinogenic transcriptional regulators of c-Met apart from SP1, tumor protein P53 (p53), V-ets avian erythroblastosis virus E26 oncogene homolog 1 (Ets1), paired box 3 (Pax3), AP1, death domain-associated protein (Daxx), and hypoxia-inducible factor 1 (HIF-1), which plays a crucial role in different biological aspects of c-Met [1, 50–56]. Apart from our study, other studies screened for c-Met and its transcriptional regulator MACC1 in patient specimens of different tumor types. Among these reports, a positive correlation of MACC1 and c-Met expression in patient tumor specimens has been described, e.g., for CRC even

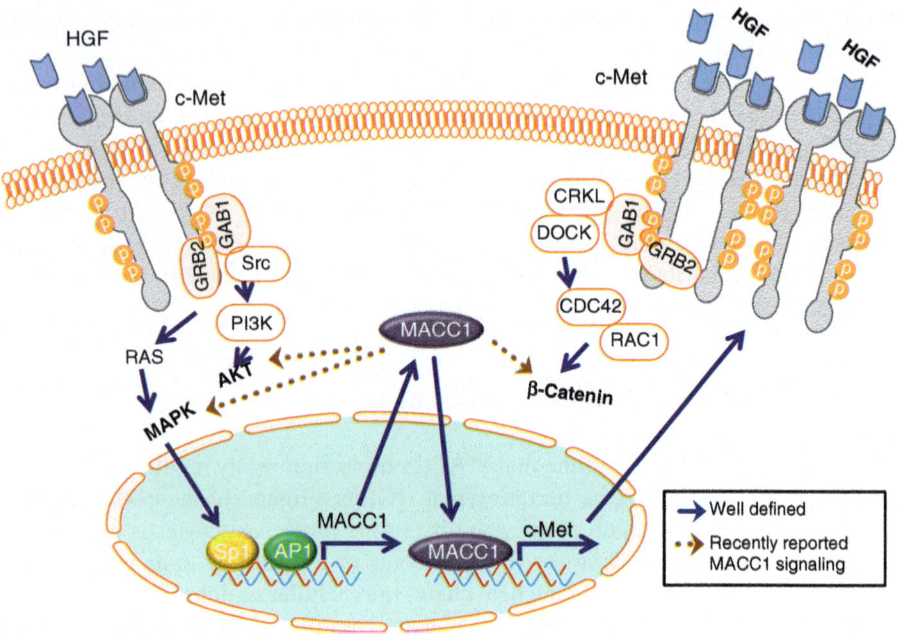

Fig. 2.2 MACC1 signaling axes: overexpression of MACC1 induced the c-Met and β-catenin expression and activation of AKT and MAPK signaling. *GRB2* growth factor receptor-bound protein 2, *GAB1* Grb2-associated adaptor protein, *Src* SRC proto-oncogene, *PI3K* phosphatidylinositol-3-kinase, *RAS* rat sarcoma viral oncogene homolog, *MAPK* mitogen-activated protein kinase, *AKT* V-Akt murine thymoma viral oncogene homolog, *CRKL* V-Crk avian sarcoma virus CT10 oncogene homolog-like, *DOCK* dedicator of cytokinesis, *CDC42* cell division cycle 42, *RAC1* Ras-related C3 botulinum toxin substrate 1

at distant metastasized organs, for rectal, hepatocellular, nasopharyngeal, and renal carcinoma [7, 57–59]. HGF/c-Met signaling axis induces MACC1 expression transcriptionally through RAS/MAPK signaling. Indeed, induced MACC1 protein translocation to the nucleus activates the c-Met expression. This positive feedback loop regulation of MACC1 and c-Met activation subsequently activates other carcinogenic signaling axes [1, 55, 60–63]. These pieces of evidence suggest that MACC1 might be a signaling molecule independently (upon induced expression) or dependently via the HGF/c-Met signaling axis (Fig. 2.2).

Based on sequence similarity predictions (Fig. 2.1), MACC1 can form protein-protein interactions and function as a signal transducer (Fig. 2.2) [59]. MACC1 knock-in studies in CRC cell lines showed significant upregulation of c-Met, cell motility, and proliferation. Specifically, overexpression of MACC1 mutated in the proline-rich and SH3 domains could not induce either together or independently the c-Met expression, cell motility, and proliferation. The c-Met ligand HGF-induced cell detachment is inhibited by MEK inhibitors UO126, but not by the PI3K inhibitor LY294002 [1]. These findings clearly show that MACC1 functions as a transcriptional factor either alone or along with other DNA-binding interacting proteins through mainly proline-rich and/or SH3 domains. Secondly, without HGF stimulation, induced MACC1 expression increases migration, cell detachment, and distant metastasis. With this functional evidence, one could suggest that without HGF activation of c-Met signaling, MACC1 alone could induce migration [1]. At least in part, the cancer cell phenotype effect could be independently mediated by protein-protein interaction domains of MACC1 (Fig. 2.2). In support of our hypothesis in terms of protein-protein interactions, Voss et al. showed by phage display screening that the proline-rich motif of human FasL is a SH3 domain interaction partner of MACC1 [46]. Further, Ebsen et al. showed with a similar screening approach that ADAM10 is also a SH3 domain interaction partner of MACC1, which induces cell proliferation, migration, and invasion [45, 64].

Changes in various signaling pathways associated with manipulation of endogenous MACC1 expression (knock-in and knockdown studies) allow cells to gain several hallmarks of cancer, like cell proliferation, cell survival, apoptosis, migration, EMT, invasion, and distant metastasis [65]. Meng et al. showed that MACC1 plays a major role in nasopharyngeal carcinogenesis through the AKT/β-catenin signaling axis. Small interfering (si) RNA knockdown of MACC1 inhibited the activation of AKT and reduced the gene expression of β-catenin, c-Met, and its regulator c-Myc. Furthermore, knockdown of MACC1 increased the amount of cleaved caspase 3, which is an apoptosis inducer. They observed a significant positive co-expression of MACC1, p-AKT, and β-catenin in nasopharyngeal carcinoma specimens [7]. A similar study in CRC by Zhen et al. showed MACC1-induced expression of c-Met, β-catenin, and its downstream signaling molecules c-Myc, Cyclin D1, and matrix metallopeptidase 9 (MMP9). Further, silencing of MACC1 with siRNA reduced the β-catenin expression and its translocation to the nucleus, whereas silencing of β-catenin with siRNA did not alter the expression or localization of MACC1 in the cell. Furthermore, they discovered a significant co-expression of MACC1 and β-catenin in CRC patient specimens [66]. Overexpression of MACC1 induced the extracellular matrix (ECM) remodeling genes fibronectin, matrix metallopeptidase 2 (MMP2), and MMP9, as well as the mesenchymal phenotype markers vimentin and CD44, and reduced the epithelial phenotype markers E-cadherin and α-catenin in gastric cancer cell lines. Importantly, apart from these molecules, MACC1 upregulated the key EMT switch initiative transcriptional factors like twist family BHLH transcription factor 1/2 (TWIST1/2). In concordance with other cancers, a significant positive correlation was observed for the expression levels of MACC1, c-Met, and β-catenin in gastric cancer patient specimens [67, 68]. MACC1 overexpression reduces cell apoptosis and induced cell growth in hepatocellular carcinoma cell lines, which is achieved by enhancing the HGF-driven activation of PI3K/AKT signaling axis and its downstream molecules like Bcl-2-associated death promoter (BAD), caspase 9, and forkhead box O3 (FOXO3) activation. MACC1 and p-AKT expression showed positive correlation in hepatocellular carcinoma tumor specimens [69]. These experiments and in vivo corroborative studies are clearly showing that MACC1 is one of the key signaling molecules in tumorigenesis allowing cells to acquire hallmarks of cancer, like cellular proliferation, colony formation, anti-apoptotic abilities, migration, invasion, EMT, distant metastasis, and angiogenesis (Fig. 2.3).

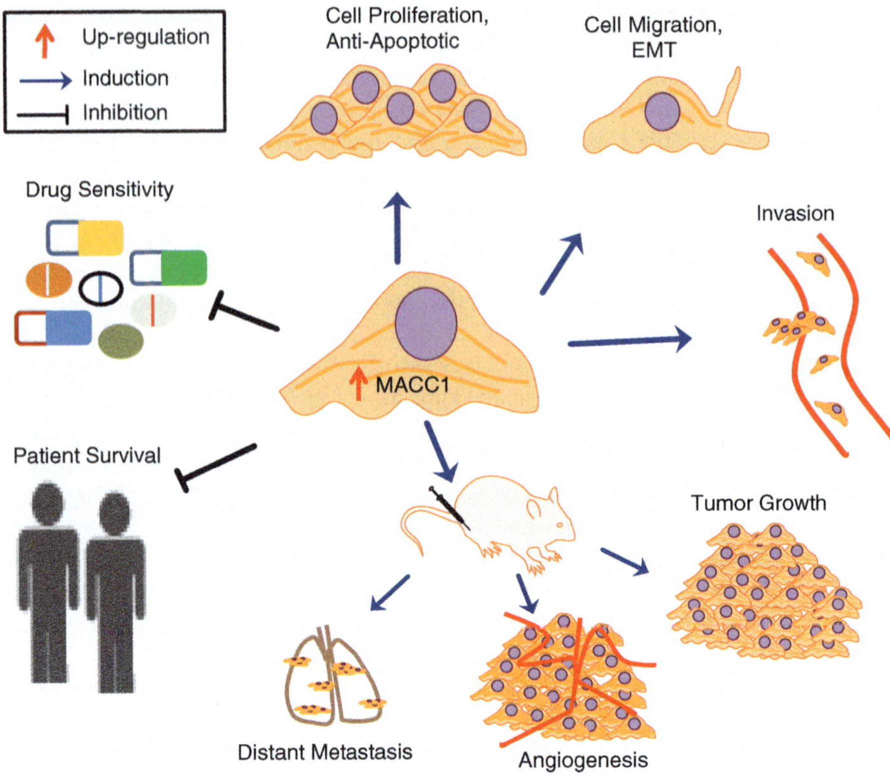

Fig. 2.3 MACC1 functional role in carcinogenesis: increased MACC1 expression in cells induces cell proliferation, EMT, anti-apoptosis, cell motility, migration, invasion, tumor growth, angiogenesis, and distant metastasis leading to reduced patient's survival and drug sensitivity

2.2 Biological Impact of MACC1

2.2.1 MACC1 as Proliferation and Anti-apoptotic Inducer

The balance of proliferation and apoptosis is crucial for physiological processes. Cancer cells are characterized by unlimited proliferation without presence of stimulating growth signals. In addition, these cells gain abilities to overcome apoptotic signals. These defects in the control mechanisms of proliferation and apoptosis promote malignant transformation [65].

MACC1 is very well characterized as metastasis-associated gene but can also induce proliferation and inhibit apoptosis. Knockdown or overexpression of MACC1 in different CRC cell lines directly correlates with the proliferation and tumor growth of these cancer cells, compared to the corresponding controls [1, 66]. In the meantime the proliferation-inducing impact of MACC1 was also shown in several different entities. To name a few, Zhang et al. showed MACC1 upregulation in osteosarcoma cell lines compared to hFOB1.19 osteoblasts and the downregulation of MACC1 expression caused inhibition of proliferation, colony formation, invasion, and tumor growth [70]. Further, decreased MACC1 expression led to cell cycle arrest in G0/G1 phase and induction of apoptosis. In salivary adenoid cystic carcinoma, MACC1 expression is significantly increased, compared to the adjacent normal salivary tissue, and knockdown of MACC1 inhibited proliferation and induced apoptosis in ACCM cells [62]. Similar correlations of MACC1 expression with proliferation and apoptosis are shown in several other entities, like gastric cancer [71], nasopharyngeal cancer [7], tongue squamous cell carcinoma [72], glioblastoma [73, 74], ovarian cancer [5] and cervical cancer [75]. The in silico predicted domain composition gives a hint that MACC1 could interact via the ZU5-UPA-death domain (DD) with signaling molecules that are directly involved in apoptosis regulation [28]. In addition, based on the domain composition of MACC1, bioinformatics studies provide additional hints for a role of MACC1 in apoptosis [27]. By using phage display screening, MACC1 was identified as possible interaction partner with the FasL, which is a known crucial molecule for apoptosis [46].

The impact of MACC1 on proliferation and apoptosis supports the possible role of MACC1 during cancer initiation and malignant transformation.

2.2.2 MACC1 as Epithelial Mesenchymal Transition (EMT) Inducer

A crucial molecular program that enables tumor cells to proliferate and establish metastases is EMT [76–78]. EMT is characterized by the transition of cells from epithelial into mesenchymal phenotype. Epithelial cells are characterized by the presence of regular cell-cell contacts and adhesion to the surrounding cellular fabric, preventing the detachment of individual cells. In contrast, mesenchymal cells do not form such intracellular contacts and are characterized by an irregular cell shape. Molecular EMT is defined by the induction of N-cadherin, vimentin, and fibronectin and the downregulation of cytokeratin and occludin. The crucial transition is the loss of the cell-cell adhesion molecule E-cadherin [79]. Mainly expressed in epithelial cells, E-cadherin plays a central role in embryonic development and mediates signals important in the morphogenesis of epithelial tissues [80]. During invasion of surrounding structures or distant metastasis, various malignant tumors show a suppression of E-cadherin expression, mediated mainly through transcriptional repressors, such as the transcription factors ZEB1 and ZEB2, survival of motor neuron protein interacting protein 1 (SIP1) [81, 82], TWIST [83–85], and snail family zinc finger (SNAIL) [86].

Recent reports using knock-in or knockdown studies showed substantial evidence for the role of MACC1 in EMT. Wang et al. examined mRNA expression of MACC1, c-Met, and EMT markers (E-cadherin, fibronectin, and vimentin) in 22 gastric cancer patients and found that MACC1 and c-Met were positively co-expressed in tumor specimens and also the expression of both of these genes was also positively correlated with fibronectin and vimentin. Furthermore, MACC1 knock-in and knockout studies clearly showed that MACC1 induces EMT by downregulating the epithelial marker E-cadherin and upregulating the mesenchymal markers fibronectin and vimentin in gastric cancer cell lines. Apart from these EMT markers, they found induced expression of MMP2, MMP3, and CD44 upon MACC1 overexpression [68]. A similar kind of study with pancreatic cancer cells showed that siRNA-mediated MACC1 knockdown increased the expression of E-cadherin [87]. Very recently, Wang et al. showed that MACC1 promotes vasculogenic mimicry in gastric cancer by upregulating one of the key EMT switch transcriptional regulators TWIST1 and TWIST2. Upon HGF stimulation, they observed increased nuclear translocation of MACC1, TWIST1, and TWIST2, while a c-Met inhibitor reduced these effects in gastric cancer cells. In parallel, they observed increased nuclear expression of MACC1, TWIST1, and TWIST2 in gastric cancer tissue specimens, compared to normal tissue samples [67]. Another study by Hung et al. in gastric cancer cell lines showed that posttranscriptional inhibition of MACC1 by miR-338-3p repressed the EMT. Secondly, they found that ZEB2, one of the crucial transcriptional regulators of the EMT switch, is a target of miR-338-3p. Either ectopic overexpression of miR-338-3p led to increased expression of the epithelial marker E-cadherin and decreased mesenchymal markers N-cadherin, fibronectin, and vimentin at both mRNA and protein level. In parallel, they observed inverse correlations between the expression of miR-338-3p and ZEB2 or MACC1 in gastric cancer patient tissue samples. Additionally, they showed a clear inhibition of the c-Met/AKT signaling axis, which is known to play a major role in EMT [20]. This evidence clearly shows that MACC1 is an EMT inducer in different cancer types.

2.2.3 MACC1 as Migration, Invasion, and Metastasis Inducer

An estimated 90% of cancer-related deaths are caused by the direct or indirect effects of metastatic dissemination [88]. Metastasis is a disease process that involves migration, local invasion, intravasation, systemic dissemination, extravasation, and establishment of secondary tumors at distant sites. Each of these steps is accomplished by an intricate interplay between tumor cells and their surrounding microenvironment and requires a complex molecular regulation at multiple levels [65]. For the first time, we have described MACC1 as migration, invasion, and metastasis inducer in CRC by mouse models [1]. SW480 and SW620 are colon cancer cell lines derived from the primary and from the lymph node metastasis tumor of the same patient, respectively. SW480 cells show less migratory and invasive phenotype compared to SW620, which is corroborating with the endogenous expression of MACC1. SW480 cells that overexpress MACC1 show a high level of cell migration, invasion, and colony formation. Similarly, wound healing assays revealed high motility and growth activity. In comparison, MACC1-siRNA-silenced SW620 cells lost significantly the ability of migration, invasion, and growth activity. Apart from knock-in and knockout studies, mutation of SH3 and proline-rich domains in MACC1 completely abrogated cell migration and invasion [1]. Similar to our findings, Zhen et al. found that siRNA-MACC1-silenced CRC cell lines showed a significant reduction of migration, invasion, colony formation, and tumor growth [66].

Apart from our initial report, other research groups also established MACC1 in different cancer models as an inducer of migration and invasion. To mention few, Meng et al. showed that MACC1 plays a major role in inducing the migration, invasion, and colony formation of nasopharyngeal carcinoma [7]. Similarly, in gallbladder cancer, ovarian carcinoma, cervical cancer, osteosarcoma, and gastric cancer cell lines, it was shown that either overexpression of MACC1 or siRNA-mediated silencing significantly mitigated cell migration, invasion, and colony formation [63, 68, 70, 89, 90]. Very recently, we identified SPON2 as a transcriptional gene of MACC1, prognostic indicator of CRC metastasis, and able to induce in vivo metastasis [91].

Intrasplenic transplantation of stable MACC1-expressing SW480 cells showed a higher number of metastatic lesions than control cells. In contrast to this, liver metastasis inducing SW620 cells showed 35% reduced metastasis when expressing short hairpin (sh)RNA that targets MACC1. Similar results were observed with cells that express shRNA against c-Met, the transcriptional target of MACC1, when compared with SW620 cells that express control shRNA. The role of MACC1 in inducing tumor cell metastasis was also shown in mice with another CRC cell line LS174T by overexpressing MACC1 [1]. In addition to this, we have performed real-time imaging to show MACC1 as a tumor progressive and metastasis inducer. We have prepared luciferase-IRES-MACC1 stably expressing SW480 and luciferase-IRES-shRNA-MACC1 stably expressing SW620 cells, along with their respective control vectors expressing cells. Intrasplenic transplantation of these cells and subsequent live imaging clearly demonstrated that MACC1-expressing SW480 cells showed induced tumor growth and liver metastasis, which is significantly reduced when using shRNA-MACC1-expressing SW620 cells [47]. Zhen et al. also showed that MACC1 significantly induced the tumor growth in CRC [66]. Similar to our findings, Wang et al. generated MACC1 and shRNA-MACC1 stably expressing gastric cancer cell line BGC823 and transplanted in the flank regions of mice legs. MACC1-overexpressing cells showed increased tumor growth and distant metastasis to the lungs and liver, whereas shRNA-MACC1-expressing cells showed significant reduction in tumor growth and distant metastasis (Fig. 2.3) [67, 68]. Very recently, we published the generation of the first transgenic MACC1 mouse models. We showed that MACC1 induces tumor progression in transgenis mice and CRC patients via increased pluripotency markers Nanog and Oct4 [92].

These pieces of evidence clearly demonstrate that MACC1 has a key role in migration, invasion, and distant metastasis.

2.2.4 MACC1 as Angiogenesis Inducer

The generation of tumor-associated vasculature is a process referred to as tumor angiogenesis, which is essential for tumor growth and progression. Tumor-associated vessels promote tumor progression by providing nutrients, oxygen, and favor tumor metastasis by facilitating tumor cell entry into the blood circulation. In this process vascular endothelial growth factor (VEGF), angiopoietin-1, angiopoietin-2, and their cognate receptors (VEGFR-1, VEGFR-2, Tie-2) play a major role in differentiation, proliferation, and migration of cells to form new vessels [93]. TWIST1 is known to induce EMT and tumor progression and is also a key regulator of VEGF and VEGFR1 [94, 95]. Very recent findings suggest that MACC1 plays a role in tumor angiogenesis. Wang et al. showed that HGF treatment increased the translocation of MACC1, TWIST1, and TWIST2 to the nucleus. Moreover, overexpression of MACC1 leads to induction of TWIST1 and TWIST2 expression and upregulates TWIST downstream signaling pathway molecules, such as vascular E-cadherin and VEGFR2, and downregulates E-cadherin in gastric cancer cells. Therefore, either through activation of HGF/c-Met signaling or independent overexpression of MACC1, vasculogenic mimicry (tube formation) was induced in 3D cultures of BGC/823 cells. They also found in vivo corroborative results with resected gastric cancer patient specimens that induced MACC1 expression positively correlated with increased vasculogenic mimicry density in patient tumors. Moreover, the 3-year survival rate was only 8.6% in patients who are double positive for MACC1 and vasculogenic mimicry, whereas it was 41.7% in negative patients. Nuclear expression of MACC1, TWIST1, and TWIST2 was higher in gastric tumor tissues when compared with the matched non-tumor tissues [67]. In another study by Sun et al., it was shown that MACC1 induces the formation of lymphatic vessels (lymphangiogenesis), a process believed to be similar to angiogenesis by upregulating VEGF-C and VEGF-D in gastric cells. Analogous results were observed with mouse xenografts [96]. These two pieces of evidence strongly suggest a role of MACC1 in angiogenesis. However, more studies have to be carried out with different cancer types to show its broader functional role in angiogenesis and tumor progression.

2.2.5 MACC1 Role in Metabolomics

Prognostic assessment of metastasis and intervention in metastatic disease still represent critical clinical problems. Given the fact that metastatic disease is characterized by and dependent on distinct metabolites, metabolic profiles of metastatic vs. localized solid tumors are identified in specimens of different tumor types.

Metastatic tumor cells need to process ECM, escape anoikis, and shift between epithelial and mesenchymal phenotype. Although these features have all been individually linked to metabolism, a systematic survey of the metabolic changes required for MACC1-mediated metastasis is still lacking. Metabolic profiling offers a powerful approach to generate a representation of altered metabolite profiles obtained from patient tissues or blood. Although a large body of publications report on the identification of metabolic profiles in localized vs. metastatic cancer in recent years, the link of these metabolic profiles to distinct mechanisms of metastasis is still poorly understood.

Here, we address the role of MACC1 in cancer metabolism. Several studies describe MACC1 in the context of glucose metabolism. In gastric cancer, Lin and colleagues discovered significantly upregulated MACC1 expression in response to glucose deprivation-induced metabolic stress via adenosine monophosphate-activated protein kinase (AMPK) signaling [71]. In patients, MACC1 expression positively correlated with the maximum standardized uptake value of 18F-deoxyglucose. Further, MACC1 increased 18F-deoxyglucose uptake in gastric cancer cells and xenografted mice. Mechanistically, MACC1 enhanced the Warburg effect by upregulating activity and expression of glycolytic enzymes, such as hexokinase, pyruvate dehydrogenase kinase, and lactate dehydrogenase. The authors also demonstrated functional consequences of this metabolic shift, such as enhanced cell viability and resistance to apoptosis. Furthermore, they showed enhanced vulnerability to metabolic stress of those cells silenced for MACC1 or blocked for the Warburg effect. Taken together, this study provides evidence for metabolic stress-induced MACC1 upregulation, thereby promoting gastric cancer growth against metabolic stress via the Warburg effect.

The link of MACC1 and its role in glucose metabolism were also reported by Ji and colleagues for hepatocellular cancer [97]. They demonstrated higher protein expression of both, MACC1 and 6-phosphofructo-2-kinase/fructose 2,6 bisphosphatase (PFKFB2) in hepatocellular carcinoma compared to corresponding non-tumor tissues. Simultaneous high expression of MACC1 and PFKFB2 was associated with a high Edmondson classification, advanced TNM stage, and, more importantly, with a lower overall survival rate. However, expression of MACC1, but not of PFKFB, served as a prognostic factor for postoperative survival.

Very recently, Li and colleagues also reported on MACC1 as a biomarker for poor prognosis of hepatocellular cancer via promoting proliferation through enhanced glucose metabolism [98]. The authors confirmed upregulated MACC1 protein expression in hepatocellular cancer cells and demonstrated an association with the hexokinase 2 (HK2) protein expression. Knockdown of MACC1 induced reduction of glycogen consumption and lactate production, resulting in reduced proliferation in the MHCC-97H cells. The authors conclude that MACC1 might promote proliferation in part via enhancement of glucose metabolism by HK2.

Besides glucose metabolism, MACC1 was also found to modulate lipogenesis [99]. Duan and colleagues demonstrated significant correlation of MACC1 levels with major enzymes of lipogenesis including fatty acid synthase (FASN)/ATP citrate lyase (ACLY) and acetyl-CoA carboxylase (ACC) in MACC1 genetically engineered in vitro models of gastric cancer, BGC/823, and MKN-28. In gastric cancer patients, high FASN expression predicted more advanced disease, more frequent postoperative recurrence, more metastases, higher mortality rate, and correlated with MACC1 expression. Thus, MACC1 promotes cancer cell lipogenesis by activation of lipogenic enzymes.

In summary, the decisive contribution of MACC1 to tumor initiation, progression, and metastasis also has to be contemplated in the light of MACC1-induced modulation of metabolic pathways, such as glucose and lipid metabolism.

2.3 MACC1 in Human Cancers

2.3.1 MACC1 as Cancer Inducer

Besides the role of MACC1 as metastasis-inducing gene, there is increasing evidence that MACC1 is also involved in early stages of carcinogenesis. Increased MACC1 expression levels are associated with steps which are crucial for the transition from adenoma to carcinoma. The first description of MACC1 expression as novel prognostic marker for metastasis in CRC showed already increased MACC1 expression in early stages of primary tumors and metastasis [1]. MACC1 expression was increased in all stages of adenocarcinomas, compared to normal colon mucosa, liver, or adenoma tissue samples. Also, in early stages, MACC1 was significantly increased in primary tumors which later on developed metachronous metastasis, compared to primary tumors without metastasis. In a further study it was a stepwise increase of MACC1 expression detectable in different neoplastic lesions [100]. From adenoma to high-grade dysplasia and intramucosal adenocarcinoma, the MACC1 expression was significantly increased. Further, the MACC1 expression was elevated during tumor progression and also to early-stage invasive carcinomas. The MACC1 and c-Met expression was following different expression patterns. c-Met expression was not significantly different among adenoma, high-grade dysplasia, intramucosal adenocarcinoma, and T1 tumors, but significantly increased in later invasive stages (T2) [100]. These findings suggest that MACC1 might contribute to CRC carcinogenesis in a c-Met independent manner. Further, MACC1 has high potential to be a prognostic marker for early diagnosis and poor prediction. Interestingly, when analyzing samples from Afro-American patients, increased MACC1 levels in tissues and blood identified colon adenoma patients at high risk [101]. MACC1 was shown to be higher expressed also in several other cancer entities [59]. Xie and colleagues also identified altered MACC1 expression levels in different stages of hepatocellular cancer (HCC) [102]. Interestingly, the inverse correlation between MACC1 expression and patient survival is significant in patients with early-stage HCC or normal AFP (alpha-fetoprotein) level. In contrast, analysis of all patients in all stages did not show any significant correlations of MACC1 and survival. MACC1 protein expression is a promising biomarker for early-stage diagnosis in CRC and other cancer entities like HCC.

Especially in the context of inflammatory bowel disease (IBD)-associated dysplasia was the MACC1 expression level significantly increased compared to corresponding inflammatory or normal colon tissue [103]. Further increase in MACC1 expression levels was detected from dysplasia to conventional colitis-associated cancer (CAC). MACC1 overexpression was detectable in 67% of conventional CAC but not in dysplasia, inflammation, or normal mucosa. MACC1 itself was not differently expressed in several differentiation grades of CAC. There was no correlation between MACC1 and c-Met expression detectable. MACC1 expression is stepwise increasing from IBD-associated colitis to dysplasia. These expression patterns point out the c-Met independent relevance of MACC1 during the conventional CAC carcinogenesis. MACC1 could be a potential prognostic marker for early detection of CAC and might allow for intervention strategies based on MACC1-specific inhibitors.

2.3.2 MACC1 as Prognostic Marker

Risk estimation by using genetic factors for early identification of high-risk patients is highly demanded but still limited in modern clinical diagnostics. Therefore, there is a need for biomarkers that allow for an individual diagnosis and prognosis of cancer patients. One candidate is the MACC1 gene, which we identified as novel prognostic biomarker for CRC. MACC1 expression levels were significantly increased in primary tumors compared to normal mucosa. Further, MACC1 was shown to be a prognostic marker for metachronous metastasis in CRC patients [1]. Apart from this, there are several studies confirming altered MACC1 expression levels in CRC to be prognostic for patient survival [57, 61, 66, 100, 104–110].

In 2012, we analyzed 179 UICC stage II CRC patient samples and correlated the expression of MACC1, sterile alpha motif (SAM) and SH3 domain containing 1 (SASH1) or osteopontin, the microsatellite stability (MSI) status, as well as KRAS and v-raf murine sarcoma viral oncogene homolog B1 (BRAF) mutation status with respect to patient survival [111]. The most common KRAS mutations of exon 2 in codon 12 (KRAS G12) or codon 13 (KRAS G13) were not separately compared to KRAS wild-type (wt) tumors with regard to patient survival. We confirmed increased expression of MACC1 ($P < 0.001$) as the only independent prognostic marker for recurrence prediction. Furthermore, integrative two-step cluster analysis allocated patients according to their tumor genetics. Beside the MSI status and BRAF mutation status, mutated KRAS and high MACC1 expression defined the group with the highest risk of recurrence.

Recently, we showed in a patient-based study of 99 CRC patients that high MACC1 expression (hazard ratio (HR): 5.02, 95% confidence interval (CI): 2.19–11.53, $P < 0.001$) and KRAS G13 mutation (HR: 4.48, 95% CI: 1.01–19.79, $P = 0.018$) were independent prognostic markers for shorter MFS [112]. Patients with high MACC1 expression and KRAS G13 mutation exhibited the worst prognosis (HR: 14.48, 95% CI: 3.37–62.18, $P < 0.001$). Based on their molecular characteristics, we categorized the patients into four clusters by using the SPSS two-step cluster function. These clusters reflect the three different described pathways for colorectal carcinogenesis: "traditional pathway," "serrated pathway," and "alternate pathway" [113–115]. The "traditional pathway," which reflects the adenoma carcinoma sequence [116, 117], can be further subdivided into low- and high-risk groups for metachronous metastases by assessing MACC1 expression.

We also performed a heterogeneity analysis of MACC1 in a retrospective study of whole tissue sections from 187 CRCs [118]. MACC1 expression was analyzed at different sites within the colorectal adenocarcinomas. MACC1 was significantly increased from the tumor center to the invasive front of the tumors and prognostic for metastasis and poor survival. In addition, high MACC1 expression was detectable in 55% of tumor budding cells. Independent of the distribution within the tumor, increased MACC1 expression also correlated with higher pT and pN stages and venous and lymphatic invasion. An additional study confirmed the correlation of increased MACC1 expression and the invasive behavior of CRC cells and further showed MACC1 as a marker for prediction of postoperative liver metastasis [105].

In the meantime, there are patient-based studies performed in nearly all solid cancer types showing the increased MACC1 expression levels in patient tumors and the prognostic role of MACC1 concerning patient survival. The detailed description about the performed studies were described by our group in the review "MACC1 as novel target for solid cancers" [59]. In addition to the already described cancer entities like colorectal, gastric, pancreatic, hepatocellular, lung, ovarian, breast, nasopharyngeal, kidney, esophageal cancer, and glioblastoma, MACC1 is also identified as prognostic marker for osteosarcoma [119], cervical cancer [120], gallbladder cancer [121] and tongue squamous cell carcinoma [72].

Apart from the previously well-described prognostic marker studies, we would like to point out the very recently published patient-based clinical study about the prognostic value of MACC1 in intrahepatic cholangiocarcinoma (ICC) and hilar cholangiocarcinoma (Klatskin tumors) [122]. The incidence rate of ICC is low, but besides the hepatocellular carcinomas the second-most common primary tumor of the liver. The treatment possibilities for these tumors are very limited, and the response to chemotherapy is very poor. So far, no promising targeted therapy could be established. In this patient-based cohort of Klatskin tumors ($n = 76$) and ICC ($n = 80$), we analyzed the mRNA expression levels of MACC1, c-Met, and HGF of tumors compared to normal liver tissue using qRT-PCR. The results of the mRNA expression were also validated by IHC. We identified significantly higher MACC1 expression levels in both tumor entities. Patients with high MACC1 expression in the Klatskin tumors had shorter overall survival and disease-free survival ($P < 0001$ and $P < 0.001$, respectively). Multivariate Cox regression analysis confirmed MACC1 as an independent prognostic marker for overall survival of patients with Klatskin tumors (HR: 2.777; 95% CI: 1.389–5.555; $P < 0.004$). MACC1 expression levels could be a promising candidate as a biomarker for these tumor entities, which are rare. Curative treatment possibilities for these patients are very limited.

Recently, there are several meta-analyses available confirming the prognostic value of MACC1 in solid tumors [123], in digestive system neoplasms [124] and in hepatocellular carcinomas [125].

The role of MACC1 as prognostic marker is not limited to expression level analysis. There are studies about single nucleotide polymorphism (SNP) in the intronic regions [126] or exons [127] of MACC1 in CRC. Lang et al. performed a study with CRC patients for six SNPs in the intronic regions of MACC1. MACC1 tagging SNP rs1990172 positively correlated with patient survival. Our group identified three SNPs in the coding exons of 154 CRC patients [127]. Patients younger than 60 years with UICC stage I or II tumors harboring the SNP rs975263 were characterized by increased risk for metachronous metastasis. Apart from CRC, MACC1 SNPs were also analyzed in 187 hepatocellular carcinoma patients treated with liver transplantation concerning risk of recurrence. The SNPs rs1990172 and rs975263 showed significantly higher risk for relapse after liver transplantation [128]. Further, the SNPs rs1990172 and rs975263 were shown to significantly correlate with increased risk for progression or death in a retrospective study, including 164 HER-2-positive breast cancer patients [129]. The third analyzed SNP rs3735615 showed a significantly better event-free survival and overall survival. In silico analysis revealed that this protective impact could be due to negative functional alteration of the MACC1 protein.

In addition to SNPs, also the impact of MACC1 gene amplification was studied in 103 mCRCs patients [61]. MACC1 expression levels and increased copy numbers correlated with unfavorable tumor characteristics, like lesion size, multiple metastases, and detection of intravascular metastatic cells.

All these previously described analyses are based on extracted RNA and DNA of patient biopsies. Therefore, we tried to establish a more useful detection method for clinical diagnostic and monitoring overcoming the limitations of "snapshot" analyses by using tumor tissues with limited availability. We screened CRC patient plasma samples to detect circulating mRNA transcripts in CRC patient blood [130]. High MACC1 transcript levels in the patient samples correlated with poorer patient survival. The highest MACC1 levels were measured in patients with metastases and reflect the same pattern as was shown in patient biopsies. This method could allow for monitoring the disease by taking blood samples without the need of taking biopsies. Therefore, it is not surprising that two other studies also used the detection of circulating MACC1 transcripts to demonstrate the prognostic relevance of MACC1 in small cell lung cancer [131] and gastric cancer [132]. These studies confirmed the previously described reports about the increased MACC1 expression and correlation of MACC1 expression levels and survival in small cell lung cancer and gastric cancer [68, 133, 134]. Wang and colleagues analyzed the MACC1 mRNA expression in plasma samples of 272 patients with NSCLC, 61 patients with benign lung disease, and 80 healthy humans by qRT-PCR [131]. There is a significant increase in a stepwise manner of the MACC1 mRNA levels from healthy volunteers over benign lung disease ($P < 0.001$) to NSCLC ($P < 0.001$) detectable. High MACC1 expression is an independent prognostic factor for poor overall survival and disease-free survival. For gastric cancer, several retrospective studies were performed, examining increased MACC1 expression in tumor tissue compared to normal tissue. High MACC1 expression significantly correlated with peritoneal dissemination [133], tumor progression, recurrence, metastasis, and mortality [68]. Interestingly, our group published recently the first blood-based prospective study about the MACC1 mRNA expression in plasma samples for diagnostic and prognostic purpose in gastric cancer patients [132]. Therefore, we analyzed 76 samples of gastric cancer patients and 54 samples from 54 healthy volunteers for the MACC1 mRNA expression. In all gastric cancer patients, the MACC1 expression was increased compared to the tumor-free control samples. High MACC1 transcript levels in the plasma of gastric cancer patients correlated with poorer survival compared to gastric cancer patients with low MACC1 transcripts.

This noninvasive blood-based method allows for diagnostic and prognostic analyses of patient plasma samples in clinical routine. Thus, even the success of curative therapy might be monitored over longer time with this cheap and quick method. Therefore, it is interesting to analyze the relevance of MACC1 not only as a promising prognostic marker but also as a predictive marker for therapy response.

2.3.3 MACC1 as Predictive Marker

The usefulness of MACC1 as diagnostic and prognostic biomarker has meanwhile been demonstrated in a large body of reports for many solid cancer types. However, there are also some reports published, demonstrating that MACC1 is of predictive value for therapy response. So far, reports on rectal, hepatocellular, and pancreatic cancer, on glioblastoma and tongue squamous cell carcinoma, are available linking MACC1 to chemoresistance.

Kawamura and colleagues analyzed the expression of MACC1 and of its transcriptional target c-Met in rectal cancer treated with neoadjuvant chemoradiotherapy (CRT) followed by surgery [57]. The CRT regimen included four cycles of 5-fluorouracil (5-FU; 600 mg/m^2 for 24 h by continuous intravenous route), and tegafur-uracil (UFT; 400 mg/m^2 orally for 5 days), concurrent with 20–45 Gy radiation. First, the authors observed a positive correlation of MACC1 and c-Met by qRT-PCR as well as by immunohistochemistry in the patient samples. Furthermore, high expression of both the genes MACC1 or c-Met was associated with reduced relapse-free survival of the patients. The authors conclude that expression of both genes may be useful for predicting survival of patients with rectal cancer treated with preoperative CRT followed by surgery.

For advanced HCC, Yang and colleagues evaluated MACC1 for outcome prediction after cryoablation therapy, which is a local therapy for HCC [135]. MACC1 mRNA and nuclear protein expression was significantly increased in tumors of these patients compared to normal liver tissue controls. Higher expression of MACC1 mRNA and nuclear protein in tumors correlated with shorter post cryoablation median time-to-progression and overall survival, compared to patients with low MACC1 expression. Thus, the authors summarize that in addition to the already described association of MACC1 with a poor prognosis of HCC, a higher intratumoral expression of MACC1 or nuclear translocation is predictive of poor outcomes of cryotherapy in these patients.

Besides these evaluations of the predictive value of MACC1 in the context of multimodal treatment regimens, several groups tested the impact of MACC1 to predict therapy response for defined chemotherapies. Wang and colleagues analyzed MACC1 protein expression in the serum of patients with pancreatic cancer by using a costumer-made MACC1-ELISA [87]. High MACC1 expression correlated with the formation of lymph node metastasis and distant metastasis, as well as a later TNM stage. Furthermore, they tested pancreatic cancer cell lines for chemosensitivity against gemcitabine, a cytotoxic nucleoside analogue, which is the current therapeutic strategy for advanced pancreatic cancer. They observed in CFPAC-1 pancreatic cancer cells that downregulation of MACC1 by RNA interference (RNAi) resulted in sensitization toward gemcitabine treatment through the inhibition of the RAS/ERK signaling pathway. They conclude that MACC1 is not only associated with metastasis development but also with chemoresistance to gemcitabine.

Duan and colleagues investigated the impact of MACC1 in the context of Herceptin resistance in gastric cancer cells [99]. They used the gastric cancer cell lines BGC-823 and MKN-28, with ectopic MACC1 overexpression or RNAi-mediated downregulation of MACC1. They demonstrated that cells with MACC1 overexpression were more likely resistant to Herceptin. High MACC1 expression promoted the axis of FASN/ACC/ACLY expression and FASN activity ($P < 0.05$). In gastric cancer patients, stages I–IV, high FASN expression predicted more advanced disease, more frequent postoperative recurrence, more metastases, a higher mortality rate, and correlated with MACC1 expression. The proliferation effect of MACC1 acting on GC was attenuated by FASN blockade with different schedules. Thus, the authors linked MACC1 to Herceptin resistance via activation of the lipogenic enzymes in gastric cancer.

Two further studies evaluated MACC1 in the context of chemoresistance to cisplatin. Shang and colleagues determined the influence of MACC1 expression on cisplatin sensitivity in human U251 glioblastoma cells [136]. Knockdown of MACC1 by siRNA did not only reduce MACC1 expression but was also associated with an increase in apoptosis rate and an elevation of the growth inhibitory rate in these glioblastoma cells, U251 cells. After treatment with cisplatin, the apoptosis rate as well as the growth inhibitory rate of MACC1-silenced U251 cells increased. The authors summarize the link of MACC1 expression and sensitivity of glioblastoma cells to cisplatin chemotherapy.

Another report evaluating MACC1 for chemoresistance toward cisplatin was published by Li and colleagues for tongue squamous cell carcinoma [137]. The authors found a positive correlation of MACC1 with lymphatic metastasis, extracellular matrix metalloproteinase inducer (EMMPRIN) expression, and poor overall patient survival. By downregulation of MACC1 in a tongue squamous cell carcinoma cell line TSCCA, cisplatin resistance was attenuated, apoptosis was increased, migration and invasion were decreased, and secretion of urokinase-type plasminogen activator system (uPA) in the supernatant of the culture medium and uPA expression was diminished.

Taken together, these first reports may hint to the use of MACC1 as a predictive marker for therapy response in several solid cancer types. However, the employment of MACC1 for prediction of response to entire treatment regimens or to defined drugs requires more experimental and clinical studies to be exploited routinely in the clinic.

2.3.4 MACC1 as Druggable Target

MACC1 is known to regulate survival, proliferation, apoptosis, colony formation, migration, invasion, EMT, and distant metastasis either by transcriptional and post-transcriptional regulation or by means of other carcinogenic signaling axis molecules. Moreover, it is shown to be overexpressed in a multitude of cancers screened until today and also its induced gene expression correlated with poor patient prognosis [59]. MACC1 was discovered as an important transcriptional regulator of c-Met, which is known to induce cancer and metastasis. In addition, MACC1-based knock-in and knockdown studies showed that it mitigates the important carcinogenic signaling axes, like PI3K/AKT, MAPK, and β-catenin [1, 7, 66]. These pieces of evidence suggest that MACC1 might be a potential druggable target.

To support this hypothesis, MACC1 knockdown studies clearly demonstrated that downregulation of c-Met, β-catenin, ECM remodeling genes like fibronectin, MMP2, MMP9, TWIST1, TWIST2, and mesenchymal phenotype markers vimentin, CD44, inhibition of p-AKT, p-ERK1/2 activation, and mitigating the metabolic stress, which are crucial pathways involved in carcinogenesis and metastasis [20, 67, 68, 87].

Secondly, posttranscriptional downregulation of MACC1 by miRNAs (miR-143, miR-574, miR-200a, miR-218 and miR-338-3p) significantly inhibited tumorigenesis, tumor growth, and distant metastasis in different cancer types [16, 19–21, 138].

MACC1-overexpressing gastric cancer cell lines showed drug resistance to Herceptin [99]. A siRNA-MACC1-silenced glioblastoma cell line showed an increase in apoptosis. After treatment with cisplatin at various concentrations, the percentage of apoptosis significantly increased, and the rate of growth was inhibited in a dose-dependent manner [136]. Similarly, MACC1-silenced pancreatic cancer cells were more sensitive to gemcitabine [87]. Higher MACC1 and c-Met-expressing rectal cancer patients showed significantly worse relapse-free survival after neoadjuvant chemoradiotherapy [57]. The effectiveness of chemotherapy is limited by drug resistance due to the differential expression of molecules in different patients. With respect to MACC1, very little is known about its role in different treatment methods and drug response. However, with all the above-mentioned evidence of MACC1's role in tumorigenesis, metastasis, and drug resistance, it is mandatory to develop new MACC1-based targeting therapeutic approaches. To name a few:

1. Specific MACC1 transcriptional inhibitors
2. shRNA-MACC1-/miRNAs-based therapeutic approaches using miRNA mimics
3. Therapeutic targeting of MACC1 domains (e.g., SH3 and proline-rich domain) using specific peptides
4. Combinatorial therapeutic approaches involving other cancer-specific markers to fight MACC1-induced cancer efficiently and effectively for better treatment outcomes

2.4 Outlook

The ultimate goal in translational oncology today is the development of personalized tailored interventions for inhibition or even prevention of cancer progression and metastasis resulting in improved patient survival. Therefore, a better understanding of the biological processes of tumor initiation, progression, and metastasis on improved prognosis and prediction, is desperately needed. However, the currently used molecular markers are not sufficient for an early identification of patients at high risk for tumor progression and metastasis.

This chapter summarizes the current knowledge on the prognostic and predictive importance for tumorigenesis as well as tumor progression including metastasis formation of MACC1, a newly discovered gene by our group in 2009. We were the first to identify MACC1 and its role as a tumor stage-independent predictor for CRC metastasis linked to metastasis-free survival when determined in the primary tumor or in patient blood. Following our initial publication, evidence was provided by more than 140 published reports generated by scientists around the globe. They confirmed our data on CRC and further solid cancer entities and concluded that MACC1 is a remarkable biomarker for disease prognosis and treatment response prediction for a variety of solid cancers.

Furthermore, the capabilities and involvements of MACC1 in cancer biology are highlighted comprehensively in this chapter. The impact of MACC1 is discussed in detail for signaling cascades known to be activated in cancer, which are regulating MACC1 or which are regulated by this molecule. Moreover, the contribution of MACC1 for biological processes such as cell proliferation and apoptosis, cell motility including migration, EMT, and invasion, as well as for angiogenesis is underlined. This knowledge, accumulated within the last 7 years, underlines the decisive role of this molecule in fundamental processes leading to cancer when deregulated.

Taken together, MACC1 represents a promising therapeutic target for prevention and/or restriction of tumor progression and metastasis, as well as for tumor sensitization toward different treatment regimens eventually resulting in better response rates of a variety of tumor types. So far, no MACC1-specific inhibitors targeting MACC1 on the transcriptional, the translational, or the posttranslational levels or designed to intervene in protein-protein interactions of MACC1 with potential binding partners are published. However, upon availability of specific inhibitors targeting MACC1, their beneficial exploitation have to be demonstrated in clinical trials, to reach the ultimate goal of personalized tailored interventions—improved patient survival.

Acknowledgments This work was supported by the German Cancer Consortium (DKTK) (to G.M. and U.S.).

References

1. Stein U, Walther W, et al. MACC1, a newly identified key regulator of HGF-MET signaling, predicts colon cancer metastasis. Nat Med. 2009;15(1):59–67.
2. Juneja M, et al. Promoter identification and transcriptional regulation of the metastasis gene MACC1 in colorectal cancer. Mol Oncol. 2013;7(5):929–43.
3. Kuo IY, et al. Low SOX17 expression is a prognostic factor and drives transcriptional dysregulation and esophageal cancer progression. Int J Cancer. 2014;135(3):563–73.
4. Zhang R, Ren F, Shi H. Expression of metastasis-associated in colon cancer-1 in different stages of epithelial ovarian cancer. Zhongguo Yi Xue Ke Xue Yuan Xue Bao. 2014;36(1):47–51.
5. Zhuang H, et al. Aberrant expression of pim-3 promotes proliferation and migration of ovarian cancer cells. Asian Pac J Cancer Prev. 2015;16(8):3325–31.
6. Li Y-Y, Mukaida N. Pathophysiological roles of Pim-3 kinase in pancreatic cancer development and progression. World J Gastroenterol. 2014;20(28):9392–404.
7. Meng F, et al. MACC1 down-regulation inhibits proliferation and tumourigenicity of nasopharyngeal carcinoma cells through Akt/β-catenin signaling pathway. PLoS One. 2013;8(4):e60821.
8. Morris MA, et al. Role of the Epstein-Barr virus-encoded latent membrane protein-1, LMP1, in the pathogenesis of nasopharyngeal carcinoma. Future Oncol. 2009;5(6):811–25.
9. Zhang Y, Wang Z, Chen M, Peng L, Wang X, Ma Q, Ma F, Jiang B. MicroRNA-143 targets MACC1 to inhibit cell invasion and migration in colorectal cancer. Mol Cancer. 2012;11:23. doi:10.1186/1476-4598-11-23.
10. Chen X, et al. Role of miR-143 targeting KRAS in colorectal tumorigenesis. Oncogene. 2009;28(10):1385–92.

11. Akao Y, et al. Role of microRNA-143 in Fas-mediated apoptosis in human T-cell leukemia Jurkat cells. Leuk Res. 2009;33(11):1530–8.
12. Ng EKO, et al. MicroRNA-143 targets DNA methyltransferases 3A in colorectal cancer. Br J Cancer. 2009;101(4):699–706.
13. Cordes KR, et al. miR-145 and miR-143 regulate smooth muscle cell fate and plasticity. Nature. 2009;460(7256):705–10.
14. Wang L, et al. MiR-143 acts as a tumor suppressor by targeting N-RAS and enhances temozolomide-induced apoptosis in glioma. Oncotarget. 2014;5(14):5416–27.
15. Zhang H, et al. microRNA-143, down-regulated in osteosarcoma, promotes apoptosis and suppresses tumorigenicity by targeting Bcl-2. Oncol Rep. 2010;24(5):1363–9.
16. Cui Z, et al. Hsa-miR-574-5p negatively regulates MACC-1 expression to suppress colorectal cancer liver metastasis. Cancer Cell Int. 2014;14(1):47.
17. Kalabus JL, Cheng Q, Blanco JG. MicroRNAs differentially regulate carbonyl reductase 1 (CBR1) gene expression dependent on the allele status of the common polymorphic variant rs9024. PloS One. 2012;7(11):e48622.
18. Ji S, et al. miR-574-5p negatively regulates Qki6/7 to impact β-catenin/Wnt signalling and the development of colorectal cancer. Gut. 2013;62(5):716–26.
19. Feng J, et al. miR-200a suppresses cell growth and migration by targeting MACC1 and predicts prognosis in hepatocellular carcinoma. Oncol Rep. 2014;33(2):713–20.
20. Huang N, et al. MiR-338-3p inhibits epithelial-mesenchymal transition in gastric cancer cells by targeting ZEB2 and MACC1/Met/Akt signaling. Oncotarget. 2015;6(17):15222–34.
21. Ilm K, et al. MACC1 is post-transcriptionally regulated by miR-218 in colorectal cancer. Oncotarget. 2016;7(33):53443–58.
22. Ceppi P, et al. Loss of miR-200c expression induces an aggressive, invasive, and chemoresistant phenotype in non-small cell lung cancer. Mol Cancer Res. 2010;8(9):1207–16.
23. Park S-M, et al. The miR-200 family determines the epithelial phenotype of cancer cells by targeting the E-cadherin repressors ZEB1 and ZEB2. Genes Dev. 2008;22(7):894–907.
24. Dinkel H, et al. The eukaryotic linear motif resource ELM: 10 years and counting. Nucleic Acids Res. 2014;42(D1):D259–66.
25. Buchan DWA, et al. Scalable web services for the PSIPRED Protein Analysis Workbench. Nucleic Acids Res. 2013;41(Web Server issue):W349–57.
26. Yachdav G, et al. PredictProtein-an open resource for online prediction of protein structural and functional features. Nucleic Acids Res. 2014;42(Web Server issue):W337–43.
27. Kokoszyńska K, et al. Unexpected domain composition of MACC1 links MET signaling and apoptosis. Acta Biochim Pol. 2009;56(2):317–24.
28. Stein U, Dahlmann M, Walther W. MACC1—more than metastasis? Facts and predictions about a novel gene. J Mol Med. 2010;88(1):11–8.
29. Dell'Angelica EC, et al. Association of the AP-3 adaptor complex with clathrin. Science (New York, NY). 1998;280(5362):431–4.
30. Braun A, et al. EHD proteins associate with syndapin I and II and such interactions play a crucial role in endosomal recycling. Mol Biol Cell. 2005;16(8):3642–58.
31. Kieken F, et al. Structural insight into the interaction of proteins containing NPF, DPF, and GPF motifs with the C-terminal EH-domain of EHD1. Protein Sci. 2009;18(12):2471–9.
32. Kirchhausen T, Owen D, Harrison SC. Molecular structure, function, and dynamics of clathrin-mediated membrane traffic. Cold Spring Harb Perspect Biol. 2014;6(5):a016725.
33. Grant BD, Caplan S. Mechanisms of EHD/RME-1 protein function in endocytic transport. Traffic. 2008;9(12):2043–52.
34. Miliaras NB, Wendland B. EH proteins: multivalent regulators of endocytosis (and other pathways). Cell Biochem Biophys. 2004;41(2):295–318.
35. Naslavsky N, Caplan S. EHD proteins: key conductors of endocytic transport. Trends Cell Biol. 2011;21(2):122–31.
36. Liao Y, et al. An ancient autoproteolytic domain found in GAIN, ZU5 and nucleoporin98. J Mol Biol. 2014;426(24):3935–45.
37. D'Osualdo A, et al. CARD8 and NLRP1 undergo autoproteolytic processing through a ZU5-like domain. PLoS One. 2011;6(11):e27396.

38. Bock FJ, et al. P53-induced protein with a death domain (PIDD): master of puppets? Oncogene. 2012;31(45):4733–9.
39. Heinz LX, et al. The death domain-containing protein Unc5CL is a novel MyD88-independent activator of the pro-inflammatory IRAK signaling cascade. Cell Death Differ. 2012;19(4):722–31.
40. Huo L, et al. Cdc42-dependent formation of the ZO-1/MRCKβ complex at the leading edge controls cell migration. EMBO J. 2011;30(4):665–78.
41. Wang C, et al. Structure of the ZU5-ZU5-UPA-DD tandem of ankyrin-B reveals interaction surfaces necessary for ankyrin function. Proc Natl Acad Sci. 2012;109(13):4822–7.
42. Yasunaga M, Ipsaro JJ, Mondragón A. Structurally similar but functionally diverse ZU5 domains in human erythrocyte ankyrin. J Mol Biol. 2012;417(4):336–50.
43. Wang R, et al. Autoinhibition of UNC5b revealed by the cytoplasmic domain structure of the receptor. Mol Cell. 2009;33(6):692–703.
44. Kaneko T, Li L, Li SS-C. The SH3 domain—a family of versatile peptide- and protein-recognition module. Front Biosci. 2008;13:4938–52.
45. Ebsen H, et al. Identification of SH3 domain proteins interacting with the cytoplasmic tail of the a Disintegrin and Metalloprotease 10 (ADAM10). PLoS One. 2014;9(7):e102899.
46. Voss M, Lettau M, Janssen O. Identification of SH3 domain interaction partners of human FasL (CD178) by phage display screening. BMC Immunol. 2009;10:53.
47. Pichorner A, et al. In vivo imaging of colorectal cancer growth and metastasis by targeting MACC1 with shRNA in xenografted mice. Clin Exp Metastasis. 2012;29(6):573–83.
48. Gherardi E, et al. Targeting MET in cancer: rationale and progress. Nat Rev Cancer. 2012;12(2):89–103.
49. Trusolino L, Bertotti A, Comoglio PM. MET signalling: principles and functions in development, organ regeneration and cancer. Nat Rev. Mol Cell Biol. 2010;11(12):834–48.
50. Epstein JA, et al. Pax3 modulates expression of the c-Met receptor during limb muscle development. Proc Natl Acad Sci U S A. 1996;93(9):4213–8.
51. Gambarotta G, et al. Ets up-regulates MET transcription. Oncogene. 1996;13(9):1911–7.
52. Morozov VM, et al. Regulation of c-met expression by transcription repressor Daxx. Oncogene. 2008;27(15):2177–86.
53. Pennacchietti S, et al. Hypoxia promotes invasive growth by transcriptional activation of the met protooncogene. Cancer Cell. 2003;3(4):347–61.
54. Seol DW, et al. Regulation of the c-met proto-oncogene promoter by p53. J Biol Chem. 1999;274(6):3565–72.
55. Stein U, Smith J, et al. MACC1 controls met: what a difference an Sp1 site makes. Cell Cycle. 2009;8(15):2467–9.
56. Zhang X, et al. Both Sp1 and Smad participate in mediating TGF-beta1-induced HGF receptor expression in renal epithelial cells. Am J Physiol Ren Physiol. 2005;288(1):F16–26.
57. Kawamura M, et al. Correlation of MACC1 and MET expression in rectal cancer after neoadjuvant chemoradiotherapy. Anticancer Res. 2012;32(4):1527–31.
58. Qiu J, et al. Identification of MACC1 as a novel prognostic marker in hepatocellular carcinoma. J Transl Med. 2011;9(1):166.
59. Stein U. MACC1—a novel target for solid cancers. Expert Opin Ther Targets. 2013;17(9):1039–52.
60. Arlt F, Stein U. Colon cancer metastasis: MACC1 and Met as metastatic pacemakers. Int J Biochem Cell Biol. 2009;41(12):2356–9.
61. Galimi F, et al. Genetic and expression analysis of MET, MACC1, and HGF in metastatic colorectal cancer: response to Met inhibition in patient xenografts and pathologic correlations. Clin Cancer Res. 2011;17(10):3146–56.
62. Li H, et al. Overexpression of MACC1 and the association with hepatocyte growth factor/c-Met in epithelial ovarian cancer. Oncol Lett. 2015;9(5):1989–96.
63. Sheng XJ, et al. MACC1 induces metastasis in ovarian carcinoma by upregulating hepatocyte growth factor receptor c-MET. Oncol Lett. 2014;8(2):891–7.
64. Shao Y, et al. Effect of A disintegrin and metalloproteinase 10 gene silencing on the proliferation, invasion and migration of the human tongue squamous cell carcinoma cell line TCA8113. Mol Med Rep. 2015;11(1):212–8.

65. Hanahan D, Weinberg RA. Hallmarks of cancer: the next generation. Cell. 2011;144(5):646–74.
66. Zhen T, et al. MACC1 promotes carcinogenesis of colorectal cancer via β-catenin signaling pathway. Oncotarget. 2014;5(11):3756–69.
67. Wang L, et al. Metastasis-associated in colon cancer-1 promotes vasculogenic mimicry in gastric cancer by upregulating TWIST1/2. Oncotarget. 2015;6(13):11492–506.
68. Wang L, et al. Metastasis-associated in colon cancer-1 upregulation predicts a poor prognosis of gastric cancer, and promotes tumor cell proliferation and invasion. Int J Cancer. 2013;133(6):1419–30.
69. Yao Y, et al. MACC1 suppresses cell apoptosis in hepatocellular carcinoma by targeting the HGF/c-MET/AKT pathway. Cell Physiol Biochem. 2015;710061:983–96.
70. Zhang K, et al. MACC1 is involved in the regulation of proliferation, colony formation, invasion ability, cell cycle distribution, apoptosis and tumorigenicity by altering Akt signaling pathway in human osteosarcoma. Tumour Biol. 2014;35(3):2537–48.
71. Lin L, et al. MACC1 supports human gastric cancer growth under metabolic stress by enhancing the Warburg effect. Oncogene. 2014;34(21):2700–10.
72. Li H-F, et al. Downregulation of MACC1 inhibits invasion, migration and proliferation, attenuates cisplatin resistance and induces apoptosis in tongue squamous cell carcinoma. Oncol Rep. 2015;33(2):651–60.
73. Hagemann C, et al. Impact of MACC1 on human malignant glioma progression and patients' unfavorable prognosis. Neuro Oncol. 2013;15(12):1696–709.
74. Sun L, et al. Silence of MACC1 expression by RNA interference inhibits proliferation, invasion and metastasis, and promotes apoptosis in U251 human malignant glioma cells. Mol Med Rep. 2015;12(3):3423–31.
75. Hua F, et al. Effects of small interfering RNA silencing MACC-1 expression on cell proliferation, cell cycle and invasion ability of cervical cancer SiHa cells. Zhonghua zhong liu za zhi [Chin J Oncol]. 2014;36(7):496–500.
76. Kalluri R. EMT: When epithelial cells decide to become mesenchymal-like cells. J Clin Investig. 2009;119(6):1417–9.
77. Kalluri R, Weinberg RA. The basics of epithelial-mesenchymal transition. J Clin Investig. 2009;119(6):1420–8.
78. Tsuji T, Ibaragi S, Hu GF. Epithelial-mesenchymal transition and cell cooperativity in metastasis. Cancer Res. 2009;69(18):7135–9.
79. Heimann R, et al. Separating favorable from unfavorable prognostic markers in breast cancer: the role of E-cadherin. Cancer Res. 2000;60(2):298–304.
80. Wheelock MJ, et al. Cadherin switching. J Cell Sci. 2008;121(Pt 6):727–35.
81. Bolós V, et al. The transcription factor Slug represses E-cadherin expression and induces epithelial to mesenchymal transitions: a comparison with Snail and E47 repressors. J Cell Sci. 2003;116(Pt 3):499–511.
82. Conacci-Sorrell M, et al. Autoregulation of E-cadherin expression by cadherin-cadherin interactions: the roles of β-catenin signaling, Slug, and MAPK. J Cell Biol. 2003;163(4):847–57.
83. Kang Y, Massagué J. Epithelial-mesenchymal transitions: twist in development and metastasis. Cell. 2004;118(3):277–9.
84. Karreth F, Tuveson DA. Twist induces an epithelial-mesenchymal transition to facilitate tumor metastasis. Cancer Biol Ther. 2004;3(11):1058–9.
85. Lee TK, et al. Twist overexpression correlates with hepatocellular carcinoma metastasis through induction of epithelial-mesenchymal transition. Clin Cancer Res. 2006;12(18):5369–76.
86. Cano A, et al. The transcription factor snail controls epithelial-mesenchymal transitions by repressing E-cadherin expression. Nat Cell Biol. 2000;2(2):76–83.
87. Wang G, et al. MACC1: a potential molecule associated with pancreatic cancer metastasis and chemoresistance. Oncol Lett. 2012;4(4):783–91.
88. Sporn MB. The war on cancer. Lancet. 1996;347(9012):1377–81.
89. Chai H, Yang Y. Effects of MACC1 siRNA on biological behaviors of HeLa. Arch Gynecol Obstet. 2014;289(6):1271–80.

90. Wang Y, et al. Downregulated expression of metastasis associated in colon cancer 1 (MACC1) reduces gallbladder cancer cell proliferation and invasion. Tumour Biol. 2014;35(4):3771–8.
91. Schmid F, Wang Q, Huska MR, Andrade-Navarro MA, Lemm M, Fichtner I, Dahlmann M, Kobelt D, Walther W, Smith J, Schlag PM, Stein U. SPON2, a newly identified target gene of MACC1, drives colorectal cancer metastasis in mice and is prognostic for colorectal cancer patient survival. Oncogene. 2016;35(46):5942–52.
92. Lemos C, Hardt M, Juneja M, Voss C, Förster S, Jerchow B, Haider W, Bläker H, Stein U. MACC1 induces tumor progression in transgenic mice and colorectal cancer patients via increased pluripotency markers Nanog and Oct4. Clin Cancer Res. 2016;22:2812–23.
93. Fabris L, et al. Epithelial expression of angiogenic growth factors modulate arterial vasculogenesis in human liver development. Hepatology. 2008;47(2):719–28.
94. Singh S, et al. The role of TWIST in angiogenesis and cell migration in giant cell tumor of bone. Adv Biol. 2014;2014:1–8.
95. Yang J, et al. Twist, a master regulator of morphogenesis, plays an essential role in tumor metastasis. Cell. 2004;117(7):927–39.
96. Sun L, et al. Metastasis-associated in colon cancer-1 upregulates vascular endothelial growth factor-C/D to promote lymphangiogenesis in human gastric cancer. Cancer Lett. 2015;357(1):242–53.
97. Ji D, et al. MACC1 expression correlates with PFKFB2 and survival in hepatocellular carcinoma. Asian Pac J Cancer Prev. 2014;15(2):999–1003.
98. Li Y, et al. Metastasis-associated in colon cancer-1 is associated with poor prognosis in hepatocellular carcinoma, partly by promoting proliferation through enhanced glucose metabolism. Mol Med Rep. 2015;12(1):426–34.
99. Duan J, et al. Participation of metastasis-associated in colon cancer-1 gene on lipogenesis and chemoresistance of gastric cancer. J Clin Oncol. 2014;32(15 suppl):e15026.
100. Ren B, et al. MACC1 is related to colorectal cancer initiation and early-stage invasive growth. Am J Clin Pathol. 2013;140(5):701–7.
101. Ashktorab H, Hermann P, Nouraie M, Shokrani B, Lee E, Haidary T, Brim H, Stein U. Increased MACC1 levels in tissues and blood identify colon adenoma patients at high risk. J Transl Med. 2016;14:215.
102. Xie C, et al. MACC1 as a prognostic biomarker for early-stage and AFP-normal hepatocellular carcinoma. PloS One. 2013;8(5):e64235.
103. Harpaz N, et al. Expression of MACC1 and MET in inflammatory bowel disease-associated colonic neoplasia. Inflamm Bowel Dis. 2014;20(4):703–11.
104. Boardman LA. Overexpression of MACC1 leads to downstream activation of HGF/MET and potentiates metastasis and recurrence of colorectal cancer. Genome Med. 2009;1(4):36.
105. Ge Y, et al. Positive MACC1 expression correlates with invasive behaviors and postoperative liver metastasis in colon cancer. Int J Clin Exp Med. 2015;8(1):1094–100.
106. Isella C, et al. MACC1 mRNA levels predict cancer recurrence after resection of colorectal cancer liver metastases. Ann Surg. 2013;257(6):1089–95.
107. Shirahata A, Shinmura K, et al. MACC1 as a marker for advanced colorectal carcinoma. Anticancer Res. 2010;30(7):2689–92.
108. Sueta A, et al. Differential role of MACC1 expression and its regulation of the HGF/c-Met pathway between breast and colorectal cancer. Int J Oncol. 2015;46(5):2143–53.
109. Yamamoto H, et al. MACC1 expression levels as a novel prognostic marker for colorectal cancer. Oncol Lett. 2014;8(5):2305–9.
110. Zlobec I. Novel biomarkers for the prediction of metastasis in colorectal cancer. Exp Opin Med Diagn. 2013;7(2):137–46.
111. Nitsche U, et al. Integrative marker analysis allows risk assessment for metastasis in stage II colon cancer. Ann Surg. 2012;256(5):763–71.
112. Ilm K, et al. High MACC1 expression in combination with mutated KRAS G13 indicates poor survival of colorectal cancer patients. Mol Cancer. 2015;14(1):1–7.
113. Leggett B, Whitehall V. Role of the serrated pathway in colorectal cancer pathogenesis. Gastroenterology. 2010;138(6):2088–100.

114. Markowitz SD, Bertagnolli MM. Molecular origins of cancer: molecular basis of colorectal cancer. N Engl J Med. 2009;361(25):2449–60.
115. Worthley D-L, et al. Colorectal carcinogenesis: road maps to cancer. World J Gastroenterol. 2007;13(28):3784–91.
116. Fearon ER, Vogelstein B. A genetic model for colorectal tumorigenesis. Cell. 1990;61(5):759–67.
117. Vogelstein B, et al. Genetic alterations during colorectal-tumor development. N Engl J Med. 1988;319(9):525–32.
118. Koelzer VH, et al. Heterogeneity analysis of Metastasis Associated in Colon Cancer 1 (MACC1) for survival prognosis of colorectal cancer patients: a retrospective cohort study. BMC Cancer. 2015;15(1):1–11.
119. Zhang K, et al. High expression of MACC1 predicts poor prognosis in patients with osteosarcoma. Tumour Biol. 2013;35(2):1343–50.
120. Guo L, et al. Metastasis-associated colon cancer-1 is a novel prognostic marker for cervical cancer. Int J Clin Exp Pathol. 2014;7(7):4150–5.
121. Chen L, et al. Prognostic significance of metastasis associated in colon cancer 1 (MACC1) expression in patients with gallbladder cancer. J Cancer Res Ther. 2015;10(4):1052–6.
122. Lederer A, et al. Metastasis-associated in colon cancer 1 is an independent prognostic biomarker for survival in klatskin tumor patients. Hepatology (Baltimore, Md.). 2015;62(3):841–50.
123. Wang G, Fu Z, Li D. MACC1 overexpression and survival in solid tumors: a meta-analysis. Tumour Biol. 2014;36(2):1055–65.
124. Wu Z, et al. Prognostic value of MACC1 in digestive system neoplasms: a systematic review and meta-analysis. Biomed Res Int. 2015;2015:252043.
125. Sun D-W, et al. Prognostic and clinicopathological significance of MACC1 expression in hepatocellular carcinoma patients: a meta-analysis. Int J Clin Exp Med. 2015;8(4):4769–77.
126. Lang AH, et al. A common variant of the MACC1 gene is significantly associated with overall survival in colorectal cancer patients. BMC Cancer. 2012;12:20.
127. Schmid F, et al. SNPs in the coding region of the metastasis-inducing gene MACC1 and clinical outcome in colorectal cancer. Mol Cancer. 2012;11(1):49.
128. Zheng Z, et al. Single nucleotide polymorphisms in the metastasisassociated in colon cancer-1 gene predict the recurrence of hepatocellular carcinoma after transplantation. Int J Med Sci. 2014;11(2):142–50.
129. Muendlein A, et al. Significant survival impact of MACC1 polymorphisms in HER2 positive breast cancer patients. Eur J Cancer. 2014;50(12):2134–41.
130. Stein U, et al. Circulating MACC1 transcripts in colorectal cancer patient plasma predict metastasis and prognosis. PloS One. 2012;7(11):e49249.
131. Wang Z, et al. Circulating MACC1 as a novel diagnostic and prognostic biomarker for nonsmall cell lung cancer. J Cancer Res Clin Oncol. 2015;141(8):1353–61.
132. Burock S, et al. Circulating Metastasis Associated in Colon Cancer 1 transcripts in gastric cancer patient plasma as diagnostic and prognostic biomarker. World J Gastroenterol. 2015;21(1):333–41.
133. Shirahata A, Sakata M, et al. MACC 1 as a marker for peritoneal-disseminated gastric carcinoma. Anticancer Res. 2010;30(9):3441–4.
134. Wang Z, et al. MACC1 overexpression predicts a poor prognosis for non-small cell lung cancer. Med Oncol (Northwood, London, England). 2014;31(1):790.
135. Yang Y-P, et al. High intratumoral metastasis-associated in colon cancer-1 expression predicts poor outcomes of cryoablation therapy for advanced hepatocellular carcinoma. J Transl Med. 2013;11(1):41.
136. Shang C, et al. Influence of the MACC1 gene on sensitivity to chemotherapy in human U251 glioblastoma cells. Asian Pac J Cancer Prev. 2015;16:195–9.
137. Li HH, et al. The expression of MACC1 and its role in the proliferation and apoptosis of salivary adenoid cystic carcinoma. J Oral Pathol Med. 2015;44(10):810–7.
138. Zhang Y, et al. MicroRNA-143 targets MACC1 to inhibit cell invasion and migration in colorectal cancer. Mol Cancer. 2012;11(1):23.

Axl and Its Mediated Signaling Axis in Cancer

3

Giridhar Mudduluru

Contents

G. Mudduluru
Max-Delbrück-Center for Molecular Medicine in the Helmholtz Association, Robert-Rössle-Str. 10, 13125 Berlin, Germany

German Cancer Consortium (DKTK), Heidelberg, Germany

Experimental and Clinical Research Center, Charité University Medicine, Berlin, Berlin, Germany
e-mail: giridhar.mudduluru@mdc-berlin.de

J. Haybaeck (ed.), *Mechanisms of Molecular Carcinogenesis – Volume 1*,
DOI 10.1007/978-3-319-53659-0_3

Abstract

AXL is a receptor tyrosine kinase activated by growth arrest-specific 6 (GAS6) or by ligand-independent homophilic and/or heterophilic interactions that regulate cancer cell proliferation, survival, migration, invasion, distant metastasis, the epithelial to mesenchymal transition (EMT), angiogenesis, and drug resistance. Axl belongs to the Tyro-3, AXL, and Mer (TAM) family of receptor molecules, known to be expressed in a number of organs and cell lines with a few exceptions such as lymphocytes and granulocytes. However, inappropriate Axl upregulation leads to uncontrolled cell growth, and its abundant expression is detected in a number of cancers such as colorectal and breast tumors. The transcriptional regulation of Axl is epigenetically inhibited by CpG hyper-methylation. Furthermore, the zinc finger transcriptional factor family members Sp1 and Sp3 are the constitutive regulators of Axl. Under oncogenic conditions, AP-1 family members mainly enhance its expression. Moreover, an overexpression of MZF1 induces Axl expression and mediates the migratory and invasive behavior of cells. Axl is also posttranscriptionally regulated by the small noncoding tumor suppressor microRNAs (miRNAs) miR-34 and miR-199. A malfunction of these different regulatory mechanisms in controlling Axl expression can induce Axl expression in cancer phenotypes. In addition to aspects of its regulation, this chapter will cover details of Axl structure, its expression in diverse cancer entities, and its signaling axis in the mediation of functions related to cancer phenotypes, including cell proliferation, antiapoptotic effects, EMT, cancer metastasis, angiogenesis, and drug resistance.

3.1 Axl Structure, Stimuli, and Expression

3.1.1 Discovery and Structure of Axl

Axl (also called UFO, ARK, and Tyro7) is a receptor tyrosine kinase which belongs to the TAM subfamily. This group comprises the three receptor tyrosine kinases Axl, Mer, and Tyro-3, which share a common structure. Axl was originally identified as a transforming gene in human leukemia and subsequently cloned and named as Axl from the Greek word "anexelekto," meaning uncontrolled [1, 2]. The *Axl* gene is evolutionarily conserved between vertebrate species. The amino acid sequence of Axl revealed it to be a novel type I transmembrane protein with an intracellular tyrosine kinase domain. This gene is ubiquitously expressed, being detectable in a wide variety of organs and cell lines of epithelial, mesenchymal, and hematopoietic origin, as well as non-transformed cells, although it is absent in lymphocytes and granulocytes [2]. *Axl* is located on chromosome 19q13.1-q13.2 and yields two forms of transcripts derived through alternative splicing. Variant 1 encodes the full-length isoform (5014 bases), and variant 2 (4987 bases) lacks exon 10. The protein that results from variant 2 lacks an internal 9-amino acid sequence; apart from this, no functional differences between these isoforms have yet been reported. The full-length 894-amino acid protein has a mass of 140 kDa and exhibits a roughly equal distribution of amino acids on either side of the plasma membrane. The structure of the extracellular domain of Axl makes it unique among receptor tyrosine kinases because it consists of a juxtaposition of two immunoglobulin-like repeats and two fibronectin type III repeats [2]. This pattern of structural elements is reminiscent of many cell adhesion molecules, and Axl has been implicated in cell–cell interactions [3]. Like other receptor tyrosine kinase molecules, the intracellular signaling functions of Axl cytoplasmic moieties (tyrosine kinase domain) have been elucidated through the cloning of receptor targets (CORT) method. Tyrosine residues at 779, 821, and 866 are active and function as docking sites for a range of intracellular tyrosine molecules including phosphoinositide phospholipase Cγ (PLCγ), growth factor receptor-bound protein 2 (GRB2), c-Src, and lymphocyte-specific protein tyrosine kinase (Lck) [4].

3.1.2 Activation of Axl Signaling

In 1995, an Axl stimulatory factor was purified from a conditioned medium of the Wi38 cell line and identified by N-terminal sequencing as growth arrest-specific gene 6 (GAS6) [5]. The 678-amino acid Gas6 protein is the latest addition to the vitamin K-dependent family of proteins. Gas6 shows a 43% amino acid sequence identity with protein S, an abundant serum protein [6]. Gas6 has the same domain organization as protein S: an N-terminal region containing 11 c-carboxyglutamic acid residues (Gla), a loop region, four EGF-like repeats, and a C-terminal sex hormone-binding globulin (SHBG)-like structure that is composed of two globular laminin G-like (LG) domains [7]. Several studies utilizing either site-specific blocking antibodies or partial protein constructs have established the SHBG region of both Gas6 and protein S as being the receptor-binding site. More detailed molecular studies revealed that the first LG domain in the Gas6 SHBG region is necessary for Axl binding [8]. The existence of a minimal Gas6–Axl complex, derived from its crystal structure, has provided a detailed view of the regions involved in their interactions [9]. In this complex, the two Ig-like domains of an Axl monomer are cross-linked by the first LG domain of a Gas6 molecule in a first high-affinity interaction. The lateral diffusion of such 1:1 complexes leads to dimerization and the formation of a circular 2:2 assembly. Two distinct sites of Gas6–Axl contact were revealed, one major and one minor; the minor site is conserved only within the Axl subfamily. No direct Axl–Axl or Gas6–Gas6 contacts were apparent in the complex (Fig. 3.1). In the major contact site, several charged residues in both Axl and Gas6 were identified to form part of polar β-sheet surfaces which interact with each other. It is interesting that protein S does not possess a distribution of charged residues similar to that found in Gas6, which may explain its inability to bind to Axl. Gas6 not only binds and activates the Axl signaling axis; it behaves the same way toward the other TAM family member receptors, with nanomolar binding affinities (0.4, 2.7, and 29 nM) in the order Axl → Tyro-3 → Mer [10].

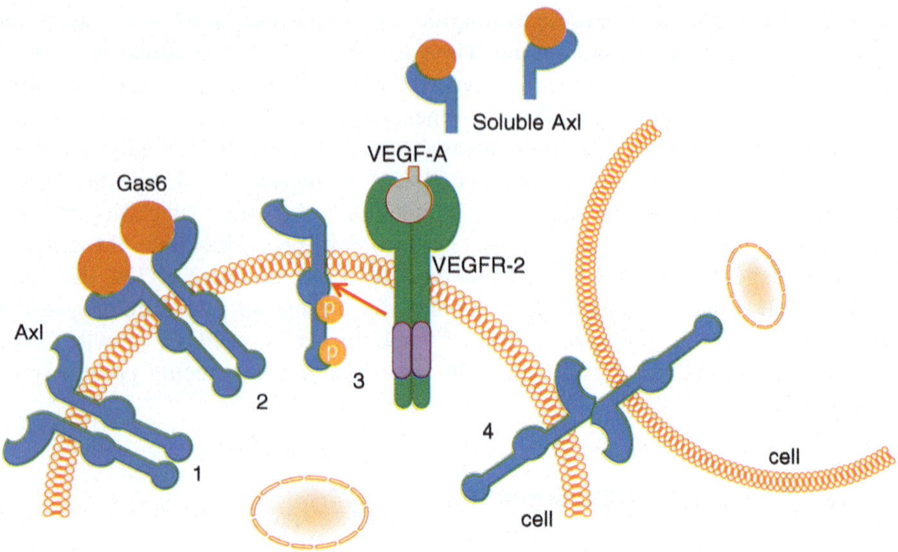

Fig. 3.1 Models for Axl signaling axis activation: (*1* and *4*) Direct, ligand-independent homophilic or heterophilic interaction between two Axl/Tyro-3 monomers in a single cell or between neighboring cells. (*2*) Ligand-induced dimerization of Axl monomers from two 1:1 (ligand–receptor) complexes to one 2:2 (2 ligand–receptor) complex. (*3*) Activation of VEGFR activates the Axl intracellular tyrosine kinase. (*4*) Hypothetical model for interaction between two Axl monomers on neighboring cells

Apart from the previously described conventional activation of the Axl signaling axis, other experimental models and a hypothetical model were proposed for Axl signaling. The first one is a ligand-independent homophilic interaction between two Axl monomers under overexpressed conditions in the same cell [11, 12]. Similar phenomena may also occur between two neighboring Axl expressing cells. The ligand-independent homophilic interactions were well described for Tyro-3, a TAM family member [13]. Axl also harbors a similar structure as Tyro-3; one could speculate that similar interactions might be possible between Axl and Tyro-3 [14]. The second way of signaling is new in its kind, which is a heterophilic interaction between two receptor signaling cascades. Binding of vascular endothelial growth factor A (VEGF-A) activates vascular endothelial growth factor receptor 2 (VEGFR2), which subsequently activates the T cell-specific adaptor protein (TSAd) and Src family kinases (SFKs), which engage Axl at its juxtamembrane domain and activates signaling [15] (Fig. 3.2). Yet hypotheses about heterodimeric interactions between TAM family members are based purely on structural similarities and have yet to be demonstrated experimentally (Fig. 3.1).

The extracellular regions of several transmembrane proteins such as adhesion molecules and growth factors and cytokine receptors have been found in circulating forms in human plasma [16]. These soluble ectodomains are shed from the full-length protein and thereby may limit the accessibility of the cell-bound receptor to the ligand.

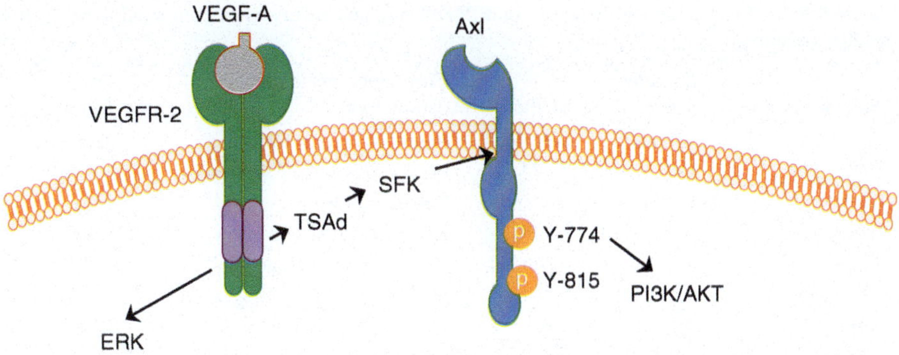

Fig. 3.2 Vascular endothelial growth factor (VEGF) signaling axis activates Axl pathway. VEGF-A activates several intracellular events that includes VEGFR2. This indeed activates TSAd, further SFK, which engages Axl and activates the Axl-mediated PI3K/AKT pathway (without Axl ligand binding or homophilic intracations). *TSAd* T cell-specific adaptor protein, *SFK* Src family kinase

Therefore they may represent an important posttranslational mechanism for controlling ligand efficacy under certain clinical conditions. The soluble Axl ectodomain is released as a result of proteolytic cleavage in conditioned media of various cell lines [17, 18]. A mouse Axl ectodomain was detected in tumor cell and dendritic cell medium and in serum. It has been suggested that proteolytic cleavage through the disintegrin-like metalloproteinase ADAM 10 is involved in its generation [19]. Furthermore, a significant amount of soluble Axl, but not Tyro-3 or Mer, was found to be in complex with Gas6 in mouse serum [19]. These observations indicate the potential value of investigating the presence of soluble Axl ectodomain in human plasma. The detection and quantitation of plasma Axl might reflect an altered regulation of Gas6–Axl system components under various clinical conditions and may therefore be of diagnostic value. In conclusion, the soluble form of Axl could reduce Axl signaling-mediated cancer phenotypes by depleting the availability of Gas6 (Fig. 3.1).

3.1.3 Axl Is Upregulated in Cancer

After the identification of Axl and structural and function determinations establishing it as a receptor tyrosine kinase and transforming gene, a number of screens have been carried out to detect its presence in a number of normal and cancer cell lines and resected patient tumor specimens. These studies have demonstrated that Axl is overexpressed in several cancer entities (Table 3.1). Furthermore, it has been shown that Axl is overexpressed in most cancer cell lines such as non-small cell lung cancer (NSCLC), breast cancer (BRC), and colorectal cancer (CRC), and its expression positively correlates with cell adhesion or invasive potential [18, 36, 37]. Increased Axl expression in cancer entities positively correlated with the cancer progression and poor prognosis and has also been identified as a potential druggable target in patient treatment.

Table 3.1 Axl expression is increased in different human cancers when compared to respective healthy specimens

Tumor entity	Sample size	Method	Correlations with clinical parameters	References
Colon	17 and 223	WB, IHC, RT-PCR	Potential marker for cancer progression and an oncotarget	[20–22]
Gastric	96	IHC	Poor prognosis	[23]
Breast	569	RT-PCR	Lymphovascular invasion	[24]
Lung	88	RT-PCR, IHC	Worse clinicopathological features and prognosis	[25]
Thyroid	27	IHC	Cancer progression	[26]
Liver	137	IHC	Lymph node metastasis	[27]
Prostate	96	RT-PCR	Cancer progression	[28]
Osteosarcoma	62	IHC	Poor prognosis	[29]
Renal Cell	221	ELISA	Cancer progression	[30]
Glioma	121	RT-PCR, IHC	Poor prognosis	[31]
Leukemia				
Myeloid	112	RT-PCR, WB	Poor prognosis	[32]
Chronic Lymphocytic	22	WB	Unique target molecule for treatment	[33]
Ovarian	72	RT-PCR, IHC	Poor prognosis	[34]
Esophageal	92	IHC	Cancer progression	[35]

IHC immunohistochemistry, *IFC* immunofluorescence, *RT-PCR* real-time polymerase chain reaction, *WB* Western blot, *ELISA* enzyme-linked immunosorbent assay, *TMA* tissue microarrays

3.2 Axl-Mediated Hallmarks of Cancer

3.2.1 Axl Signaling in Cell Proliferation

Cell proliferation is a normal physiological process which occurs in all types of cells and organs. Under normal physiological circumstances, cell proliferation is tightly controlled. Changes at cellular, genetic, and/or epigenetic levels lead to a reprogramming in which cell proliferation occurs and may result in uncontrolled cell division. Janssen et al. were the first to report that Axl is a transforming and cell-proliferating gene under induced expression conditions [1]. After this initial report, studies in more cancer entities buttressed a functional role for Axl in inducing cell proliferation. Varnum et al. identified Gas6 as a stimulus of an Axl signaling axis which mediates mouse fibroblast cell transformation and increased proliferation [5]. Before the identification of Gas6 as an Axl ligand, a study with interleukin-3-dependent 32D cells using a chimeric receptor containing the recombinant epidermal growth factor (EGF) receptor extracellular and transmembrane domains and the Axl kinase domain showed that PI3K is dispensable for Ras/ERK-mediated

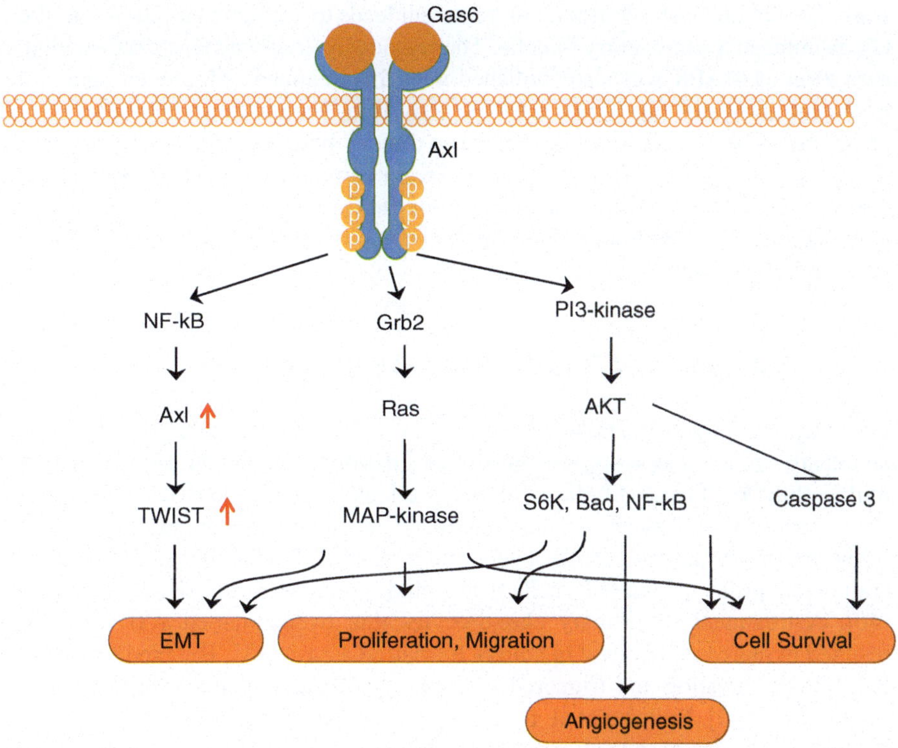

Fig. 3.3 Axl signaling axis: Activation of different intracellular molecules, mediating cell proliferation, survival, migration, invasion, metastasis, EMT, and angiogenesis (*red upward arrow*: Gas6/Axl signaling activates the transcriptional regulation of the gene) (color figure online)

cell proliferation and also that different threshold levels are needed for Ras/ERK activation. This study gives a demonstration that different extracellular domains dramatically alter the intracellular responses of receptor kinases to different stimuli, and this is especially true for Axl [38]. Later studies with different cell types revealed that the Ras/ERK pathway contributes to Axl-mediated cell survival [39]. In addition to mechanisms involving Gas6 activation, the induction of Axl expression can stimulate progression through the cell cycle and the division of NIH3T3 cells through the mitogen-activated protein kinase (MAPK) pathway [40]. Depending on the availability of Gas6 ligand and Axl protein in various cell types, this means that the MAPK/ERK pathway might be important in Gas6/Axl-mediated cell signaling in some contexts [41].

Apart from the MAPK/ERK pathway, many Axl-mediated downstream pathways have been linked to PI3K/AKT and the ribosomal protein S6 kinase (S6K)-mediated signaling. Goruppi et al. used specific PI3K and S6K inhibitors to show that these two kinases are essential mediators in Gas6/Axl-mediated cell proliferation (Fig. 3.3). Studies involved pretreatments of serum-started NIH3T3 cells with the specific PI3K inhibitor Wortmannin and S6K inhibitor rapamycin before stimulation with Gas6. Wortmannin treatment leads to a significant inhibition of the Gas 6-induced S-phase entry of cells. The previously described Gas6/Axl-mediated activation of MAPK was also abolished with this treatment. Moreover, rapamycin treatment abrogated Gas6-induced S6K activation, leading to an inhibition of cell proliferation. While Axl does not appear to directly bind Src, Gas6 activation nonetheless induced the phosphorylation of Src in serum-starved NIH3T3 cells [42]. These studies clearly demonstrated a role for the Gas6/Axl-mediated signaling axis as an inducer of cell proliferation through distinct intermediary molecules including PI3K, Ras/ERK, and Src.

3.2.2 Antiapoptotic Functions of Axl Signaling

Under normal physiological conditions, most cells generally undergo programmed cell death events (apoptosis) to control cell growth or remove damaged cells from multicellular organisms. Studies with the interleukin-3-dependent 32D cell line first demonstrated that Gas6/Axl signaling reduced the induction of apoptosis [38]. Even in the absence of a stimulus, the overexpression of Axl alone increased cell proliferation and protected NIH3T3 cells from cell death under serum-starved conditions [40]. A treatment of primary chondrocytes that overexpressed Axl with Gas6 resulted in increased survival in colony formation assays [43]. The binding of PI3K led to Axl activation and triggered multiple downstream pathways and increased cell survival [42]. Bellosta et al. used fibroblasts from Axl knockout mice in a further demonstration of the antiapoptotic activities of Axl. Serum-starved Axl knockout cells showed increased levels of apoptosis which could not even be rescued by the addition of Gas6. In addition, Axl protected these fibroblast cells from apoptosis even after treatment with TNF alpha or an overexpression of c-Myc [39]. These initial reports revealed that Gas6/Axl has antiapoptotic signaling functions both in vitro and in vivo.

The expression of Ark, the mouse homologue of Axl, was detected in Gn10 GnRH cells derived from migrating tumors in olfactory cells, but not in GT1-7 cells derived from the post-migratory tumor in the forebrain. Comparatively in these lines, Gn10 are more resistant to serum-stimulated apoptosis. Gas6/Axl signaling activated the PI3K and ERK pathways in Gn10 cells, and the effects were blocked by the ERK-specific inhibitor PD98059 and PI3K inhibitor Wortmannin [41]. Gas6/Axl signaling similarly induced antiapoptotic effects in human umbilical vein endothelial cells and human pulmonary artery endothelial cells [44, 45]. Further its shown that Gas6/Axl mediated antiapoptotic signaling is mediated through AKT activation [46]. These studies reveal a clear and general role for antiapoptotic processes mediated by Gas6/Axl in different cells under conditions of Axl overexpression.

Nuclear factor kappa B (NF-κB) is known to translocate into the nucleus upon activation of the PI3K/AKT cell survival pathway. Under normal conditions, NF-κB is found in the cytoplasm as a homodimer or a heterodimer including members of a family of structurally related proteins. Five members of the family have been identified: RelA (p65), RelB, cRel, NF-κB1 (p50/p105), and NF-κB2 (p52/p100). All are inactive in association with inhibitory proteins of the IκB family (IκBα, β, ε, and Bcl3) or as precursors of NF-κB1 (p105) and NF-κB2 (p100), whose nuclear localization signals are masked until they are further processed. Upon stimulation, the inhibitors IκB and p105 are activated, a process which triggers their proteolytic degradation via the ubiquitin-proteasome pathway. This leaves activated forms of NF-κB that can translocate to the nucleus and induce the transcriptional regulation of antiapoptotic genes like BCl-X_L and BCL2 [47, 48]. Demarchi et al. showed NF-κB as one of the key downstream molecules in mediating the Gas6/Axl signaling antiapoptotic property. As described above, Gas6 stimulation increased the nuclear translocation of active NF-κB and its binding to BCl-X_L promoter and induced the BCl-X_L expression in serum-starved NIH3T3 cells [49]. These authors also observed glycogen synthase kinase 3 (GSK3) activation, which is known to induce cell proliferation and antiapoptotic functions [50]. A similar functional role of NF-κB was shown under the Gas6/Axl signaling axis in endothelial cells. Gas6 treatment led to the activation of AKT and NF-κB and increased the expression of the antiapoptotic gene BCL2—one mechanism by which Axl-stimulated pathways protect cells from apoptosis. The treatment had the supportive effect of decreasing the expression of the caspase 3 activation products p12 and p20 [51], which are proapoptotic. In vascular smooth muscle cells, Son et al. showed that Gas6/Axl signaling inhibited the function of the proapoptotic molecules BAD and caspase 3 [52]. An shRNA-mediated Axl knockdown in NSCLC cell lines showed increases of apoptosis over controls [53]. In OE33 and OE19 esophageal cancer cells, Axl overexpression attenuated cellular and molecular markers of apoptosis induced by tumor necrosis factor-related apoptosis-inducing ligand (TRAIL). In support of this finding, Axl knockdown FLO-1 cells were sensitive to TRAIL-induced apoptosis. However, another mechanism seemed to be involved: Axl expression did not alter DR4 or DR5 expression, but associations of Axl with DR5 were detected. This blocked the recruitment of caspase 8 to the death-inducing signaling complex. This confirms that Axl mediates TRAIL-induced apoptosis by mitigating the death-inducing signaling complex [54].

All of these studies provide clear evidence for the importance of Gas6/Axl signaling or Axl hemophilic activation through an increased activity of downstream signaling pathways (PI3K/AKT, MAPK, GSK3, TRAIL) and specific molecules (NF-κB, BAD, BCl-X_L, BCL2) that are involved in cell survival/antiapoptosis in a range of cell types.

3.2.3 Axl Signaling Induces Angiogenesis

Angiogenesis is important in sustaining tumor growth and survival. A tumor mass stimulates the formation of new blood vessels to permit tumor expansion, local invasion, and metastasis and mainly to deliver oxygen, nutrients, and survival and growth factors. Several factors are known to be important for angiogenesis, including the vascular endothelial growth factor receptor (VEGFR) and corresponding receptor-associated signaling that promotes endothelial proliferation and migration.

Other angiogenesis regulators include EGF-like domain-containing protein 7 (EGFL7) and alpha-5 beta-1 (α5β1) integrin, which contribute to the formation of new blood vessels. Angiogenesis processes have been further extrapolated through studies of interactions between endothelial cell interactions and the extracellular matrix [55–58].

A functional role for Axl in vasculature was first demonstrated by O'Donnell et al. [45]. Axl expression was detectable in capillary endothelium, in vascular smooth muscle cells of arterioles and veins, and in a subset of synovial cells in the synovial tissue of a patient suffering from rheumatoid arthritis (RA), a disease whose key features are abnormalities in angiogenesis and synovial cell hyperplasia. In vitro studies showed that Gas6/Axl signaling protected human umbilical vein endothelial cells (HUVECs) from tumor necrosis factor alpha (TNFα)-mediated cytotoxicity [45]. The hint that Axl might be involved in angiogenesis came from observing its high expression in vascular cells.

This role was confirmed by Axl gain- or loss-of-function experiments in vitro and in vivo. Axl antisense screening reduced the proliferation of HUVECs and their directed migration to vitronectin haptotaxis. Furthermore, in co-culture branching morphogenesis/tube formation VGFR-dependent assays, HUVEC/primary pulmonary artery smooth muscle cells showed a functional role for Axl in angiogenesis. Knockdowns of Axl mitigated tube formation, fiber length, and branching, which are main functional events in endothelial tube morphogenesis. Similar results were pronounced in an SCID mouse angiogenesis model. shRNA-Axl- and shRNA-VEGFR2-silenced HUVECs were mixed with Matrigel, seeded into poly-L-lactic acid scaffolds and subsequently implanted into SCID mice. After 14 days, a significant reduction in human Tie2 expression levels was observed in both shRNA-Axl and shRNA-VEGFR2 cells, which indicates a role for Axl in neovascularization [59]. Similarly, Axl-siRNA-silencing reduced the expression of Dickkopf-homologue 3 (DKK3) and Tie2, which play a role in tube formation. However, silencing these two genes had no effect on Axl expression in HUVECs. Anti-VEGF treatment attenuated the reduction of tube formation introduced by siRNA-Axl or siRNA-DKK3 [60]. Implanting cells that stably expressed Axl-dominant negative constructs in nude mice using the dorsal skinfold chamber model demonstrated that Axl-dominant negative-expressing cells reduced the density and diameter of tumor vessels to a degree comparable to that of Axl wild-type-expressing cells [12].

Angiogenesis is promoted by high levels of lactate, which are also essential for wound healing and tumorigenesis. Lactate activates the PI3K/AKT pathway in primary human endothelial cells and is essential for their organization into tubes, as well as for vessel formation in mouse aortic explants. This activation is mediated by the ligand-mediated activation of the three receptor tyrosine kinases Axl, Tie2, and VEGF receptor 2 and has been confirmed through pharmacological inhibitions of their kinase activity or by suppressing their expression [61]. VEGF-A generally activates a series of intracellular events through the activation of VEGFR2 and subsequent downstream molecules such as Src family kinase (SFK) in a T cell-specific adaptor protein (TSAd)-dependent manner. It is also known that VEGFR2 activates the PI3K/AKT pathway in endothelial cells. An interesting and novel aspect of these findings is that activated SFKs engage Axl via its JM domain and thereby promote autophosphorylation at Y773 and Y815 in the absence of any external mediated activation of Axl. The activated tyrosine residues within the optimal motif bind to the SH2 domains of p85, and this further activates PI3K, which produces lipids that are essential for the activation of AKT [15]. This is one of the most important aspects of the Axl signaling, responding for other receptor signaling through intracellular kinase domain activation, without its receptor or ligand (Figs. 3.2 and 3.3). These studies clearly demonstrate that Axl has a signaling function in angiogenesis and by consequence a role in the angiogenic mechanisms related to metastases.

3.2.4 Axl Signaling Induces Epithelial–Mesenchymal Transition

Epithelial–mesenchymal transition (EMT) is a process whereby epithelial cells lose their cell polarity and cell–cell adhesion and gain migratory and invasive properties to become mesenchymal stem cells. As cells undergo EMT, they gain increased resistance to apoptosis and alter their production of extracellular matrix (ECM) components and ECM-degrading enzymes. The switch that occurs in EMT initiation is also accompanied by changes in a number of key molecules including the expression and activity of specific transcription factors and specific cell-surface proteins, a reorganization and expression of cytoskeletal proteins, and changes in microRNA expression patterns [62]. Kalluri et al. provided a summary of many genes that function as EMT markers, including key transcriptional regulators such as Snail, Slug, and Twist, in a recent review [62].

A role for Axl functions in EMT in pancreatic cancer was first demonstrated [63]. This work showed that shRNA-Axl silencing in MIAPaCa-2 cells led to a significant reduction in signaling in MAPK/ERK and PI3K/AKT pathways. It also revealed a significant downregulation in the major transcriptional factors that initiate the mesenchymal switch: Snail, Slug, and Twist [63]. Interestingly, the overexpression of these three factors in immortalized mammary epithelial cells enhances Axl expression and autocrine signaling loop with Gas6 ligand (Fig. 3.3). This also enhanced the expression of the mesenchymal markers N-cadherin and vimentin and reduced the expression of the epithelial markers E-cadherin and β-catenin. In parallel, shRNA-Axl silencing in highly metastatic breast cancer cells MDA-MB-231 reduced the threshold of mesenchymal-like features and mediated cancer progression events [64, 65]. Another study showed that EMT induced by Slug and H-Ras is mediated through Axl via vimentin in breast cancer cells [66]. Furthermore, Axl expression positively correlated with the vimentin expression in EGFR-mutant NSCLC tumors in vivo [67]. This evidence confirms increased Axl expression induction by mesenchymal cells and that its expression is induced by the mesenchymal transcriptional factors Slug, Snail, and Twist through vimentin.

Axl is posttranscriptionally downregulated by the tumor suppressor and EMT inhibitor miRNA miR-34a [36, 68]. miR-34a is also known to downregulate other

EMT-inducing genes including axin-like protein (AXIN2), carbonic anhydrase 9 (CA9), C-X-C motif chemokine 10 (CXCL10), FOS-like antigen 1 (FOSL1), fucosyltransferase 8 (FUT8), growth arrest-specific protein 1 (GAS1), Kruppel-like factor 6 (KLF6), and podocalyxin-like protein 1 (PODXL) [68]. As discussed above, Gas6/Axl signaling activates the AKT and NF-κB signaling cascades, which are known to induce the EMT phenotype in a range of cancers.

3.2.5 Axl Signaling in Cancer Metastasis

In 2012 the World Health Organization (WHO) registered more than 8.2 million cancer-related death cases mainly lung, liver, stomach, colorectal, breast, and esophageal cancer. More than 90% of these cancer-related deaths are due to cancer metastasis, which is induced by a loss of the normal self-control of signaling cascades [69]. Cancer metastasis is a multistep process accompanied by a number of changes in morphology and molecular functions. The way cancer cells spread from a primary site to different local or distant organs in a patient depends on the tumor type. This process begins after gaining self-sustainability at the tumor site through angiogenesis and extracellular degradation and a gain in migratory properties, followed by the steps of migration, local invasion, intravasation, and transport through the circulatory system to different organs, extravasation, and formation as micrometastasis or macrometastasis, which require local angiogenesis for establishment [69, 70].

Axl is capable of inducing all aspects of cancer metastasis events including migration, invasion, angiogenesis, and EMT in different cancer entities. An overexpression of Axl has been reported using differential display PCR methods in the highly metastasizing prostate cancer cell line DU145 [71]. However, for the first time, the role of Axl in terms of cancer metastasis, explained through adenovirus type V E1A protein (E1A), is known to reverse the transformed phenotype, to inhibit metastasis, and to induce apoptosis. Overexpression of E1A inhibited the expression of Axl and prevented the Gas6/Axl signaling axis which induces signaling cascades required for cell survival, including AKT and NF-κB; it also inactivated or downregulated apoptosis-inducing genes such as BAD and Fas-ligand, thus inhibiting cancer progression [72]. Nakano et al. derived low and highly metastatic cell lines from highly metastatic cell lines through the dilution plating method [73]. A cDNA array analysis by the authors revealed that five genes, including Axl, were significantly upregulated in highly metastatic cell lines compared to the lines with low metastatic activity. Similarly, suppression subtractive hybridization screening of lung cancer cell lines revealed an induction of Axl expression in the highly metastatic cell line PLA-801D compared to the low metastatic cell line PLA-801C [74]. Axl overexpression induced the expression of matrix metalloproteinase 9 (MMP-9) and activated the ERK pathway, enhanced the transactivation of NF-κB, and induced the translocation of brahma-related gene 1 (Brg-1) to the nucleus. Axl-mediated MMP-9 expression and the invasiveness of cancer cells were also significantly inhibited by interfering with the pathway: either through a dominant negative overexpression of ERK, IkB, or Brg-1 or a specific inhibition of ERK and NF-κB [75].

Axl staining was positive in 54% of pancreatic cancer specimens and significantly associated with lymph node metastases. Here, too, a specific inhibition of the ERK and P13K/AKT pathway showed a loss of Axl function in in vitro studies. Axl silencing decreased the amount of the GTPase proteins Rho and Rac and mirrored the migration and invasion of MIAPaCa-2 cells [63]. Even a dominant negative inhibition of just the Axl kinase domain activity had similar effects by inhibiting tumor cell invasion in fetal rat brain aggregates [12]. Similarly, inhibition of Axl expression and/or functions inhibited cell migrations, invasions, and distant metastasis to the lung in an orthotopic breast cancer model [60, 64].

Interfering with other molecules functionally associated with Axl has similar effects in multipole cancers. Yes-associated protein 1 (YAP1) plays a major role in tumorigenesis and tumor progression in multiple cancers. A shRNA-YAP1-mediated knockdown of YAP1 significantly inhibited the expression of Axl, the proliferating cell nuclear antigen (PCNA), and MMP-9 and mediated the invasive potential of LAC A549 and GAC SGC-7901 cell lines [76]. In clear cell renal cell carcinoma (ccRCC), Rankin et al. demonstrated cross talk between von Hippel–Lindau/hypoxia-inducible transcription factor (HIF) and Gas6/Axl signaling [77]. HIF1 and HIF2 transactivate Axl expression by binding the hypoxia-response element in the Axl proximal promoter. In parallel, the authors determined that Gas6/Axl signaling uses lateral activation of met proto-oncogene (MET) through SRC to maximize cellular invasion abilities [77]. Similarly, Axl activates transforming growth factor-β (TGF-β) signaling and the tumor progression events it mediates in hepatocellular carcinoma (HCC) [78, 79].

Myeloid zinc finger 1 (MZF1) is a transcription factor known to induce cell proliferation, antiapoptotic properties, and transformation of cells. siRNA-MZF1 silencing reduced the number of tumors and prolonged the time it took them to form [80–82]. The overexpression of MZF1 induced migration and invasion by inducing Axl expression in colorectal and cervical cancer cell lines. Furthermore, an in vivo chicken embryo metastasis assay showed that the overexpression of MZF1 induced tumor growth and the formation of distant metastases in an Axl-dependent manner. A positive correlation was found between Axl and MZF1 expression in colorectal cancer tumor specimens [22].

Axl is transcriptionally induced by AP-1 family members through the MAPK pathway, under oncogenically stimulated conditions [83]. As described above, the MAPK/ERK pathways are one of the key downstream pathways activated by different oncogenic stimuli-like phorbol esters, which activates the AP-1 family. Interestingly, the invasive capacity of a panel of NSCLC, breast cancer, and CRC cells correlated positively with Axl mRNA and protein expression [36]. Less invasive cells, on the other hand, exhibited low levels of Axl expression due to epigenetic regulation [84]. Additionally, the Axl posttranscriptional regulators miR-34 and miR-199 significantly reduced Axl-mediated migration, invasion, and distant metastasis in NSCLC, BRC, and CRC cell lines. miR-34 and miR-199 expression negatively correlated with Axl expression in NSCLC patient tumor specimens as compared with levels in normal specimens [36].

Further evidence that Axl is an important cancer-inducing receptor tyrosine kinase comes from loss- or gain-of-function studies of key molecules in the pathway, including Axl and its transcriptional or posttranscriptional regulators. Studies based on a range of drugs that specifically inhibited the functions of Axl, or are in combination with other relevant molecules such as VEGFR- and EGFR-specific inhibitors, have supported this finding [85–90].

3.2.6 Axl in Cancer Stemness

Cancer stem cells (CSC) can be divided into types that divide or differentiate rapidly; others develop into mature cells without any further divisions. Two explanations have been offered. The first (cancer stem cell theory) is a more systematic model proposing that cells divide and feed tumor growth with self-renewal ability. These cells divide certain number of times and then differentiate as specialized mature tumor cells. The second (stochastic) model proposes that each cancer cell possesses the same potential to self-renewal and differentiate [91–97]. Till today, a number of cancer stem cell markers were reported generally as well as cancer entity specific [98].

Ahtiainen et al. reported Axl CSC functional properties in breast cancer cell lines. The authors checked the innate immunity to oncolytic adenovirus Ad5/3-Delta24 in conventional treatment-resistant, non-cancer-initiating cells (CIC) with $CD44^{+}/CD24^{-/low}$ population and normal breast tissue $CD44^{+}/CD24^{-/low}$ stem cells. Under these conditions, the authors observed that normal breast tissue cells have intact type I INF signaling compared to the breast cancer CIC $CD44^{+}/CD24^{-/low}$ population, which showed dysregulated innate immune response due to the dysfunctional virus recognition caused by impaired trafficking of Toll-like receptors (TLR) 9 and cofactor MyD88 and the absence of TLR2, having a deleterious impact on TLR pattern recognition receptor signaling. Further, they increased inhibitory signaling via the suppression of cytokine signaling of Axl/Tyro-3 and Mer. The presence of these CIC stem cells has been put forward as an explanation for the resistance of relapsed or metastatic cancers to treatments [99]. CD44 and ALDH1 are known and well characterized CSC markers for many cancer entities. shRNA-Axl silencing significantly altered the expression of intercellular junction molecules increasing cell–cell adhesion with downregulation of Wnt and TGFβR signaling and negatively correlated with CD44 and ALDH1 expression in squamous cell carcinoma (SCC) cell lines [100]. Specific inhibition of Axl with amuvatinib inhibited the breast cancer stem cell self-renewal and restored chemosensitivity for the drug [101]. Further studies showed that the overexpression of Axl and $CD44^{+}$ positively correlated with drug resistance against metformin and imatinib [102, 103].

3.3 Axl as Regulator of Drug Resistance in Cancer

Resistance to chemo and molecular targeted therapies is a major problem in present cancer treatments, which has stimulated research into the mechanisms that cause resistance and strategies to overcome them. Drug resistance can be divided into two types. One is intrinsic (an inbuilt system that arises in a cancer mass or cells due to the molecules that are expressed or genetic modifications that it has experienced). The second is acquired resistance, which is a change in the molecular profiles of cells or the tumor mass that has been stimulated by the treatment itself [104]. A number of studies have reported that Axl-induced expression increases drug resistance in different cancer entities.

Macleod et al. developed an ovarian carcinoma cell line that was resistant to 20-fold dosages of cisplatin ($PE01^{CDDP}$) by exposing the parental cell line PE01 to increasing concentrations of cisplatin. Under these conditions, Axl is overexpressed 2.9 times in the resistant cell line compared to the original parental cell line PE01 [105]. Cisplatin is used as a first-line treatment for esophageal adenocarcinoma (EAC), known for a high rate of chemotherapy resistance and poor outcome. Inducing Axl expression increased the IC^{50} value twofold and increased cell survival threefold. The inhibition of Axl expression reduced the cell survival twofold when treated with cisplatin. Cisplatin is known to activate endogenous p73β and increase expression of p-c-ABL(Y412) and p-p73β(Y99). In general, this molecular mechanism plays a role in inducing apoptosis, which is blocked by Axl expression [106]. Kurokawa et al. reported that acquired resistance to cisplatin is due to the EMT-like changes that Axl induces in NSCLC cells [107]. Bladder cancer cell line and patient derived xenografts after treatment with PI3K/mTOR (PF-04691502) and MEK (PD-0325901)-specific inhibitors reduced tumor growth and decreased the secretion of the vascular endothelial growth factor. However, this increased Axl expression [108]. Another study based on NSCLC cells confirmed Axl's role in acquired resistance under treatment with drugs like cisplatin [109].

R428 is a selective, small molecule inhibitor of Axl that blocks its catalytic and precancerous activities. R428 treatment reduced Axl-induced AKT phosphorylation, cancer cell invasion, angiogenesis, and the production of pro-inflammatory cytokines. It also reduced the expression of the cytokine granulocyte macrophage colony-stimulating factor and Snail in a dosage-dependent manner. Interestingly, using R428 to inhibit Axl-mediated cellular and molecular functions during cisplatin treatments achieved an enhanced suppression of liver metastases [110]. Axl knockdowns in RAC cell lines reduced migration, invasion, and in vivo engraftment, accompanied by a downregulation in the activity of the Ral GTPase proteins (RalA and RalB). Similar effects were obtained using an A428 inhibitor. Blocking Axl functions also abrogated the phosphorylation of ERBB2 (Her-2/neu) at the Tyr877 residue, which reveals the cross-functional effects of R428 on different receptor signaling axes [35].

Induced EGFR expression has been associated with the development of head and neck cancer (HNC) and a poor prognosis for patients. Clinical trials based on the EGFR inhibitor erlotinib were not successful in HNC patients. Glies et al. developed a cell line, which is resistant to erlotinib. Compared to the parental cell line, the HN5-erlotinib resistant cell line exhibited an EMT phenotype and affected migrations. Surprisingly, Axl exhibited a higher degree of phosphorylation in this line [68]. Similarly, cases of NSCLC with activating mutations of EGFR respond to EGFR-targeted tyrosine kinase inhibitors such as erlotinib, but the sensitivity is short and these cells acquire resistance to the drug. These resistant cells are characterized by an increased activation of Axl and EMT [67]. Specifically inhibiting Axl with R428 induced the cells to become sensitive to erlotinib treatment [68], suggesting that the expression or activation of Axl might be the cause of resistance in the patient's samples. Another study demonstrated that inhibiting Axl activity increased the sensitivity of head and neck squamous cell carcinoma to chemotherapy, cetuximab, and radiation [111]. Axl is upregulated in metformin resistant prostate cancer cell lines and inhibits Axl with R428 sensitized the cells to metformin treatment [102]. In support of these findings, it has been shown that Axl expression is increased in myeloid leukemia cell lines and also Gas6/Axl signaling for chemotherapy. The cells also exhibited increased Bcl-2 and Twist expression [112]. More evidence of Axl's role in cancer stemness comes from the different responses of metastasized lesions to small molecule inhibitors. R428 effectively inhibited liver and lung metastasic lesions more effectively than when applied to peritoneal metastasis-derived cells [113].

An overexpression of Axl was also found in HER2-positive and estrogen receptor (ER)-positive lapatinib-resistant breast cancer clones derived from lapatinib-sensitive BT474 cells by chronic exposure to lapatinib. The authors found that Axl overexpression is the cause for this drug resistance; treatment with foretinib (a multikinase inhibitor of Axl, MET, and VEGFR) restores the sensitivity of these cells to lapatinib [114]. Metastatic renal cell carcinoma (RCC) patients show high rates of resistance to antiangiogenic therapy. To understand this, Zhou et al. generated RCC cell lines with resistance to sunitinib (an antiangiogenic small molecule). Gene profiling assays revealed an upregulation and activation of Axl, MET, and EMT genes. Angiogenesis was also enhanced by co-culturing RCC with human umbilical vein endothelial cells. Further, the authors stained tissues from the RCC patients that had been treated with sunitinib, which revealed that Axl and MET are mediators of sunitinib resistance. Pretreatments with specific inhibitors of Axl and MET or a suppression of these genes inhibited the metastatic behavior of RCC cell lines and rescued the acquired resistance to sunitinib in a xenograft model [115]. Similarly, Axl knockdown in glioblastoma cell lines led to higher sunitinib sensitivity, reduced migration, and increased apoptosis [116]. Many other studies have confirmed the importance of Axl expression and its activated signaling in drug resistance. All of this evidence clearly demonstrates that Axl pays an important role in drug resistance and acquired resistance in different treatment strategies.

Conclusions

Axl is known to be overexpressed in a number of cancer entities and induces cell survival, proliferation, antiapoptosis, colony formation, migration, invasion, EMT, and distant metastasis formation. These hallmarks of cancer can be mediated either through the overexpression of Axl or Gas6 stimulation, through their different effects on cancer-associated signaling cascades. Moreover, Axl overexpression is associated with poor patient survival in several cancer entities. Drug resistance is a major problem in cancer therapies, and Axl is known for inducing drug resistance and cancer stemness. A number of gain- or loss-of-function experiments have demonstrated that Axl has potential as a druggable target both in vitro and in vivo. The expression status of Axl and availability of its ligand Gas6 play major roles in Axl-mediated signaling. As discussed above, under normal conditions, the regulation of Axl at transcriptional and posttranscriptional stages is tightly controlled. However, during cancer progression, Axl expression is induced in cancer cell lines and tissues through a loss of regulation. The inhibition of Axl expression—either through inhibitors at the transcriptional level or miRs at the posttranscriptional level—would control its expression. A second way to inhibit Axl-mediated cancer progression and metastasis would be to block Axl signaling either at its Gas6 binding site and/or through active intracellular motifs and specific inhibitors. This would inhibit cross talk signaling between Axl and enhance other signaling cascades such as VEGFR2 receptor tyrosine molecules. Recent advances in technology and research into cancer genotypes have thus produced an enormous amount of data on Axl and its effects and are encouraging signs that a personalized approach based on Axl might be an effective way to treat patients. All of this evidences the importance of Axl as a potential therapeutic target for mitigating many of the hallmarks of cancer.

Acknowledgments This work was supported by the German Cancer Consortium (DKTK). I thank Prof. Ulrike Stein for her suggestions, critical reading and encouragement in writing this chapter and Harikrishna Raddhakrishna, Steffen Fuchs, and Kathrina Ilm for their critical reading and discussion of this chapter.

References

1. Janssen JW, et al. A novel putative tyrosine kinase receptor with oncogenic potential. Oncogene. 1991;6(11):2113–20.
2. O'Bryan JP, et al. axl, a transforming gene isolated from primary human myeloid leukemia cells, encodes a novel receptor tyrosine kinase. Mol Cell Biol. 1991;11(10):5016–31.
3. McCloskey P, et al. GAS6 mediates adhesion of cells expressing the receptor tyrosine kinase Axl. J Biol Chem. 1997;272(37):23285–91.
4. Braunger J, et al. Intracellular signaling of the Ufo/Axl receptor tyrosine kinase is mediated mainly by a multi-substrate docking-site. Oncogene. 1997;14(22):2619–31.
5. Varnum BC, et al. Axl receptor tyrosine kinase stimulated by the vitamin K-dependent protein encoded by growth-arrest-specific gene 6. Nature. 1995;373(6515):623–6.
6. Dahlbäck B, Villoutreix BO. Regulation of blood coagulation by the protein C anticoagulant pathway: novel insights into structure-function relationships and molecular recognition. Arterioscler Thromb Vasc Biol. 2005;25:1311–20.
7. Villoutreix BO, et al. SHBG region of the anticoagulant cofactor protein S: secondary structure prediction, circular dichroism spectroscopy, and analysis of naturally occurring mutations. Proteins. 1997;29(4):478–91.
8. Fisher PW, et al. A novel site contributing to growth-arrest-specific gene 6 binding to its receptors as revealed by a human monoclonal antibody. Biochem J. 2005;387(Pt 3):727–35.
9. Sasaki T, et al. Structural basis for Gas6-Axl signalling. EMBO J. 2006;25(1):80–7.
10. Nagata K, et al. Identification of the product of growth arrest-specific gene 6 as a common ligand for Axl, Sky, and Mer receptor tyrosine kinases. J Biol Chem. 1996;271(47):30022–7.

11. Bellosta P, et al. The receptor tyrosine kinase ARK mediates cell aggregation by homophilic binding. Mol Cell Biol. 1995;15(2):614–25.
12. Vajkoczy P, et al. Dominant-negative inhibition of the Axl receptor tyrosine kinase suppresses brain tumor cell growth and invasion and prolongs survival. Proc Natl Acad Sci U S A. 2006;103(15):5799–804.
13. Heiring C, Dahlbäck B, Muller YA. Ligand recognition and homophilic interactions in Tyro3: structural insights into the Axl/Tyro3 receptor tyrosine kinase family. J Biol Chem. 2004;279(8):6952–8.
14. Toshima J, et al. Autophosphorylation activity and association with Src family kinase of Sky receptor tyrosine kinase. Biochem Biophys Res Commun. 1995;209(2):656–63.
15. Ruan G-X, Kazlauskas A. Axl is essential for VEGF-A-dependent activation of PI3K/Akt. EMBO J. 2012;31(7):1692–703.
16. Fernandez-Botran R. Soluble cytokine receptors: basic immunology and clinical applications. Crit Rev Clin Lab Sci. 1999;36(3):165–224.
17. O'Bryan JP, et al. The transforming receptor tyrosine kinase, Axl, is post-translationally regulated by proteolytic cleavage. J Biol Chem. 1995;270:551–7.
18. Wimmel A, et al. Axl receptor tyrosine kinase expression in human lung cancer cell lines correlates with cellular adhesion. Eur J Cancer. 2001;37(17):2264–74.
19. Budagian V, et al. A promiscuous liaison between IL-15 receptor and Axl receptor tyrosine kinase in cell death control. EMBO J. 2005;24(24):4260–70.
20. Craven RJ, et al. Receptor tyrosine kinases expressed in metastatic colon cancer. Int J Cancer. 1995;60(6):791–7.
21. Martinelli E, et al. AXL is an oncotarget in human colorectal cancer. Oncotarget. 2015;6(27):23281–96.
22. Mudduluru G, Vajkoczy P, Allgayer H. Myeloid zinc finger 1 induces migration, invasion, and in vivo metastasis through Axl gene expression in solid cancer. Mol Cancer Res. 2010;8(2):159–69.
23. Wu CW, et al. Clinical significance of AXL kinase family in gastric cancer. Anticancer Res. 2002;22:1071–8.
24. D'Alfonso TM, et al. Axl receptor tyrosine kinase expression in breast cancer. J Clin Pathol. 2014;67(8):690–6.
25. Ishikawa M, et al. Higher expression of receptor tyrosine kinase Axl, and differential expression of its ligand, Gas6, predict poor survival in lung adenocarcinoma patients. Ann Surg Oncol. 2013;20:S467–76.
26. Avilla E, et al. Activation of TYRO3/AXL tyrosine kinase receptors in thyroid cancer. Cancer Res. 2011;71(5):1792–804.
27. Xu J, et al. Axl gene knockdown inhibits the metastasis properties of hepatocellular carcinoma via PI3K/Akt-PAK1 signal pathway. Tumour Biol. 2014;35(4):3809–17.
28. Paccez JD, et al. The receptor tyrosine kinase Axl is an essential regulator of prostate cancer proliferation and tumor growth and represents a new therapeutic target. Oncogene. 2013;32(6):689–98.
29. Han J, et al. Gas6/Axl mediates tumor cell apoptosis, migration and invasion and predicts the clinical outcome of osteosarcoma patients. Biochem Biophys Res Commun. 2013;435(3):493–500.
30. Gustafsson A, et al. Differential expression of Axl and Gas6 in renal cell carcinoma reflecting tumor advancement and survival. Clin Cancer Res. 2009;15(14):4742–9.
31. Hutterer M, et al. Axl and growth arrest-specific gene 6 are frequently overexpressed in human gliomas and predict poor prognosis in patients with glioblastoma multiforme. Clin Cancer Res. 2008;14(1):130–8.
32. Ben-Batalla I, et al. Axl, a prognostic and therapeutic target in acute myeloid leukemia mediates paracrine crosstalk of leukemia cells with bone marrow stroma. Blood. 2013;122(14):2443–52.
33. Ghosh AK, et al. The novel receptor tyrosine kinase Axl is constitutively active in B-cell chronic lymphocytic leukemia and acts as a docking site of nonreceptor kinases: implications for therapy. Blood. 2011;117(6):1928–37.

34. Rea K, et al. Novel Axl-driven signaling pathway and molecular signature characterize high-grade ovarian cancer patients with poor clinical outcome. Oncotarget. 2015; 6(13):30859–75.
35. Hector A, et al. The Axl receptor tyrosine kinase is an adverse prognostic factor and a therapeutic target in esophageal adenocarcinoma. Cancer Biol Ther. 2010;10(10):1009–18.
36. Mudduluru G, et al. Regulation of Axl receptor tyrosine kinase expression by miR-34a and miR-199a/b in solid cancer. Oncogene. 2011;30(25):2888–99.
37. Zhang Y, et al. AXL is a potential target for therapeutic intervention in breast cancer progression. Cancer Res. 2008;68(6):1905–15.
38. Fridell YW, et al. Differential activation of the Ras/extracellular-signal-regulated protein kinase pathway is responsible for the biological consequences induced by the Axl receptor tyrosine kinase. Mol Cell Biol. 1996;16(1):135–45.
39. Bellosta P, et al. Signaling through the ARK tyrosine kinase receptor protects from apoptosis in the absence of growth stimulation. Oncogene. 1997;15:2387–97.
40. Goruppi S, Ruaro E, Schneider C. Gas6, the ligand of Axl tyrosine kinase receptor, has mitogenic and survival activities for serum starved NIH3T3 fibroblasts. Oncogene. 1996;12(3):471–80.
41. Allen MP, et al. Growth arrest-specific gene 6 (Gas6)/adhesion related kinase (Ark) signaling promotes gonadotropin-releasing hormone neuronal survival via extracellular signal-regulated kinase (ERK) and Akt. Mol Endocrinol. 1999;13(2):191–201.
42. Goruppi S, et al. Requirement of phosphatidylinositol 3-kinase-dependent pathway and Src for Gas6-Axl mitogenic and survival activities in NIH 3T3 fibroblasts. Mol Cell Biol. 1997;17(8):4442–53.
43. Loeser RF, et al. Human chondrocyte expression of growth-arrest-specific gene 6 and the tyrosine kinase receptor axl. Potential role in autocrine signaling in cartilage. Arthritis Rheum. 1997;40(8):1455–65.
44. Healy AM, et al. Gas 6 promotes Axl-mediated survival in pulmonary endothelial cells. Am J Physiol Lung Cell Mol Physiol. 2001;280(6):L1273–81.
45. O'Donnell K, et al. Expression of receptor tyrosine kinase Axl and its ligand Gas6 in rheumatoid arthritis: evidence for a novel endothelial cell survival pathway. Am J Pathol. 1999;154(4):1171–80.
46. Sawabu T, et al. Growth arrest-specific gene 6 and Axl signaling enhances gastric cancer cell survival via Akt pathway. Mol Carcinog. 2007;46(2):155–64.
47. Chen C, Edelstein LC, Gélinas C. The Rel/NF-kappaB family directly activates expression of the apoptosis inhibitor Bcl-x(L). Mol Cell Biol. 2000;20(8):2687–95.
48. Zong WX, et al. The prosurvival Bcl-2 homolog Bfl-1/A1 is a direct transcriptional target of NF-kappaB that blocks TNFalpha-induced apoptosis. Genes Dev. 1999; 13(4):382 7.
49. Demarchi F, et al. Gas6 anti-apoptotic signaling requires NF-kappa B activation. J Biol Chem. 2001;276(34):31738–44.
50. Marchand B, et al. Inhibition of glycogen synthase kinase-3 activity triggers an apoptotic response in pancreatic cancer cells through JNK-dependent mechanisms. Carcinogenesis. 2012;33(3):529–37.
51. Hasanbasic I, et al. Intracellular signaling pathways involved in Gas6-Axl-mediated survival of endothelial cells. Am J Physiol Heart Circ Physiol. 2004;287(3):H1207–13.
52. Son B-K, et al. Gas6/Axl-PI3K/Akt pathway plays a central role in the effect of statins on inorganic phosphate-induced calcification of vascular smooth muscle cells. Eur J Pharmacol. 2007;556(1–3):1–8.
53. Linger RMA, et al. Mer or Axl receptor tyrosine kinase inhibition promotes apoptosis, blocks growth and enhances chemosensitivity of human non-small cell lung cancer. Oncogene. 2013;32(29):3420–31.
54. Hong J, Belkhiri A. AXL Mediates TRAIL Resistance in Esophageal Adenocarcinoma. Neoplasia. 2013;15(3):296–304.
55. Bergers G, Benjamin LE. Tumorigenesis and the angiogenic switch. Nature reviews. Cancer. 2003;3(6):401–10.

56. Delfortrie S, et al. Egfl7 promotes tumor escape from immunity by repressing endothelial cell activation. Cancer Res. 2011;71(23):7176–86.
57. Hicklin DJ, Ellis LM. Role of the vascular endothelial growth factor pathway in tumor growth and angiogenesis. J Clin Oncol Off J Am Soc Clin Oncol. 2005;23(5):1011–27.
58. Parker LH, et al. The endothelial-cell-derived secreted factor Egfl7 regulates vascular tube formation. Nature. 2004;428(6984):754–8.
59. Holland SJ, et al. Multiple roles for the receptor tyrosine kinase Axl in tumor formation. Cancer Res. 2005;65(20):9294–303.
60. Li Y, et al. Axl as a potential therapeutic target in cancer: role of Axl in tumor growth, metastasis and angiogenesis. Oncogene. 2009;28(39):3442–55.
61. Ruan G-X, Kazlauskas A. Lactate engages receptor tyrosine kinases Axl, Tie2, and vascular endothelial growth factor receptor 2 to activate phosphoinositide 3-kinase/Akt and promote angiogenesis. J Biol Chem. 2013;288:21161–2172.
62. Kalluri R, Weinberg RA. The basics of epithelial-mesenchymal transition. J Clin Investig. 2009;119(6):1420–8.
63. Koorstra JBM, et al. The Axl receptor tyrosine kinase confers an adverse prognostic influence in pancreatic cancer and represents a new therapeutic target. Cancer Biol Ther. 2009;8(7):618–26.
64. Gjerdrum C, et al. Axl is an essential epithelial-to-mesenchymal transition-induced regulator of breast cancer metastasis and patient survival. Proc Natl Acad Sci U S A. 2010;107(3):1124–9.
65. Thiery JP, Sleeman JP. Complex networks orchestrate epithelial-mesenchymal transitions. Nature reviews. Mol Cell Biol. 2006;7(2):131–42.
66. Vuoriluoto K, et al. Vimentin regulates EMT induction by Slug and oncogenic H-Ras and migration by governing Axl expression in breast cancer. Oncogene. 2011;30(12):1436–48.
67. Zhang Z, et al. Activation of the AXL kinase causes resistance to EGFR-targeted therapy in lung cancer. Nat Genet. 2012;44(8):852–60.
68. Giles KM, et al. Axl mediates acquired resistance of head and neck cancer cells to the epidermal growth factor receptor inhibitor erlotinib. Mol Cancer Ther. 2013;12(11):2541–58.
69. Hanahan D, Weinberg RA. Hallmarks of cancer: the next generation. Cell. 2011;144(5):646–74.
70. Laufs S, Schumacher J, Allgayer H. Urokinase-receptor (u-PAR): an essential player in multiple games of cancer: a review on its role in tumor progression, invasion, metastasis, proliferation/dormancy, clinical outcome and minimal residual disease. Cell Cycle. 2006;5(16):1760–71.
71. Jacob AN, et al. A receptor tyrosine kinase, UFO/Axl, and other genes isolated by a modified differential display PCR are overexpressed in metastatic prostatic carcinoma cell line DU145. Cancer Detect Prev. 1999;23(4):325–32.
72. Lee WP, et al. Akt is required for Axl-Gas6 signaling to protect cells from E1A-mediated apoptosis. Oncogene. 2002;21(3):329–36.
73. Nakano T, et al. Biological properties and gene expression associated with metastatic potential of human osteosarcoma. Clin Exp Metastasis. 2003;20(7):665–74.
74. Zhang JQ, et al. Differentially expressed genes in human giant-cell lung cancer lines with different metastatic potentials. Zhonghua Zhong Liu Za Zhi. 2004;26(10):590–3.
75. Tai K-Y, et al. Axl promotes cell invasion by inducing MMP-9 activity through activation of NF-kappaB and Brg-1. Oncogene. 2008;27(29):4044–55.
76. Cui Z-L, et al. YES-associated protein 1 promotes adenocarcinoma growth and metastasis through activation of the receptor tyrosine kinase Axl. Int J Immunopathol Pharmacol. 2012;25(4):989–1001.
77. Rankin EB, et al. Direct regulation of GAS6/AXL signaling by HIF promotes renal metastasis through SRC and MET. Proc Natl Acad Sci U S A. 2014;111(37):13373–8.
78. Li Y, et al. Axl as a downstream effector of TGF-β1 via PI3K/Akt-PAK1 signaling pathway promotes tumor invasion and chemoresistance in breast carcinoma. Tumour Biol. 2015;36(2):1115–27.

79. Reichl P, et al. Axl activates autocrine transforming growth factor-β signaling in hepatocellular carcinoma. Hepatology. 2015;61(3):930–41.
80. Hromas R, et al. Forced over-expression of the myeloid zinc finger gene MZF-1 inhibits apoptosis and promotes oncogenesis in interleukin-3-dependent FDCP.1 cells. Leukemia. 1996;10(6):1049–50.
81. Hromas R, Morris J, Cornetta K. Aberrant expression of the myeloid zinc finger gene, MZF-1, is oncogenic. Cancer Res. 1995;55:3610–4.
82. Hsieh Y-H, et al. Suppression of tumorigenicity of human hepatocellular carcinoma cells by antisense oligonucleotide MZF-1. Chin J Physiol. 2007;50(1):9–15.
83. Mudduluru G, Leupold JH, et al. PMA up-regulates the transcription of Axl by AP-1 transcription factor binding to TRE sequences via the MAPK cascade in leukaemia cells. Biol Cell. 2010;103(1):21–33.
84. Mudduluru G, Allgayer H. The human receptor tyrosine kinase Axl gene--promoter characterization and regulation of constitutive expression by Sp1, Sp3 and CpG methylation. Biosci Rep. 2008;28(3):161–76.
85. Dunne PD, et al. AXL is a key regulator of inherent and chemotherapy-induced invasion and predicts a poor clinical outcome in early-stage colon cancer. Clin Cancer Res. 2014;20(1):164–75.
86. Huang JS, et al. Oxidative stress enhances Axl-mediated cell migration through an Akt1/Rac1-dependent mechanism. Free Radic Biol Med. 2013;65:1246–56.
87. Kirane A, et al. Warfarin blocks Gas6-mediated Axl activation required for pancreatic cancer epithelial plasticity and metastasis. Cancer Res. 2015;75(18):3699–705.
88. Liu S, et al. Design, synthesis, and validation of Axl-targeted monoclonal antibody probe for microPET imaging in human lung cancer xenograft. Mol Pharm. 2014;11(11):3974–9.
89. Tokarz P, Blasiak J. The role of microRNA in metastatic colorectal cancer and its significance in cancer prognosis and treatment. Acta Biochim Pol. 2012;59(4):467–74.
90. Yakes FM, et al. Cabozantinib (XL184), a novel MET and VEGFR2 inhibitor, simultaneously suppresses metastasis, angiogenesis, and tumor growth. Mol Cancer Ther. 2011;10(12):2298–308.
91. Al-Hajj M, et al. Therapeutic implications of cancer stem cells. Curr Opin Genet Dev. 2004;14(1):43–7.
92. Dick JE. Stem cell concepts renew cancer research. Blood. 2008;112(13):4793–807.
93. Knudson AG, Strong LC, Anderson DE. Heredity and cancer in man. Prog Med Genet. 1973;9:113–58.
94. Morrison SJ, et al. A genetic determinant that specifically regulates the frequency of hematopoietic stem cells. J Immunol. 2002;168(2):635–42.
95. Saadatpour A, et al. Single-cell analysis in cancer genomics. Trends Genet. 2015;31(10):576–86.
96. Sell S. Stem cell origin of cancer and differentiation therapy. Crit Rev Oncol Hematol. 2004;51(1):1–28.
97. Visvader JE, Lindeman GJ. Cancer stem cells in solid tumours: accumulating evidence and unresolved questions. Nature reviews. Cancer. 2008;8(10):755–68.
98. Bao B. et al. (2013) Overview of cancer stem cells (CSCs) and mechanisms of their regulation: implications for cancer therapy. Current protocols in pharmacology/editorial board, S.J. Enna (editor-in-chief). [et al.], Chapter 14, Unit 14.25
99. Ahtiainen L, et al. Defects in innate immunity render breast cancer initiating cells permissive to oncolytic adenovirus. PLoS One. 2010;5(11):e13859.
100. Cichoń MA, et al. The receptor tyrosine kinase Axl regulates cell-cell adhesion and stemness in cutaneous squamous cell carcinoma. Oncogene. 2014;33(July 2013):4185–92.
101. Asiedu MK, et al. AXL induces epithelial-to-mesenchymal transition and regulates the function of breast cancer stem cells. Oncogene. 2014;33(10):1316–24.
102. Bansal N, et al. Axl receptor tyrosine kinase is up-regulated in metformin resistant prostate cancer cells. Oncotarget. 2015;6(17):15321–31.

103. Grosso S, et al. Gene expression profiling of imatinib and PD166326-resistant CML cell lines identifies Fyn as a gene associated with resistance to BCR-ABL inhibitors. Mol Cancer Ther. 2009;8:1924–33.
104. Longley DB, Johnston PG. Molecular mechanisms of drug resistance. J Pathol. 2005;205(2):275–92.
105. Macleod K, et al. Altered ErbB receptor signaling and gene expression in cisplatin-resistant ovarian cancer. Cancer Res. 2005;65(15):6789–800.
106. Hong J, et al. ABL regulation by AXL promotes cisplatin resistance in esophageal cancer. Cancer Res. 2013;73(1):331–40.
107. Kurokawa M, et al. Cisplatin influences acquisition of resistance to molecular-targeted agents through epithelial-mesenchymal transition-like changes. Cancer Sci. 2013;104(7):904–11.
108. Cirone P, et al. Patient-derived xenografts reveal limits to PI3K/mTOR- and MEK-mediated inhibition of bladder cancer. Cancer Chemother Pharmacol. 2014;73(3):525–38.
109. Kim K-C, Lee C. Reversal of Cisplatin resistance by epigallocatechin gallate is mediated by downregulation of axl and tyro 3 expression in human lung cancer cells. Korean J Physiol Pharmacol. 2014;18(1):61–6.
110. Holland SJ, et al. R428, a selective small molecule inhibitor of Axl kinase, blocks tumor spread and prolongs survival in models of metastatic breast cancer. Cancer Res. 2010;70(4):1544–54.
111. Brand TM, et al. AXL is a logical molecular target in head and neck squamous cell carcinoma. Clin Cancer Res. 2015;21(11):2601–12.
112. Hong CC, et al. Receptor tyrosine kinase AXL is induced by chemotherapy drugs and overexpression of AXL confers drug resistance in acute myeloid leukemia. Cancer Lett. 2008;268(2):314–24.
113. Kim M-S, et al. Heterogeneity of pancreatic cancer metastases in a single patient revealed by quantitative proteomics. Mol Cell Proteomics. 2014;13(11):2803–11.
114. Liu L, et al. Novel mechanism of lapatinib resistance in HER2-positive breast tumor cells: activation of AXL. Cancer Res. 2009;69(17):6871–8.
115. Zhou L, et al. Targeting MET and AXL overcomes resistance to sunitinib therapy in renal cell carcinoma. Oncogene. 2016;35(21):2687–97. doi:10.1038/onc.2015.343.
116. Martinho O, Zucca LE, Reis RM. AXL as a modulator of sunitinib response in glioblastoma cell lines. Exp Cell Res. 2015;332(1):1–10.

Insulin Signaling Linking Metabolism and Malignancy

4

Sonja M. Kessler and Alexandra K. Kiemer

Contents

Abstract

Dysregulation of insulin/insulin-like growth factor (IGF) pathways is a major feature of both the metabolic syndrome (MetS) and cancer. This chapter explains the molecular events linking MetS to carcinogenesis, thereby focusing on the insulin/IGF signaling. Specific differences in receptor expression, ligand affinity, and substrate activation enabling differential signaling of insulin and IGFs are summarized.

S.M. Kessler (✉) • A.K. Kiemer
Department of Pharmacy, Pharmaceutical Biology, Saarland University, Saarbrücken, Germany
e-mail: s.kessler@mx.uni-saarland.de; pharm.bio.kiemer@mx.uni-saarland.de

J. Haybaeck (ed.), *Mechanisms of Molecular Carcinogenesis – Volume 1*,
DOI 10.1007/978-3-319-53659-0_4

4.1 Introduction

Both the metabolic syndrome (MetS) and cancer constitute a growing health problem worldwide. In the last decades, MetS as a risk factor for cancer has become apparent [1]. The MetS is a cluster of risk factors for both cardiovascular disease and type 2 diabetes and includes glucose intolerance or insulin resistance together with two or more of the following components: raised arterial pressure, raised plasma triglyceride and/or low HDL-C, central obesity, and microalbuminuria. Jaggers and colleagues demonstrated in a study with more than 30,000 patients that the MetS is associated with an increased risk of all-cause cancer mortality in men [2]. Also other studies reported that the individual components of the MetS independently increase the risk for the development of certain cancer types [3–5]. For example, MetS was described to be associated with increased incidences of colorectal and prostate cancer, and with the recurrence of breast cancer [6–8]. A meta-analysis reported an association of MetS with liver, colorectal, bladder, endometrial, pancreatic, and breast cancers [9].

The mechanisms linking MetS and cancer risk are not completely understood. MetS may be only concomitant with other cancer risk factors, such as decreased physical activity, consumption of high calorie foods, high dietary fat intake, low-fiber intake, and oxidative stress [9]. Still, adiposity, in particular visceral obesity, results in a chronic inflammatory state, in which adipocytes and infiltrating immune cells create a pro-tumorigenic environment by producing inflammatory cytokines and chemokines [10]. The obesity-driven altered balance between proinflammatory and antiinflammatory cytokines influences insulin sensitivity [11]. Increased concentrations of inflammatory cytokines suppress insulin signal transduction, which, in turn, promotes inflammation [12, 13]. Chronic inflammation is commonly known to promote tumorigenesis [14].

Also other symptoms of MetS have been linked to insulin resistance and type 2 diabetes, i.e., high blood pressure and hypertriglyceridemia [15]. Insulin resistance can predict microalbuminuria [16].

This chapter focuses on the link between type 2 diabetes and cancer, thereby omitting other symptoms of MetS. Especially alterations in the insulin metabolism seem to increase cancer risk [17–19]. Patients with type 2 diabetes were reported to show increased cancer risk, which may be caused by hyperinsulinemia, elevated IGF1, or potentially both factors [20]. While normal cells often show little responsiveness toward insulin and IGF-dependent growth stimulation, tumor cells highly express both insulin and IGF1 receptors [20] (Fig. 4.1). Insulin resistance is characterized by a defective classical metabolic signaling. At the same time, altered signaling is induced due to increased levels of insulin, IGFs, and other factors as discussed below. Low insulin, IGF1, and IGF2 levels appear to protect from tumorigenesis [21].

Noteworthy, insulin induces the generation of reactive oxygen species (ROS) [22] (Fig. 4.1). Also hyperglycemia is known to increase oxidative stress [23], leading to increased DNA damage in diabetic individuals compared to healthy subjects [24] (Fig. 4.1). ROS can lead to downregulation of the tumor suppressor phosphatase and tensin homolog (PTEN) [25], a process known to promote insulin

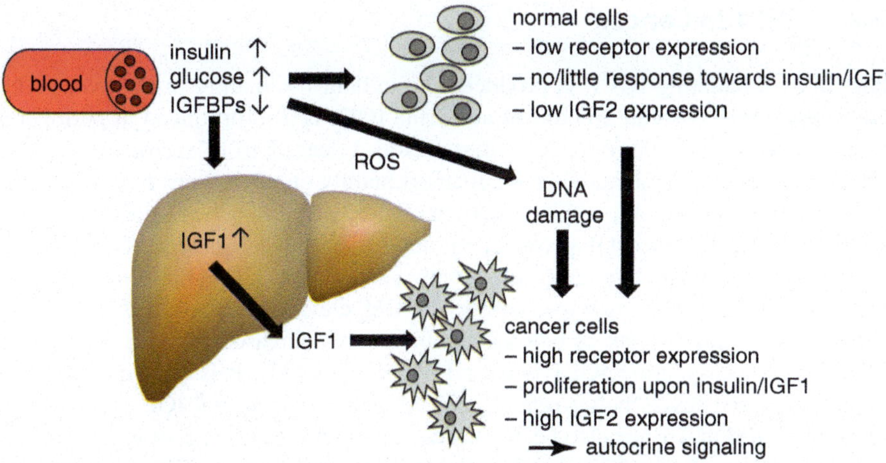

Fig. 4.1 Cancer promoting insulin/IGF signaling during insulin resistance. In normal cells of insulin target tissues, high glucose and insulin levels lead to glucose uptake and metabolic actions such as glucogen synthesis and lipid synthesis. Nonclassical insulin target tissues lack mechanisms which regulate mitogenic actions of insulin. In insulin resistance, increased systemic levels of insulin and glucose induce hepatic IGF1 production, which can lead to tumorigenesis due to the growth and survival-promoting effects of IGF1, especially in nonclassical insulin target organs. Elevated insulin and glucose levels can elicit an elevated generation of reactive oxygen species (ROS), which induce DNA damage, thereby facilitating tumor initiation

signaling. ROS generation, in general, is regarded as a hallmark of inflammation and can lead to carcinogenesis due to DNA damage [26].

Insulin/insulin-like growth factor (IGF) signaling is mediated by binding of insulin or IGFs to insulin and/or IGF receptors. IGF levels can be regulated by IGF-binding proteins (IGFBPs), which can inhibit and potentiate IGF actions by ligand binding. High circulating insulin levels decrease levels of IGFBP1 and IGFBP2, thereby increasing the bioavailability of IGF1 and concomitant changes in the cellular environment facilitating tumor formation (Fig. 4.1). In insulin resistance, nonclassical insulin target tissues which express insulin receptors are exposed to the elevated plasma levels of insulin, triglycerides, free fatty acids, and glucose [27] (Fig. 4.1). In contrast to classical insulin target tissues, such as skeletal muscle, adipose tissue, and liver, these tissues may lack a specific mechanism regulating the mitogenic actions of insulin [27]. Additional changes in signaling pathways may be induced by the increased availability of energy substrates, such as glucose, triglycerides, and free fatty acids, which also ensure energy substrates for already transformed cells [27]. High insulin levels as found in insulin resistance enhance growth hormone (GH) receptor signaling and hepatic IGF1 production [28], both of which can contribute to carcinogenesis. Concordantly, in vitro, animal, and human epidemiological studies demonstrate that despite suppressed classical metabolic insulin signaling, high concentrations of insulin and insulin-like growth factors (IGFs) promote cancer development by acting through the insulin/IGF axis [29] (Fig. 4.1).

4.2 IGF1 in Cancer

IGF1, i.e., circulating IGF1, is produced throughout life mainly in the liver under GH stimulation. A small amount of autocrine IGF1 is also produced in peripheral tissues and can be controlled by other factors released from surrounding cells. Cancer epidemiological studies have focused mainly on circulating total IGF1 and its major binding protein, IGFBP3. Circulating IGF1 is associated positively with the risk of breast, colorectal, prostate, and lung cancer, whereas total IGFBP3 concentrations are negatively associated with cancer risk [30–32]. In acromegaly patients, typically showing hypersecretion of GH, elevated levels of total IGF1, and hyperinsulinemia, the risk of colorectal cancer was increased [33]. In the healthy state, 99% of circulating IGF1 is bound by IGFBPs [34]. It is believed that free circulating IGF1 levels better reflect IGF1 bioactivity than total IGF1 levels [35]. Free circulating IGF1 has also been correlated to an increased risk of breast cancer, but independent of total IGF1 levels. In contrast to total IGF1 levels, free IGF1 was not related to tumor development in prostate cancer [36]. In addition to a hyperinsulinemia-induced increase in circulating levels of IGF1, prostate cancer cells in rodents were suggested to upregulate their intrinsic IGF1 production, thereby enabling independence from growth-promoting, circulating IGF1 [37]. In contrast, knockout mice with liver-specific IGF1 deficiency had decreased growth and metastasis of transplanted colonic adenocarcinomas and mammary tumors [38–40]. Administration of IGF1 abrogated the protective effect of IGF1 deficiency on tumor progression and resulted in neovascularization due to vascular endothelial growth factor (VEGF) induction [38, 40]. Angiogenesis is further promoted by IGF1-induced expression of hypoxia-inducible factor 1α (HIF1α) [41, 42]. Moreover, IGF1-induced metastatic tumor spread was suggested to be related to the relocation of integrins to the edge of migrating cells and the extension of lamellipodia [43, 44].

4.3 IGF2 in Cancer

IGF2 is expressed in the embryonic and neonatal state and its expression strongly drops after birth. IGF2 was reported to be reexpressed in several cancer types [45–51], defining IGF2 as an oncofetal protein [52]. Tumors take advantage of the proliferative [53, 54] and antiapoptotic properties of IGF2 by increasing IGF2 expression in tumor cells [55]. IGF2 expression was associated with the tumor grade in hepatocellular carcinoma [56, 57]. Furthermore, IGF2 expression was observed to correlate with tumor grade and lymph node metastasis in breast cancer [58]. In adrenocortical carcinoma and osteosarcoma, IGF2 expression was described to correlate with microvessel density [59, 60], to influence taxol resistance, and to be linked to a shortened disease-free survival [61]. *Igf2* transgenic mice are more susceptible to diverse malignancies [62]. Mouse models of colon cancer showing overexpression of IGF2 had a doubled tumor incidence in the presence of the adenomatous polyposis coli gene mutation [63]. Also enhanced sensitivity to IGF2 signaling led to elevated expression of proliferation-related genes and enhanced tumor development [64].

4.4 Insulin and IGF Signaling and Its Implication in Carcinogenesis

The insulin/IGF signaling network impresses through its complexity. In the following section, we point out important links between insulin/IGF signaling and carcinogenesis.

4.4.1 Insulin Receptors, IGF Receptors, and Hybrid Receptors

The three ligands insulin, IGF1, and IGF2 can act via five different receptors, namely, insulin receptors (IR) A and B, IGF1 receptor (IGF1R), and two hybrid receptors IRA/IGF1R and IRB/IGF1R. Insulin displays highest affinity for the two IRs, whereas IGF1 and 2 rather bind to the IGF1R and the hybrid receptors. IRB/IGF1R is exclusively bound by IGF1 but not by IGF2 (Fig. 4.2). The activation of the respective receptor by the different ligands can induce distinct downstream effects. Interestingly, binding of IGF2 to IRA results in a different gene expression pattern compared to binding of insulin [65], which is of relevance for tumors showing elevated IGF2 expression. However, the exact mechanisms of the different consequences of ligand binding to the insulin/IGF receptors are still unknown.

The different receptors mediate their effects through recruitment, phosphorylation, and finally activation of insulin receptor substrates (IRS), Src homology 2 domain containing transforming protein (SHC), and Janus kinase (JAK) 1/2,

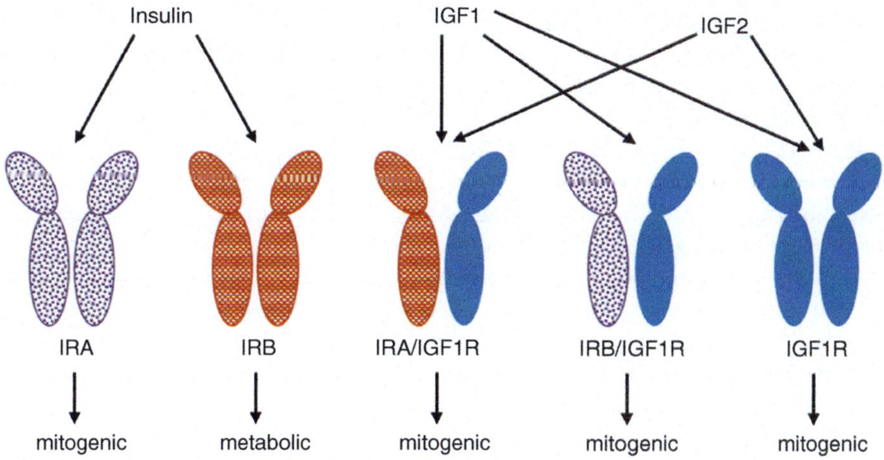

Fig. 4.2 Binding affinities of IR and IGF1R receptor ligands. Insulin preferentially binds to insulin receptors IRA and IRB. IGF1 rather activates the hybrid receptors and IGF1R. IGF1R and the hybrid receptor variant IRA/IGF1R are also bound by IGF2. IRA, the hybrid receptors, and IGF1R tend to a more mitogenic signaling, whereas IRB rather activates metabolic pathways. In cancer IRA, IGF1R, and the hybrid receptors are overexpressed, resulting in a mitogenic signaling

leading to an activation of phosphoinositide 3-kinase (PI3K), protein kinase B (PKB/AKT), and mammalian target of rapamycin (mTOR), mitogen-activated protein kinase/extracellular signal-regulated kinase (MAPK/ERK), or JAK/signal transducer and activator of transcription (STAT). Although all five receptors share the same signaling pathways, it is known that IRA and IGF1R favor mitogenic actions, whereas IRB rather induces metabolic effects (Fig. 4.2) [66–68]. Insulin resistance is caused by defects in the metabolic signaling pathways, favoring a mitogenic and growth-promoting signaling [27]. Concordantly, insulin induces transcription of a set of genes involved in metabolism, whereas insulin-like ligands increase expression of mitogenic genes [69]. Thus, differential expression of the respective receptors or their ligands in cancer, as well as in development, can implicate distinct consequences, i.e., metabolic and/or mitogenic or growth-related signaling. For example, overexpression of IGF2 in tumor cells also leads to increased mitogenic signaling via IRA [70].

IGF2 can also interact with a sixth receptor, IGF2R, which degrades IGF2 protein and therefore decreases IGF2 bioavailability. Thus, inhibitory IGF2R is often mutated or downregulated in cancer [71, 72].

4.4.2 Insulin Receptor Substrates

Autophosphorylation of the five signaling receptors mentioned above leads to the recruitment of different proteins, mainly IRS1, IRS2, and SHC, resulting in PI3K or MAPK pathway activation. Although IRS1 and IRS2 share biological effects, they exert tissue-specific roles [73]. PI3K can be activated by both IRS1 and IRS2. Besides antiapoptotic signaling, the PI3K/AKT pathway regulates metabolic pathways in tumors which promote aerobic glycolysis, a hallmark of cancer [74, 75]. Cancer cells depend rather on glycolysis than oxidative phosphorylation for energy production, even in high oxygen states, a phenomenon called the "Warburg" effect [76]. IRS2 signaling preferentially regulates tumor cell metabolism, i.e., aerobic glycolysis by inhibition of GSK-3β [77]. In line with this finding, aerobic glycolysis is diminished in IRS2 knockout cells compared to IRS1 knockout cells. Moreover, IRS2 may be required for glucose transporter (GLUT) 1 to localize to the cell surface where it can facilitate glucose uptake [78].

MAPK signaling seems to be preferentially induced by IRS1 (Fig. 4.3) [79]. Indeed, several studies suggest that IRS1 distinctly mediates the insulin/IGF1-induced mitogenic effects, whereas IRS2 appears to be more involved in generating the metabolic responses of insulin [80–83] and the migration-promoting potential of IGF1 (Fig. 4.3) [84]. However, metabolic stress induces specific phosphorylations of IRS1, which aggravate insulin resistance [85]. Specific responses were suggested to be altered by integrins differentially regulating IRS1 and IRS2 expression (Fig. 4.3) [86]. While IRS2 promotes aggressive tumor behavior, IRS1 may negatively regulate tumor progression, although IRS1 and IRS2 may play redundant roles in tumor initiation and primary tumor growth [78]. However, IRS1 was described to elevate growth and migration in breast cancer cells [87]. Different

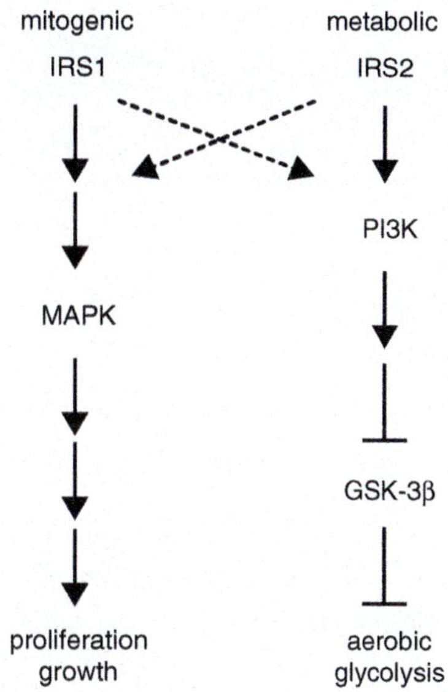

Fig. 4.3 Preferential signaling of the two insulin receptor substrates supporting tumorigenesis. IRS1 preferentially activates MAPK signaling, leading to a mitogenic signaling, thereby increasing proliferation and growth. IRS2 inhibits GSK-3βs by PI3K phosphorylation, leading to an increased aerobic glycolysis, which is essential in tumor metabolism

activation of and by IRS1 and IRS2 may be also due to the structural differences, since they share only 14 conserved sites of 21 and 23 phosphorylation sites of IRS1 and IRS2, respectively [88].

4.4.3 PI3K-Related Signaling

The PI3K/AKT pathway is the major signaling network involved in insulin/IGF signaling (Fig. 4.4). PI3K plays a central role in cancer promoting cancer cell growth, survival, motility, and metabolism [89]. By induction of several activating factors, as well as by repression of different inhibitory factors, a constitutively activated pro-survival signaling is achieved. One of these inhibitory factors is PTEN, which usually counters cell growth and cell cycle progression by inhibiting PI3K induced PIP_3 phosphorylation. PTEN displays one of the most commonly mutated tumor suppressor genes in human cancer. Loss of PTEN results in increased signaling of IGF2 through IGF1R and IRA in breast cancer cells [90]. PIP_3 activates AKT, resulting in activation of the key metabolic regulator mTOR and thereby initiating ribosomal protein synthesis and mitosis through 4E–BP1 (Fig. 4.4). Deletion of the mTOR target S6K1 in mice was shown to result in hyperinsulinemia and glucose intolerance [91]. These mTOR-induced mechanisms all favor tumor growth; thus, dysregulated mTOR signaling has been linked to numerous human cancers [92–94]. Loss of PTEN leads to constitutively activated mTOR [95]. mTOR regulation is controlled not only by PTEN but also by the tumor suppressor gene products tuberous sclerosis (TSC) 1, TSC2, and AMP-activated protein kinase (AMPK). AMPK interacts with both TSC2 and mTOR and thus directly and indirectly inhibits the activation of mTOR (Fig. 4.4) [96]. In colorectal cancer, frameshift mutations in the

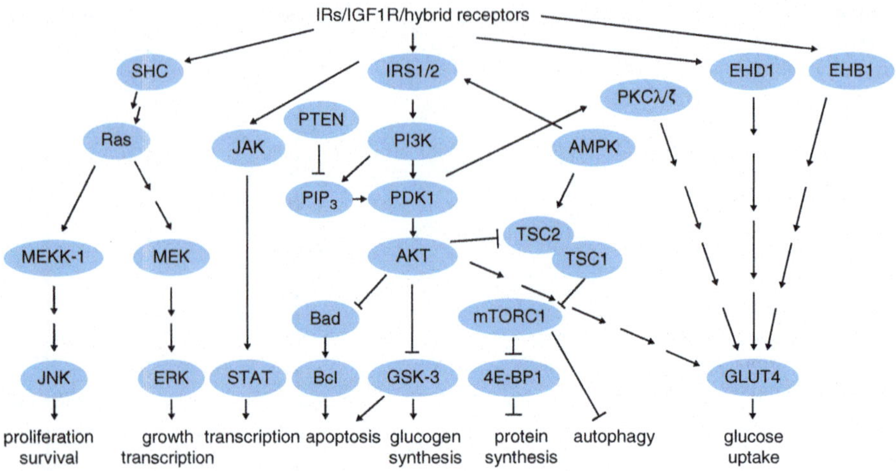

Fig. 4.4 Overview of the insulin/IGF signaling network. Central factors of the insulin/IGF signaling pathways are shown. For details see text

AMPK-encoding gene were observed [97]. mTOR itself was also shown to be mutated in several types of cancer [98–100].

Antiapoptotic insulin/IGF signaling via AKT is realized by initiating phosphorylation of the Bcl-2 family member BAD, followed by Bcl-XL leading to inhibition of apoptosis (Fig. 4.4). Moreover, multiple transcription factors, such as cAMP response element-binding protein (CREB), nuclear factor (NF)-κB, and p53, which are involved in the transcription of genes encoding apoptotic mediators, are regulated by IGFs [101]. Akt hyperactivation in cancer not only contributes to the inhibition of apoptosis but is also coupled with metabolic alterations in cancer cells, including aerobic glycolysis [102].

4.4.4 MAPK-Related Signaling

Besides PI3K activation, insulin or IGF stimulation has been shown to increase interaction with SHC [103]. SHC initiates the MAPK pathway, which represents a key promoter of cell proliferation, tumor development, tumor growth [104], as well as in the maintenance and progression of several tumors [105, 106]. The MAPK pathway involves activation of Ras, which can activate both JNK and MEK/ERK pathways (Fig. 4.4). The Ras/Raf cascade is frequently elevated in cancer, either growth factor dependently or independently, e.g., due to mutations [107, 108].

Noteworthy, ERK signaling is also implicated in metabolic alterations, such as insulin resistance. Chronic activation of ERK induces severe insulin resistance by inhibiting expressions of both GLUT4 and insulin-signaling proteins [109]. Targeting the MEK/ERK cascade normalized hyperglycemia and hyperlipidemia and improved insulin sensitivity, as well as glucose tolerance in diabetic mice [110]. Thus, the MAPK pathway displays a second important insulin/IGF-mediated pathway linking insulin resistance to cancer.

Conclusion

Insulin/IGF signaling is of particular importance in carcinogenesis, especially when tumor development is the consequence of chronic metabolic diseases. Insulin/IGF signaling mediates its effects through different signaling cascades. Not surprisingly, tumor cells activate multiple signaling pathways at once to achieve growth, protection against apoptosis, metastasis, metabolic alterations, and other features being a characteristic for cancer. Here, the activation of the insulin/IGF axis offers the advantage of activating several pathways at once for tumor development and progression. As a result from the extensive basic research, several therapeutic approaches targeting the insulin/IGF axis in cancer are currently under investigation and reviewed in detail elsewhere [111–113].

Acknowledgements The authors' work was funded, in part, by the Else Kröner-Fresenius-Stiftung (2012_A250).

References

1. Alberti KG, Eckel RH, Grundy SM, Zimmet PZ, Cleeman JI, Donato KA, Fruchart J-C, James WPT, Loria CM, Smith SC. Harmonizing the metabolic syndrome: a joint interim statement of the International Diabetes Federation Task Force on Epidemiology and Prevention; National Heart, Lung, and Blood Institute; American Heart Association; World Heart Federation; International Atherosclerosis Society; and International Association for the Study of Obesity. Circulation. 2009;120(16):1640–5. PMID: 19805654. doi:10.1161/circulationaha.109.192644.
2. Jaggers JR, Sui X, Hooker SP, LaMonte MJ, Matthews CE, Hand GA, Blair SN. Metabolic syndrome and risk of cancer mortality in men. Eur J Cancer. 2009;45(10):1831–8. PMID: 19250819. doi:10.1016/j.ejca.2009.01.031.
3. Nicolucci A. Epidemiological aspects of neoplasms in diabetes. Acta Diabetol. 2010;47(2):87–95. PMID: 20376506. doi:10.1007/s00592-010-0187-3.
4. Renehan AG, Tyson M, Egger M, Heller RF, Zwahlen M. Body-mass index and incidence of cancer: a systematic review and meta-analysis of prospective observational studies. Lancet. 2008;371(9612):569–78. PMID: 18280327. doi:10.1016/S0140-6736(08)60269-X.
5. Jafri H, Alsheikh-Ali AA, Karas RH. Baseline and on-treatment high-density lipoprotein cholesterol and the risk of cancer in randomized controlled trials of lipid-altering therapy. J Am Coll Cardiol. 2010;55(25):2846–54. PMID: 20579542. doi:10.1016/j.jacc.2009.12.069.
6. Chiu HM, Lin JT, Shun CT, Liang JT, Lee YC, Huang SP, Wu MS. Association of metabolic syndrome with proximal and synchronous colorectal neoplasm. Clin Gastroenterol Hepatol. 2007;5(2):221–9. PMID: 16931168. doi:10.1016/j.cgh.2006.06.022.
7. Laukkanen JA, Laaksonen DE, Niskanen L, Pukkala E, Hakkarainen A, Salonen JT. Metabolic syndrome and the risk of prostate cancer in Finnish men: a population-based study. Cancer Epidemiol Biomarkers Prev. 2004;13(10):1646–50. PMID: 15466982.
8. Pasanisi P, Berrino F, De Petris M, Venturelli E, Mastroianni A, Panico S. Metabolic syndrome as a prognostic factor for breast cancer recurrences. Int J Cancer. 2006;119(1):236–8. PMID: 16450399. doi:10.1002/ijc.21812.

9. Esposito K, Chiodini P, Colao A, Lenzi A, Giugliano D. Metabolic syndrome and risk of cancer: a systematic review and meta-analysis. Diabetes Care. 2012;35(11):2402–11. PMID: 23640372. doi:10.2337/dc12-0336.
10. Harvey AE, Lashinger LM, Hursting SD. The growing challenge of obesity and cancer: an inflammatory issue. Ann N Y Acad Sci. 2011;1229(1):45–52. PMID: 21793838. doi:10.1111/j.1749-6632.2011.06096.x.
11. Lumeng CN, Saltiel AR. Inflammatory links between obesity and metabolic disease. J Clin Invest. 2011;121(6):2111–7. PMID: 21633179. doi:10.1172/JCI57132.
12. Adabimohazab R, Garfinkel A, Milam EC, Frosch O, Mangone A, Convit A. Does inflammation mediate the association between obesity and insulin resistance? Inflammation. 2016;39(3):994–1003. PMID: 26956471. doi:10.1007/s10753-016-0329-z.
13. Dandona P, Aljada A, Bandyopadhyay A. Inflammation: the link between insulin resistance, obesity and diabetes. Trends Immunol. 2004;25(1):4–7. PMID: 14698276. doi:10.1016/j.it.2003.10.013.
14. Kamp DW, Shacter E, Weitzman SA. Chronic inflammation and cancer: the role of the mitochondria. Oncology (Williston Park). 2011;25(5):400–10, 413. PMID: 21710835.
15. DeFronzo RA. Insulin resistance, lipotoxicity, type 2 diabetes and atherosclerosis: the missing links. The Claude Bernard Lecture 2009. Diabetologia. 2010;53(7):1270–87. PMID: 20361178. doi:10.1007/s00125-010-1684-1.
16. Hsu C-C, Chang H-Y, Huang M-C, Hwang S-J, Yang Y-C, Tai T-Y, Yang H-J, Chang C-T, Chang C-J, Li Y-S, Shin S-J, Kuo KN. Association between insulin resistance and development of microalbuminuria in type 2 diabetes: a prospective cohort study. Diabetes Care. 2011;34(4):982–7. PMID: 21335369. doi:10.2337/dc10-1718.
17. Nilsen TIL, Vatten LJ. Prospective study of colorectal cancer risk and physical activity, diabetes, blood glucose and BMI: exploring the hyperinsulinaemia hypothesis. Br J Cancer. 2001;84(3):417–22. PMID: 11161410. doi:10.1054/bjoc.2000.1582.
18. Trevisan M, Liu J, Muti P, Misciagna G, Menotti A, Fucci F. Markers of insulin resistance and colorectal cancer mortality. Cancer Epidemiology Biomarkers Prev. 2001;10(9):937–41. PMID: 11535544.
19. Jee SH, Ohrr H, Sull JW, Yun JE, Ji M, Samet JM. Fasting serum glucose level and cancer risk in Korean men and women. JAMA. 2005;293(2):194–202. PMID: 15644546. doi:10.1001/jama.293.2.194.
20. Taubes G. Unraveling the obesity-cancer connection. Science. 2012;335(6064):28–32. PMID: 22223787. doi:10.1126/science.335.6064.28.
21. Gallagher EJ, LeRoith D. Minireview: IGF, Insulin, and Cancer. Endocrinology. 2011;152(7):2546–51. PMID: 21540285. doi:10.1210/en.2011-0231.
22. Goldstein BJ, Mahadev K, Wu X, Zhu L, Motoshima H. Role of insulin-induced reactive oxygen species in the insulin signaling pathway. Antioxid Redox Signal. 2005;7(7-8):1021–31. PMID: 15998257. doi:10.1089/ars.2005.7.1021.
23. Paolisso G, Giugliano D. Oxidative stress and insulin action: is there a relationship? Diabetologia. 1996;39(3):357–63. PMID: 8721784. doi:10.1007/s001250050454.
24. Dandona P, Thusu K, Cook S, Snyder B, Makowski J, Armstrong D, Nicotera T. Oxidative damage to DNA in diabetes mellitus. Lancet. 1996;347(8999):444–5. PMID: 8618487. doi:10.1016/S0140-6736(96)90013-6.
25. Lee S-R, Yang K-S, Kwon J, Lee C, Jeong W, Rhee SG. Reversible inactivation of the tumor suppressor PTEN by H_2O_2. J Biol Chem. 2002;277(23):20336–42. PMID: 11916965. doi: 10.1074/jbc.M111899200.
26. Block K, Gorin Y. Aiding and abetting roles of NOX oxidases in cellular transformation. Nat Rev Cancer. 2012;12(9):627. PMID: 22918415. doi:10.1038/nrc3339.
27. Gunter MJ, Leitzmann MF. Obesity and colorectal cancer: epidemiology, mechanisms and candidate genes. J Nutr Biochem. 2006;17(3):145–56. PMID: 16426829. doi:10.1016/j.jnutbio.2005.06.011.
28. Baxter RC, Bryson JM, Turtle JR. Somatogenic receptors of rat liver: regulation by insulin. Endocrinology. 1980;107(4):1176–81. PMID: 6250795. doi:10.1210/endo-107-4-1176.

29. Godsland Ian F. Insulin resistance and hyperinsulinaema in the development and progression of cancer. Clin Sci (Lond). 2010;118(5):315–32. PMID: 19922415. doi:10.1042/cs20090399.
30. Renehan AG, Frystyk J, Flyvbjerg A. Obesity and cancer risk: the role of the insulin–IGF axis. Trends Endocrinol Metab. 2006;17(8):328–36. PMID: 16956771. doi:10.1016/j.tem.2006.08.006.
31. Renehan AG, Zwahlen M, Minder C, O'Dwyer ST, Shalet SM, Egger M. Insulin-like growth factor (IGF)-I, IGF binding protein-3, and cancer risk: systematic review and meta-regression analysis. Lancet. 2004;363(9418):1346–53. PMID: 15110491. doi:10.1016/S0140-6736(04)16044-3.
32. Morris JK, George LM, Wu T, Wald NJ. Insulin-like growth factors and cancer: no role in screening. Evidence from the BUPA study and meta-analysis of prospective epidemiological studies. Br J Cancer. 2006;95(1):112–7. PMID: 16804529. doi:10.1038/sj.bjc.6603200.
33. Renehan AG, O'Connell J, O'Halloran D, Shanahan F, Potten CS, O'Dwyer ST, Shalet SM. Acromegaly and colorectal cancer: a comprehensive review of epidemiology, biological mechanisms, and clinical implications. Horm Metab Res. 2003;35(11-12):712–25. PMID: 14710350. doi:10.1055/s-2004-814150.
34. Jones JI, Clemmons DR. Insulin-like growth factors and their binding proteins: biological actions. Endocr Rev. 1995;16(1):3–34. PMID: 7758431. doi:10.1210/edrv-16-1-3.
35. Janssen JAMJL, Lely AJ, Lamberts SWJ. Circulating free insulin-like growth-factor-I (IGF-I) levels should also be measured to estimate the IGF-I bioactivity. J Endocrinol Invest. 2014;26(6):588–94. PMID: 12952376. doi:10.1007/bf03345225.
36. Janssen JAMJL, Wildhagen MF, Ito K, Blijenberg BG, Van Schaik RHN, Roobol MJ, Pols HAP, Lamberts SWJ, Schröder FH. Circulating free insulin-like growth factor (IGF)-I, total IGF-I, and IGF binding protein-3 levels do not predict the future risk to develop prostate cancer: Results of a case-control study involving 201 patients within a population-based screening with a 4-year interval. J Clin Endocrinol Metab. 2004;89(9):4391–6. PMID: 15356036. doi:10.1210/jc.2004-0232.
37. Wang YZ, Wong YC. Sex hormone-induced prostatic carcinogenesis in the noble rat: the role of insulin-like growth factor-1 (IGF-1) and vascular endothelial growth factor (VEGF) in the development of prostate cancer. Prostate. 1998;35(3):165–77. PMID: 9582085. doi:10.1002/(SICI)1097-0045(19980515)35:3<165::AID-PROS2>3.0.CO;2-G.
38. Dunn SE, Kari FW, French J, Leininger JR, Travlos G, Wilson R, Barrett JC. Dietary restriction reduces insulin-like growth factor I levels, which modulates apoptosis, cell proliferation, and tumor progression in p53- deficient mice. Cancer Res. 1997;57(21):4667–72. PMID: 9354418.
39. Wu Y, Cui K, Miyoshi K, Hennighausen L, Green JE, Setser J, LeRoith D, Yakar S. Reduced circulating insulin-like growth factor I levels delay the onset of chemically and genetically induced mammary tumors. Cancer Res. 2003;63(15):4384–8. PMID: 11861378.
40. Wu Y, Yakar S, Zhao L, Hennighausen L, LeRoith D. Circulating insulin-like growth factor-I levels regulate colon cancer growth and metastasis. Cancer Res. 2002;62(4):1030–5. PMID: 11861378.
41. Fukuda R, Hirota K, Fan F, Jung YD, Ellis LM, Semenza GL. Insulin-like growth factor 1 induces hypoxia-inducible factor 1-mediated vascular endothelial growth factor expression, which is dependent on MAP kinase and phosphatidylinositol 3-kinase signaling in colon cancer cells. J Biol Chem. 2002;277(41):38205–11. PMID: 12149254. doi:10.1074/jbc.M203781200.
42. Héron-Milhavet L, LeRoith D. Insulin-like growth factor I induces MDM2-dependent degradation of p53 via the p38 MAPK pathway in response to DNA damage. J Biol Chem. 2002;277(18):15600–6. PMID: 11877395. doi:10.1074/jbc.M111142200.
43. Canonici A, Steelant W, Rigot V, Khomitch-Baud A, Boutaghou-Cherid H, Bruyneel E, Van Roy F, Garrouste F, Pommier G, André F. Insulin-like growth factor-I receptor, E-cadherin and αv integrin form a dynamic complex under the control of α-catenin. Int J Cancer. 2008;122(3):572–82. PMID: 17955485. doi:10.1002/ijc.23164.
44. Meyer GE, Shelden E, Kim B, Feldman EL. Insulin-like growth factor I stimulates motility in human neuroblastoma cells. Oncogene. 2001;20(51):7542–50. PMID: 11709726. doi:10.1038/sj.onc.1204927.

45. Cariani E, Lasserre C, Seurin D, Hamelin B, Kemeny F, Franco D, Czech MP, Ullrich A, Brechot C. Differential expression of insulin-like growth factor II mRNA in human primary liver cancers, benign liver tumors, and liver cirrhosis. Cancer Res. 1988;48(23):6844–9. PMID: 3180092.
46. Cheng YW, Idrees K, Shattock R, Khan SA, Zeng Z, Brennan CW, Paty P, Barany F. Loss of imprinting and marked gene elevation are 2 forms of aberrant IGF2 expression in colorectal cancer. Int J Cancer. 2010;127(3):568–77. PMID: 19957330. doi:10.1002/ijc.25086.
47. Faria AM, Almeida MQ. Differences in the molecular mechanisms of adrenocortical tumorigenesis between children and adults. Mol Cell Endocrinol. 2012;351(1):52–7. PMID: 22019901. doi:10.1016/j.mce.2011.09.040.
48. Lu ZL, Luo DZ, Wen JM. Expression and significance of tumor-related genes in HCC. World J Gastroenterol. 2005;11(25):3850–4. PMID: 15991281.
49. Vasiljevic A, Champier J, Figarella-Branger D, Wierinckx A, Jouvet A, Fèvre-Montange M. Molecular characterization of central neurocytomas: potential markers for tumor typing and progression. Neuropathology. 2013;33(2):149–61. PMID: 22816789. doi:10.1109/MED.2013.6608886.
50. Wu MS, Wang HP, Lin CC, Sheu JC, Shun CT, Lee WJ, Lin JT. Loss of imprinting and overexpression of IGF2 gene in gastric adenocarcinoma. Cancer Lett. 1997;120(1):9–14. PMID: 9570380. doi:10.1016/S0304-3835(97)00279-6.
51. Zhao R, DeCoteau JF, Geyer CR, Gao M, Cui H, Casson AG. Loss of imprinting of the insulin-like growth factor II (IGF2) gene in esophageal normal and adenocarcinoma tissues. Carcinogenesis. 2009;30(12):2117–22. PMID: 19843644. doi:10.1093/carcin/bgp254.
52. Takeda S, Kondo M, Kumada T, Koshikawa T, Ueda R, Nishio M, Osada H, Suzuki H, Nagatake M, Washimi O, Takagi K, Takahashi T, Nakao A, Takahashi T. Allelic-expression imbalance of the insulin-like growth factor 2 gene in hepatocellular carcinomas and underlying disease. Oncogene. 1996;12(7):1589–92. PMID: 8622877.
53. Hartmann W, Koch A, Brune H, Waha A, Schüller U, Dani I, Denkhaus D, Langmann W, Bode U, Wiestler OD, Schilling K, Pietsch T. Insulin-like growth factor II is involved in the proliferation control of medulloblastoma and its cerebellar precursor cells. Am J Pathol. 2005;166(4):1153–62. PMID: 15793295.
54. Pacher M, Seewald MJ, Mikula M, Oehler S, Mogg M, Vinatzer U, Eger A, Schweifer N, Varecka R, Sommergruber W, Mikulits W, Schreiber M. Impact of constitutive IGF1/IGF2 stimulation on the transcriptional program of human breast cancer cells. Carcinogenesis. 2007;28(1):49–59. PMID: 16774935. doi:10.1093/carcin/bgl091.
55. Eichenmüller M, Hemmerlein B, Von Schweinitz D, Kappler R. Betulinic acid induces apoptosis and inhibits hedgehog signalling in rhabdomyosarcoma. Br J Cancer. 2010;103(1):43–51. PMID: 20517313. doi:10.1038/sj.bjc.6605715.
56. Abou-Alfa GK, Capanu M, O'Reilly EM, Ma J, Chou JF, Gansukh B, Shia J, Kalin M, Katz S, Abad L, Reidy-Lagunes DL, Kelsen DP, Chen HX, Saltz LB. A phase II study of cixutumumab (IMC-A12, NSC742460) in advanced hepatocellular carcinoma. J Hepatol. 2014;60(2):319–24. PMID: 24045151. doi:10.1016/j.jhep.2013.09.008.
57. Tovar V, Alsinet C, Villanueva A, Hoshida Y, Chiang DY, Solé M, Thung S, Moyano S, Toffanin S, Mínguez B, Cabellos L, Peix J, Schwartz M, Mazzaferro V, Bruix J, Llovet JM. IGF activation in a molecular subclass of hepatocellular carcinoma and pre-clinical efficacy of IGF-1R blockage. J Hepatol. 2010;52(4):550–9. PMID: 20206398. doi:10.1016/j.jhep.2010.01.015.
58. Qiu J, Yang R, Rao Y, Du Y, Kalembo FW. Risk factors for breast cancer and expression of insulin-like growth factor-2 (IGF-2) in women with breast cancer in Wuhan city, China. PLoS One. 2012;7(5):e36497. PMID: 22662119. doi:10.1371/journal.pone.0036497.
59. Chen P, Wang SJ, Wang HB, Ren P, Wang XQ, Liu WG, Gu WL, Li DQ, Zhang TG, Zhou CJ. The distribution of IGF2 and IMP3 in osteosarcoma and its relationship with angiogenesis. J Mol Histol. 2012;43(1):63–70. PMID: 22042095. doi:10.1007/s10735-011-9370-2.
60. Zhu Y, Xu Y, Chen D, Zhang C, Rui W, Zhao J, Zhu Q, Wu Y, Shen Z, Wang W, Ning G, Wang X. Expression of STAT3 and IGF2 in adrenocortical carcinoma and its relationship with angiogenesis. Clin Transl Oncol. 2013;16(7):644–9. PMID: 24178245. doi:10.1007/s12094-013-1130-1.

61. Huang GS, Brouwer-Visser J, Ramirez MJ, Kim CH, Hebert TM, Lin J, Arias-Pulido H, Qualls CR, Prossnitz ER, Goldberg GL, Smith HO, Horwitz SB. Insulin-like growth factor 2 expression modulates taxol resistance and is a candidate biomarker for reduced disease-free survival in ovarian cancer. Clin Cancer Res. 2010;16(11):2999–3010. PMID: 20404007. doi:10.1158/1078-0432.CCR-09-3233.
62. Rogler CE, Yang D, Rossetti L, Donohoe J, Alt E, Chang CJ, Rosenfeld R, Neely K, Hintz R. Altered body composition and increased frequency of diverse malignancies in insulin-like growth factor-II transgenic mice. J Biol Chem. 1994;269(19):13779–84. PMID:7514593.
63. Sakatani T, Kaneda A, Iacobuzio-Donahue CA, Carter MG, De Boom WS, Okano H, Ko MSH, Ohlsson R, Longo DL, Feinberg AP. Loss of imprinting of Igf2 alters intestinal maturation and tumorigenesis in mice. Science. 2005;307(5717):1976–8. PMID:15731405. doi:10.1126/science.1108080.
64. Kaneda A, Wang CJ, Cheong R, Timp W, Onyango P, Wen B, Iacobuzio-Donahue CA, Ohlsson R, Andraos R, Pearson MA, Sharov AA, Longo DL, Ko MSH, Levchenko A, Feinberg AP. Enhanced sensitivity to IGF-II signaling links loss of imprinting of IGF2 to increased cell proliferation and tumor risk. Proc Natl Acad Sci U S A. 2007;104(52):20926–31. doi:10.1073/pnas.0710359105.
65. Pandini G, Conte E, Medico E, Sciacca L, Vigneri R, Belfiore A. IGF-II binding to insulin receptor isoform A induces a partially different gene expression profile from insulin binding. Ann N Y Acad Sci. 2004;1028:450–6. PMID:18087038. doi:10.1196/annals.1322.053.
66. Belfiore A, Frasca F, Pandini G, Sciacca L, Vigneri R. Insulin receptor isoforms and insulin receptor/insulin-like growth factor receptor hybrids in physiology and disease. Endocr Rev. 2009;30(6):586–623. PMID:19752219. doi:10.1210/er.2008-0047.
67. LeRoith D, Werner H, Beitner-Johnson D, ACT R. Molecular and cellular aspects of the insulin-like growth factor I receptor. Endocr Rev. 1995;16(2):143–63. PMID:7540132. doi:10.1210/edrv-16-2-143.
68. Yakar S, Pennisi P, Kim CH, Zhao H, Toyoshima Y, Gavrilova O, LeRoith D. Studies involving the GH-IGF axis: lessons from IGF-I and IGF-I receptor gene targeting mouse models. J Endocrinol Invest. 2005;28(5 Suppl):19–22. PMID:16114270.
69. ter Braak B, Wink S, Koedoot E, Pont C, Siezen C, van der Laan JW, van de Water B. Alternative signaling network activation through different insulin receptor family members caused by pro-mitogenic antidiabetic insulin analogues in human mammary epithelial cells. Breast Cancer Res. 2015;17(1):97. PMID:26187749. doi:10.1186/s13058-015-0600-5.
70. Sciacca L, Costantino A, Pandini G, Mineo R, Frasca F, Scalia P, Sbraccia P, Goldfine ID, Vigneri R, Belfiore A. Insulin receptor activation by IGF-II in breast cancers: evidence for a new autocrine/paracrine mechanism. Oncogene. 1999;18(15):2471–9. PMID:10229198. doi:10.1038/sj.onc.1202600.
71. Kreiling JL, Montgomery MA, Wheeler JR, Kopanic JL, Connelly CM, Zavorka ME, Allison JL, MacDonald RG. Dominant negative effect of truncated mannose 6-phosphate/insulin-like growth factor II receptor species in cancer. FEBS J. 2012;279(15):2695–713. PMID: 22681933. doi:10.1111/j.1742-4658.2012.08652.x.
72. Oates AJ, Schumaker LM, Jenkins SB, Pearce AA, DaCosta SA, Arun B, Ellis MJC. The mannose 6-phosphate/insulin-like growth factor 2 receptor (M6P/IGF2R), a putative breast tumor suppressor gene. Breast Cancer Res Treat. 1998;47(3):269–81. PMID: 9516081. doi:10.1023/A:1005959218524.
73. Kido Y, Burks DJ, Withers D, Bruning JC, Kahn CR, White MF, Accili D. Tissue-specific insulin resistance in mice with mutations in the insulin receptor, IRS-1, and IRS-2. J Clin Invest. 2000;105(2):199–205. PMID:10642598. doi:10.1172/JCI7917.
74. DeBerardinis RJ, Lum JJ, Hatzivassiliou G, Thompson CB. The biology of cancer: metabolic reprogramming fuels cell growth and proliferation. Cell Metab. 2008;7(1):11–20. PMID:18177721. doi:10.1016/j.cmet.2007.10.002.
75. Hanahan D, Weinberg RA. Hallmarks of cancer: the next generation. Cell. 2011;144(5):646–74. PMID:21376230. doi:10.1016/j.cell.2011.02.013.
76. Warburg O. On the origin of cancer cells. Science. 1956;123(3191):309–14. PMID:13298683. doi:10.1126/science.123.3191.309.

77. Landis J, Shaw LM. Insulin receptor substrate 2-mediated phosphatidylinositol 3-kinase signaling selectively inhibits glycogen synthase kinase 3β to regulate aerobic glycolysis. J Biol Chem. 2014;289(26):18603–13. PMID:24811175. doi:10.1074/jbc.M114.564070.
78. Shaw LM. The insulin receptor substrate (IRS) proteins. Cell Cycle. 2011;10(11):1750–6. PMID:21597332. doi:10.4161/cc.10.11.15824.
79. Byron SA, Horwitz KB, Richer JK, Lange CA, Zhang X, Yee D. Insulin receptor substrates mediate distinct biological responses to insulin-like growth factor receptor activation in breast cancer cells. Br J Cancer. 2006;95(9):1220–8. PMID:17043687. doi:10.1038/sj.bjc.6603354.
80. Brüning JC, Winnay J, Cheatham B, Kahn CR. Differential signaling by insulin receptor substrate 1 (IRS-1) and IRS-2 in IRS-1-deficient cells. Mol Cell Biol. 1997;17(3):1513–21. PMID:9032279.
81. Miele C, Caruso M, Calleja V, Auricchio R, Oriente F, Formisano P, Condorelli G, Cafieri A, Sawka-Verhelle D, Van Obberghen E, Beguinot F. Differential role of insulin receptor substrate (IRS)-1 and IRS-2 in L6 skeletal muscle cells expressing the Arg1152 → Gln insulin receptor. J Biol Chem. 1999;274(5):3094–102. PMID:9915848. doi:10.1074/jbc.274.5.3094.
82. Myers MG, Wang LM, Sun XJ, Zhang Y, Yenush L, Schlessinger J, Pierce JH, White MF. Role of IRS-1-GRB-2 complexes in insulin signaling. Mol Cell Biol. 1994;14(6):3577–87. PMID:8196603. doi:10.1128/mcb.14.6.3577.
83. Rose DW, Saltiel AR, Majumdar M, Decker SJ, Olefsky JM. Insulin receptor substrate 1 is required for insulin-mediated mitogenic signal transduction. Proc Natl Acad Sci U S A. 1994;91(2):797–801. PMID:8290602. doi:10.1073/pnas.91.2.797.
84. Jackson JG, Zhang X, Yoneda T, Yee D. Regulation of breast cancer cell motility by insulin receptor substrate-2 (IRS-2) in metastatic variants of human breast cancer cell lines. Oncogene. 2001;20(50):7318–25. PMID:11704861. doi:10.1038/sj.onc.1204920.
85. Hançer NJ, Qiu W, Cherella C, Li Y, Copps KD, White MF. Insulin and metabolic stress stimulate multisite serine/threonine phosphorylation of insulin receptor substrate 1 and inhibit tyrosine phosphorylation. J Biol Chem. 2014;289(18):12467–84. PMID:24652289. doi:10.1074/jbc.M114.554162.
86. Lebrun P, Baron V, Hauck CR, Schlaepfer DD, Van Obberghen E. Cell adhesion and focal adhesion kinase regulate insulin receptor substrate-1 expression. J Biol Chem. 2000;275(49):38371–7. PMID:10967115. doi:10.1074/jbc.M006162200.
87. Meyer K, Albaugh B, Schoenike B, Roopra A. Type 1 insulin-like growth factor receptor/insulin receptor substrate 1 signaling confers pathogenic activity on breast tumor cells lacking REST. Mol Cell Biol. 2015;35(17):2991–3004. PMID:26100015. doi:10.1128/mcb.01149-14.
88. Sun XJ, Wang L-M, Zhang Y, Yenush L, Myers Jr MG, Glasheen E, Lane WS, Pierce JH, White MF. Role of IRS-2 in insulin and cytokine signalling. Nature. 1995;377(6545):173–7. PMID: 7675087. doi: 10.1038/377173a0.
89. Courtney KD, Corcoran RB, Engelman JA. The PI3K pathway as drug target in human cancer. J Clin Oncol. 2010;28(6):1075–83. PMID:20085938. doi:10.1200/jco.2009.25.3641.
90. Perks CM, Vernon EG, Rosendahl AH, Tonge D, Holly JMP. IGF-II and IGFBP-2 differentially regulate PTEN in human breast cancer cells. Oncogene. 2007;26(40):5966–72. PMID: 17369847. doi: 10.1038/sj.onc.1210397 .
91. Soliman GA. The mammalian target of rapamycin signaling network and gene regulation. Curr Opin Lipidol. 2005;16(3):317–23. PMID: 15891393. doi:10.1097/01.mol.0000169352.35642.06.
92. Hsieh AC, Liu Y, Edlind MP, Ingolia NT, Janes MR, Sher A, Shi EY, Stumpf CR, Christensen C, Bonham MJ, Wang S, Ren P, Martin M, Jessen K, Feldman ME, Weissman JS, Shokat KM, Rommel C, Ruggero D. The translational landscape of mTOR signalling steers cancer initiation and metastasis. Nature. 2012;485(7396):55–61. PMID: 22367541. doi:10.1038/nature10912.
93. Sabatini DM. mTOR and cancer: Insights into a complex relationship. Nat Rev Cancer. 2006;6(9):729–34. PMID: 16915295. doi:10.1038/nrc1974.
94. Zoncu R, Efeyan A, Sabatini DM. mTOR: from growth signal integration to cancer, diabetes and ageing. Nat Rev Mol Cell Biol. 2011;12(1):21–35. PMID: 21157483. doi:10.1038/nrm3025.
95. Reiling JH, Sabatini DM. Stress and mTORture signaling. Oncogene. 2006;25(48):6373–83. PMID: 17041623. doi:10.1038/sj.onc.1209889.

96. Nellist M, Burgers PC, van den Ouweland AMW, Halley DJJ, Luider TM. Phosphorylation and binding partner analysis of the TSC1–TSC2 complex. Biochem Biophys Res Commun. 2005;333(3):818–26. PMID: 15963462. doi:10.1016/j.bbrc.2005.05.175.
97. Choi MR, An CH, Yoo NJ, Lee SH. Frameshift mutations of PRKAG1 gene encoding an AMPK gamma subunit in colorectal cancers. J Gastrointest Liver Dis. 2014;23(3): 343–5. PMID: 25267969. doi:10.1543/jgld.2014.1121.
98. Ghosh AP, Marshall CB, Coric T, Shim EH, Kirkman R, Ballestas ME, Ikura M, Bjornsti MA, Sudarshan S. Point mutations of the mTOR-RHEB pathway in renal cell carcinoma. Oncotarget. 2015;6(20):17895–10. PMID: 26255626. doi:10.18632/oncotarget.4963.
99. Lim JS, Lee JH. Brain somatic mutations in MTOR leading to focal cortical dysplasia. BMB Rep. 2016;49(2):71–2. PMID: 26779999. doi:10.5483/BMBRep.2016.49.2.010.
100. Nakashima M, Saitsu H, Takei N, Tohyama J, Kato M, Kitaura H, Shiina M, Shirozu H, Masuda H, Watanabe K, Ohba C, Tsurusaki Y, Miyake N, Zheng Y, Sato T, Takebayashi H, Ogata K, Kameyama S, Kakita A, Matsumoto N. Somatic Mutations in the MTOR gene cause focal cortical dysplasia type IIb. Ann Neurol. 2015;78(3):375–86. PMID: 26018084. doi:10.1002/ana.24444.
101. Braun S, Bitton-Worms K, LeRoith D. The link between the metabolic syndrome and cancer. Int J Biol Sci. 2011;7(7):1003–15. PMID: 21912508.
102. Robey RB, Hay N. Is Akt the "Warburg kinase"?—Akt-energy metabolism interactions and oncogenesis. Semin Cancer Biol. 2009;19(1):25–31. PMID: 19130886. doi:10.1016/j.semcancer.2008.11.010.
103. Novosyadlyy R, Vijayakumar A, Lann D, Fierz Y, Kurshan N, LeRoith D. Physical and functional interaction between polyoma virus middle T antigen and insulin and IGF-I receptors is required for oncogene activation and tumour initiation. Oncogene. 2009;28(39):3477–86. PMID: 19617901. doi:10.1038/onc.2009.209.
104. Gallagher EJ, LeRoith D. The proliferating role of insulin and insulin-like growth factors in cancer. Trends Endocrinol Metab. 2010;21(10):610–8. PMID: 20663687. doi:10.1016/j.tem.2010.06.007.
105. Neuzillet C, Tijeras-Raballand A, De Mestier L, Cros J, Faivre S, Raymond E. MEK in cancer and cancer therapy. Pharmacol Ther. 2014;141(2):160–71. PMID: 24121058. doi:10.1016/j.pharmthera.2013.10.001.
106. Yumoto K, Eber MR, Berry JE, Taichman RS, Shiozawa Y. Molecular pathways: niches in metastatic dormancy. Clin Cancer Res. 2014;20(13):3384–9. PMID: 24756372. doi:10.1158/1078-0432.CCR-13-0897.
107. Caunt CJ, Sale MJ, Smith PD, Cook SJ. MEK1 and MEK2 inhibitors and cancer therapy: the long and winding road. Nat Rev Cancer. 2015;15(10):577–92. PMID: 26399658. doi:10.1038/nrc4000.
108. Pylayeva-Gupta Y, Grabocka E, Bar-Sagi D. RAS oncogenes: weaving a tumorigenic web. Nat Rev Cancer. 2011;11(11):761–74. PMID: 21993244. doi:10.1038/nrc3106.
109. Fujishiro M, Gotoh Y, Katagiri H, Sakoda H, Ogihara T, Anai M, Onishi Y, Ono H, Abe M, Shojima N, Fukushima Y, Kikuchi M, Oka Y, Asano T. Three mitogen-activated protein kinases inhibit insulin signaling by different mechanisms in 3T3-L1 adipocytes. Mol Endocrinol. 2003;17(3):487–97. PMID: 12554784. doi:10.1210/me.2002-0131.
110. Ozaki KI, Awazu M, Tamiya M, Iwasaki Y, Harada A, Kugisaki S, Tanimura S, Kohno M. Targeting the ERK signaling pathway as a potential treatment for insulin resistance and type 2 diabetes. Am J Physiol Endocrinol Metab. 2016;310(8):E643-E651. PMID: 26860984. doi:10.1152/ajpendo.00445.2015.
111. Iams WT, Lovly CM. Molecular pathways: clinical applications and future direction of insulin-like growth factor-1 receptor pathway blockade. Clin Cancer Res. 2015;21(19):4270–7. PMID: 26429980. doi:10.1158/1078-0432.ccr-14-2518.
112. Pollak M. The insulin and insulin-like growth factor receptor family in neoplasia: an update. Nat Rev Cancer. 2012;12(3):159. PMID: 22337149. doi:10.1038/nrc3215.
113. Singh P, Alex JM, Bast F. Insulin receptor (IR) and insulin-like growth factor receptor 1 (IGF-1R) signaling systems: novel treatment strategies for cancer. Med Oncol. 2014;31(1):805. PMID: 24338270. doi:10.1007/s12032-013-0805-3.

Translational Control and mTOR in Cancer

5

Rita Seeböck, Nicole Golob-Schwarzl, and Stefanie Krassnig

Contents

Abstract

The mammalian target of rapamycin (mTOR) is a key link between a cell's nutrients and energy sensors and proliferative effector molecules. Essential processes controlled by the activity of mTOR include ribosome biogenesis, protein translation, and cell cycle progression. Antagonizing mTOR activity has shown antitumor effects in preclinical studies, and first clinical trials with mTOR inhibitors gave promising results. Among the broad spectrum of tumor entities that might be sensitive to mTOR targeting, we chose glioblastoma (GBM) and colorectal cancer to report the current state of investigation. In these exemplary tumor types, we review the potential of mTOR pathway components as biomarkers as well as drug targets.

R. Seeböck (✉)
IMC University of Applied Sciences, Krems, Austria
e-mail: rita.seeboeck@fh-krems.ac.at

N. Golob-Schwarzl • S. Krassnig
Institute of Pathology, Medical University Graz, Graz, Austria
e-mail: nicole.golobschwarzl@cbmed.at; stefanie.krassnig@medunigraz.at

J. Haybaeck (ed.), *Mechanisms of Molecular Carcinogenesis – Volume 1*,
DOI 10.1007/978-3-319-53659-0_5

5.1 mTOR Signaling

The mammalian target of rapamycin (mTOR) is a serine/threonine kinase which was identified as the cellular target of rapamycin, a bioactive compound first isolated in the 1970s from soil samples collected on Easter Island [1]. Rapamycin was first used in clinics as an immunosuppressive drug predominantly following kidney transplantations to counteract graft-versus-host disease and acute transplant rejection. Furthermore, it was observed that rapamycin could decrease the frequency of tumor formation in organ transplant experiments and was hence focused on in oncologic research as potential anticancer agent [2–4].

The molecular context of rapamycin was discovered in the early 1990s, with the discovery of mTOR. This elegant genetic study in *Saccharomyces cerevisiae* revealed that FKBP (FK506 binding protein) is a critical component of a rapamycin gain-of-function inhibitory complex with TOR1 and TOR2 [5]. TOR proteins are evolutionarily conserved from yeast to human, with human, mouse, and rat TOR proteins sharing 95% identity at the amino acid level. Soon after the first discovery of TOR, also the mammalian homologue (mTOR) could be purified and ever since is one of the major pathways investigated in molecular oncology [6].

mTOR is active within the cell in two different complexes. mTOR complex 1 (mTORC1) is built up by mTOR, regulatory-associated protein of TOR (Raptor), DEP domain-containing mTOR-inactivating protein (Deptor), mLST8, and PRAS40. mTORC2 is built up by mTOR, rapamycin-insensitive companion of mTOR (Rictor), Deptor, Protor, LST8, GbL, and mSIN1 [7].

The mTOR pathway is regulated by a wide variety of cellular signals, including mitogenic growth factors, hormones, nutrients, cellular energy levels, and stress conditions. Therefore, it could be shown that mTOR is embedded within the PI3K/AKT signal transduction pathway, which is critically involved in the mediation of cell survival and proliferation. Signaling through the PI3K/AKT pathway is initiated by mitogenic signals, triggered by growth factors that bind receptors in the cell membrane. These receptors include insulin-like growth factor receptor (IGFR), platelet-derived growth factor receptor (PDGFR), and the ERBB receptor family. Receptor binding activates AKT via PI3K; this process is antagonized by PTEN activity [8]. Akt phosphorylates mTOR directly, but may also work indirectly on mTOR through the actions of the TSC1/TSC2 (tuberous sclerosis complex). The physical association of the proteins TSC1 (Hamartin) and TSC2 (Tuberin) produces a functional complex that inhibits mTOR. This inhibitory effect is understood to act via inactivation of the Ras family small GTPase Rheb by TSC2. GTP hydrolysis of Rheb blocks mTOR activation, which only occurs by Rheb-GTP [9]. Cell surface signals can, next to PI3K/Akt signaling, also trigger the mitogen-activated protein kinase (MAPK) cascade. This phosphorylation cascade of Ras-Raf-MEK and ERK leads to phosphorylation of TSC1/TSC2 and inhibition of its activity [9]. Activation of mTOR results in phosphorylation of several downstream targets. mTORC1 activation by nutrients and availability of cellular energy lead to signals inducing ribosome biogenesis and mRNA translation, all leading to cell growth and proliferation. Rapamycin-insensitive mTORC2 controls the actin cytoskeleton and thereby determines the shape of the cell [8].

The best-characterized effectors downstream of mTOR are two signaling pathways that act in parallel to control mRNA translation. Activated mTOR mediates the phosphorylation of the 4E-BP1 and the ribosomal protein S6 kinase (S6K1). 4E-BP1 represses the activity of the eIF4F complex by blocking its essential, mRNA cap-binding component eIF4e. In its unphosphorylated state, 4E-BP1 binds tightly to eIF4e. Phosphorylation of 4E-BP1 by mTOR reduces its affinity for eIF4e, and the proteins dissociate, releasing eIF4e which is then able to associate with the other components of eIF4F and act in translation initiation. Growth factor deprivation or inhibition of mTOR results in the dephosphorylation of 4E-BP1, followed by its reassociation with eIF4e and a reduction in cap-specific translation [10]. Besides 4E-BP, S6K1 is the most important downstream effector of mTOR. By phosphorylation of mTOR, S6K1 is activated. This leads to a downstream activation of the ribosomal S6 protein, essential for 40S ribosomal subunit recruitment [6].

To sum up these anabolic regulations by mTOR, mTORC1 is mainly activated by AKT, which itself can be regulated by mTORC2. AKT is activated by PI3K, which is antagonized by the tumor suppressor PTEN. Downstream of mTORC1, S6K1, and 4E-BP1 both regulate mRNA translation at the levels of translation initiation and ribosome biogenesis [4].

5.2 Eukaryotic Translation Initiation Factors (eIFs) in Cancer

Major players in translation initiation are the eukaryotic translation initiation factors (eIFs), comprising eIF1, eIF1a, eIF2, eIF2b, eIF3, eIF4a, eIF4e, eIF4g, eIF4b, eIF4h, eIF5, and eIF5b [11]. Many of these have been described to have an implication in carcinogenesis and tumor progression. We wanted to highlight this group as they are physiologically related to mTOR signaling, and their impact in cancer research has dramatically increased over the last few years. Relevant facts for eIFs in cancer are summarized below.

eIF1 is differentially expressed throughout the body and has also been described in association with genotoxic stress situations, including ionizing radiation and heat shock. This connection revealed a dependence of eIF1 on the potent tumor suppressor p53 [12].

Investigations on eIF2 subunits mainly deal with their role in stress response, but overexpression or increased activity was also linked to cancer types including various lymphoma subtypes, gastrointestinal disease, lung cancer, and melanoma [13–17].

eIF3 is the largest and most complex initiation factor with a molecular mass of 600–700 kDa and 13 described subunits, known as eIF3a-m. The subunits can form modules and complexes of varying compositions [18]. The exact contribution of individual eIF3 subunits in translation initiation is not completely refined; however, in carcinogenesis, they are individually described. The largest subunit eIF3a interacts with all other eIF3 subunits and eIF4b, which establishes a direct link to mTOR signaling. Overexpression of eIF3a was correlated to several human cancers, including breast [19], cervix [20], colon [21], lung [22], urinary bladder [23], esophagus [24], and oral squamous cell carcinoma [25]. Beyond its hypothesized interaction with mTOR [26], eIF3a was discovered as a negative modulator of the extracellular signal-regulated kinase (ERK) pathway, via interaction with SHC and Raf-1 [27].

Upregulation of other eIF3 subunits was also identified in tumors, but it is not yet known if their differential expression is a cause or consequence of carcinogenesis [11]. One eIF3 subunit that has to be dealt with separately is eIF3f, because it is the only eIF3 core subunit which was shown to be downregulated in cancer. This was shown in patients suffering from melanoma and pancreatic cancer [28, 29]. In agreement with that, overexpression of eIF3f led to proliferation inhibition and apoptosis induction in vitro [28, 29]. Similar to eIF3a, eIF3f is suggested to interact with and eventually regulate mTOR and its downstream cascade [30].

Among eIF4 proteins, there are three subunits of the eIF4F complex, namely, eIF4a, eIF4e, and eIF4g, and the independent subunit eIF4b [31]. eIF4b is a downstream target of mTOR which, when phosphorylated, binds tighter to eIF3, thus increasing translational efficiency [32]. eIF4e functions in protein translation initiation by its cap-binding activity. eIF4e and especially its phosphorylated form were intensively studied in different cancer types and found upregulated in breast [33], colon [34], head and neck [35], and ovarian carcinoma [36] and non-Hodgkin's lymphoma [17, 36]. eIF4e availability is regulated by the 4e-binding protein 1 (4E-BP1). 4E-BP1 responds to extracellular stimuli like increased insulin levels or binding of growth factors to cell surface receptors. mTOR phosphorylates 4E-BP1, thereby inactivating it and releasing eIF4e, which can consequently interact in the eIF4F complex in order to initiate translation [37].

5.3 Targeting the mTOR Pathway in Glioblastoma

Glioblastoma multiforme (GBM) is a brain tumor deriving from glial cell origin and belongs to the most common malignant brain tumors [38]. Current treatment strategies combine surgical resection, adjuvant radiotherapy, and chemotherapy [39]. Nevertheless, the outcome with a median survival of 12 months is still very poor [40]. One reason for the poor treatment response is the highly infiltrative nature of GBMs, which leads to frequent recurrences [41]. Thus, there exists an immediate need for novel treatment strategies in glioma therapy.

In glioblastoma, the PI3K/AKT/mTOR signaling has already been extensively studied as many known mutations in GBM patients lead to a constitutive activation of this important pathway. Hyperactivation of this signaling cascade can be induced by the deletion of the tumor suppressor PTEN, overexpression of EGFR, as well as mutations in PI3K [42]. Genetic alterations in the PI3K/AKT/mTOR pathway have been detected in 88% of gliomas [43]. As a result, it was demonstrated that deregulation of the AKT/mTOR signaling seems to be one of the key players driving gliomagenesis [44]. Therefore, it has become an auspicious target for potential novel GBM treatments [45, 46].

The receptor tyrosine kinase (RTK) member EGFR offered itself as promising candidate for the downstream inhibition of the whole PI3K/AKT/mTOR pathway. Nevertheless, the EGFR inhibitors gefitinib [47] and erlotinib [48] exhibited only a reasonable performance during clinical trials.

Besides the regulation of protein translation, mTOR was shown to have diverse functions in the brain such as long-term potentiation, memory formation, and synaptic plasticity [49, 50]. In GBM therapy, rapamycin (sirolimus) and rapalogues, e.g., everolimus (RAD001) and temsirolimus (CCL-779), have been evaluated in clinical trials so far [46]. Although rapamycin has been shown to effectively inhibit glioma cell growth [51], clinical trials were not successful [52]. The fact that rapamycin predominantly inhibits mTORC1, but not mTORC2, and the presence of various feedback loops might explain the failure in clinical trials. Afterward, the focus in mTOR-mediated glioma therapy shifted to combination treatments (e.g., mTOR/PI3K or EGFR) [53].

PI3K-directed therapy has improved since the first-generation inhibitor Wortmannin did not pass the preclinical phase due to clinical toxicity [54]. PX-866 revealed only minimal toxicity, reduced GBM proliferation, increased apoptosis, and prolonged survival in murine xenograft models [55]. In clinics, BKM120 successfully finished phase I and is continuing in phase II [56], whereas the PX-866 trail was completed due to a low overall response rate in phase II [57].

The protein kinase AKT, a key player in the PI3K/AKT/mTOR pathway, has also been targeted as GBM treatment approach. Perifosine inhibits the activation of AKT by preventing its phosphorylation and translocation to the plasma membrane [58]. In murine animal models, perifosine revealed promising results especially in combination with the mTOR inhibitor temsirolimus [59] and temozolomide [60]. Nevertheless, probably also due to several drawbacks (e.g., limited penetrance through the blood-brain barrier), the success in a clinical phase II trial of recurrent GBMs was only moderate [46].

Due to the rather modest results in targeting only one signaling molecule of the mTOR cascade, approaches combining two or even more targets have become improved options in glioma therapy [61]. The PI3K/mTOR inhibitor PI-103 was one of the first dual inhibitors. PI-103 induced cell cycle arrest in glioma cells without revealing neurotoxic properties [62]. However, it never endured the preclinical phase due to its weak pharmacological properties. Other dual inhibitors, e.g., NVP-BEZ235 and XL-765, have been more successful and even completed phase I trials [46]. The EGFR inhibitor BKM120 is also tested clinically in combination with radiation or the monoclonal anti-vascular endothelial growth factor (VEGF) antibody bevacizumab [63]. However, combination therapies of erlotinib and temozolomide failed to improve GBM patient prognosis [64, 65].

To conclude, the PI3K/AKT/mTOR survival pathway plays a crucial role during gliomagenesis and lends itself therefore to be investigated in more detail as potential therapeutic approach. Although mTOR signaling belongs to one of the most investigated pathways in GBM, further efforts are needed to elucidate the exact mechanism of this complex pathway during gliomagenesis and to use it as potential therapeutic target in advance.

5.4 mTOR Targeting in Colorectal Cancer

Colorectal cancer (CRCs) is the third most common cause of cancer and the fourth most cancer-related death worldwide [44]. Screening strategies and enhancements in treatment have resulted in the decrease in the morbidity and mortality associated with CRC [66]. Treatment of CRC includes a multidisciplinary approach that comprises surgery, radiation, chemotherapy, and targeted therapy [67]. Cancer cells can spread to nearby and remote lymph nodes, as well as to other organs, such as the lung and liver. The prognosis and survival rate depends on the stage of the disease and tumor location. Surgical removal of tumor tissue and nearby lymph nodes is the most common treatment strategy for early stage (stage I and II) CRC. Chemotherapy and/or combinations with radiation therapy are the treatment strategy for late stage CRC [68].

Targeted therapy strategies include monoclonal antibodies, for example, bevacizumab (Avastin, an VEGF-A inhibitor) and cetuximab (anti-EGFR), regorafenib (multiple RTK inhibitor), and aflibercept (anti-VEGF agent) [69, 70]. It is suggested to reconsider the existing examples for the selection of agents in the adjuvant treatment of CRC [71].

Genetic mutations and chromosomal instability can arise either hereditarily or sporadically. A large proportion of these aberrations involve oncogenic pathways converging on the translational machinery. These pathways are MAPK and PI3K/AKT/mTOR cascades that include components and regulators strongly associated with the CRC carcinogenesis, such as PIK3CA, K-RAS, BRAF, PTEN RTKs, and others [72–74]. Mutations of PIK3CA and decreased function of PTEN are also often found in CRC, directed to the activation of the PI3K/AKT/mTOR pathway [74, 75]. EIF4e is one of the most studied translation factors and is associated in the cancer biology in general and in CRC in particular. eIF4e is also involved in regulation by different signaling cascades, including MAPK and PI3K/AKT/mTOR. The PI3K/AKT/mTOR pathway controls 4E-BP, a tumor suppressor, which, when phosphorylated by an activated mTOR, dissociates from eIF4e and facilitates translation. Overexpression and activating phosphorylation of eIF4e, as well as inactivating phosphorylation and downregulation of 4E-BPs, are key notes in CRC. The 4E-BPs/eIF4e axis is a predictive and prognostic biomarker in the therapy of CRC.

Conclusions

Over the past few years, PI3K/AKT/mTOR signaling has been shown to be a key player driving tumorigenesis in various tumor entities [76–78]. Thus, much effort was put into targeting this major survival pathway, regrettably with moderate success. The complexity of the PI3K/AKT/mTOR signaling with its various feedback loops and cross-talk signaling seems to be one of the major challenges in a pathway-directed therapy [79, 80]. Many clinical trials have already been started to target the mTOR cascade via multiple inhibitors [53]. Future cancer therapeutic approaches will even turn more into the direction of multiple targeting and combination therapy to solve the Sisyphean task of curing cancer.

References

1. Dann SG, Selvaraj A, Thomas G. mTOR Complex1-S6K1 signaling: at the crossroads of obesity, diabetes and cancer. Trends Mol Med. 2007;13:252–9.
2. Law BK. Rapamycin: an anti-cancer immunosuppressant? Crit Rev Oncol Hematol. 2005;56:47–60.
3. Rao RD, Buckner JC, Sarkaria JN. Mammalian target of rapamycin (mTOR) inhibitors as anticancer agents. Curr Cancer Drug Targets. 2004;4:621–35.
4. Geissler EK, Schlitt HJ, Thomas G. mTOR, cancer and transplantation. Am J Transplant. 2008;8:2212–28.
5. Heitman J, Movva NR, Hall MN. Targets for cell cycle arrest by the immunosuppressant rapamycin in yeast. Science. 1991;253:905–9.
6. Gentilella A, Kozma SC, Thomas G. A liaison between mTOR signaling, ribosome biogenesis and cancer. Biochim Biophys Acta. 2015;1849:812–20.
7. Xu K, Liu P, Wei W. mTOR signaling in tumorigenesis. Biochim Biophys Acta. 2014;1846:638–54.
8. Perl A. mTOR activation is a biomarker and a central pathway to autoimmune disorders, cancer, obesity and aging. Ann N Y Acad Sci. 2015;1346:33–44.
9. Laplante M, Sabatini DM. mTOR signaling at a glance. J Cell Sci. 2009;122:3589–94.
10. Zarogoulidis P, Lampaki S, Turner JF, Huang H, Kakolyris S, Syrigos K, Zarogoulidis K. mTOR pathway: a current, up-to-date mini review. Oncol Lett. 2014;8:2367–70.
11. Spilka R, Ernst C, Mehta AK, Haybaeck J. Eukaryotic translation initiation factors in cancer development and progression. Cancer Lett. 2013;340:9–21.
12. Sheikh MS, Fornace AJ Jr. Regulation of translation initiation following stress. Oncogene. 1999;18:6121–8.
13. Lam N, Sandberg ML, Sugden B. High physiological levels of LMP1 result in phosphorylation of eIF2 alpha in Epstein-Barr virus-infected cells. J Virol. 2004;78:1657–64.
14. Lobo MV, Martin ME, Perez MI, Alonso F, Redondo C, Alvarez MI, Salinas M. Levels, phosphorylation status and cellular localization of translational factor eIF2 in gastrointestinal carcinomas. Histochem J. 2000;32:139–50.
15. Rosenwald IB, Hutzler MJ, Wang S, Savas L, Fraire AE. expression of eukaryotic translation initiation factors 4E and 2alpha is increased frequently in bronchioloalveolar but not in squamous cell carcinomas of the lung. Cancer. 2001;92:2164–71.
16. Rosenwald IB, Wang S, Savas L, Woda B, Pullman J. Expression of translation initiation factor eIF-2alpha is increased in benign and malignant melanocytic and colonic epithelial neoplasms. Cancer. 2003;98:1080–8.
17. Wang S, Rosenwald IB, Hutzler MJ, Pihan GA, Savas L, Chen JJ, Woda BA. Expression of the eukaryotic translation initiation factors 4E and 2alpha in non-Hodgkin's lymphomas. Am J Pathol. 1999;155:247–55.
18. Wagner S, Herrmannová A, Malík R, Peclinovská L, Valášek LS. Functional and biochemical characterization of human eukaryotic translation initiation factor 3 in living cells. Mol Cell Biol. 2014;34(16):3041–52. doi:10.1128/MCB.00663-14. Epub 2014 Jun 9
19. Bachmann F, Banziger R, Burger MM. Cloning of a novel protein overexpressed in human mammary carcinoma. Cancer Res. 1997;57:988–94.
20. Dellas A, Torhorst J, Bachmann F, Banziger R, Schultheiss E, Burger MM. Expression of p150 in cervical neoplasia and its potential value in predicting survival. Cancer. 1998;83:1376–83.
21. Haybaeck J, O'Connor T, Spilka R, Spizzo G, Ensinger C, Mikuz G, Brunhuber T, Vogetseder A, Theurl I, Salvenmoser W, Drax HL, Banziger R, Bachmann F, Schafer G, Burger M, Obrist

P. Overexpression of p150, a part of the large subunit of the eukaryotic translation initiation factor 3, in colon cancer. Anticancer Res. 2010;30:1047–55.
22. Pincheira R, Chen Q, Zhang JT. Identification of a 170-kDa protein over-expressed in lung cancers. Br J Cancer. 2001;84:1520–7.
23. Spilka R, Ernst C, Bergler H, Rainer J, Flechsig S, Vogetseder A, Benesch M, Brunner A, Geley S, Eger A, Bachmann F, Doppler W, Obrist P, Haybaeck J. eIF3a is over-expressed in urinary bladder cancer and influences its phenotype independent of translation initiation. Cell Oncol. 2014;37:253–67.
24. Chen G, Burger MM. p150 overexpression in gastric carcinoma: the association with p53, apoptosis and cell proliferation. Int J Cancer. 2004;112:393–8.
25. Spilka R, Laimer K, Bachmann F, Spizzo G, Vogetseder A, Wieser M, Müller H, Haybaeck J, Obrist P. Overexpression of eIF3a in squamous cell carcinoma of the oral cavity and its putative relation to chemotherapy response. J Oncol. 2012;2012:901956.
26. Harris TE, Chi A, Shabanowitz J, Hunt DF, Rhoads RE, Lawrence JC Jr. mTOR-dependent stimulation of the association of eIF4G and eIF3 by insulin. EMBO J. 2006;25:1659–68.
27. Xu TR, Lu RF, Romano D, Pitt A, Houslay MD, Milligan G, Kolch G. Eukaryotic translation initiation factor 3, subunit a, regulates the extracellular signal-regulated kinase pathway. Mol Cell Biol. 2012;32:88–95.
28. Doldan A, Chandramouli R, Shanas A, Bhattacharyya A, Leong SP, Nelson MA, Shi J. Loss of the eukaryotic initiation factor 3f in melanoma. Mol Carcinog. 2008b;47:806–13.
29. Doldan A, Chandramouli A, Shanas R, Bhattacharyya A, Cunningham JT, Nelson MA, Shi J. Loss of the eukaryotic initiation factor 3f in pancreatic cancer. Mol Carcinog. 2008a;47:235–44.
30. Csibi A, Cornille K, Leibovitch MP, Poupon A, Tintignac LA, Sanchez AM, Leibovitch SA. The translation regulatory subunit eIF3f controls the kinase-dependent mTOR signaling required for muscle differentiation and hypertrophy in mouse. PLoS One. 2010;5:e8994.
31. Lazaris-Karatzas A, Sonenberg N. The mRNA 5′ cap-binding protein, eIF-4E, cooperates with v-myc or E1A in the transformation of primary rodent fibroblasts. Mol Cell Biol. 1992;12:1234–8.
32. Kroczynska B, Kaur S, Katsoulidis E, Majchrzak-Kita B, Sassano A, Kozma SC, Fish EN, Platanias LC. Interferon-dependent engagement of eukaryotic initiation factor 4B via S6 kinase (S6K)- and ribosomal protein S6K-mediated signals. Mol Cell Biol. 2009;29:2865–75.
33. Li BD, Liu L, Dawson M, de Benedetti A. Overexpression of eukaryotic initiation factor 4E (eIF4E) in breast carcinoma. Cancer. 1997;79:2385–90.
34. Rosenwald IB, Chen JJ, Wang S, Savas L, London IM, Pullman J. Upregulation of protein synthesis initiation factor eIF-4E is an early event during colon carcinogenesis. Oncogene. 1999;18:2507–17.
35. Sorrells DL, Ghali GE, de Benedetti A, Nathan CO, Li BD. Progressive amplification and overexpression of the eukaryotic initiation factor 4E gene in different zones of head and neck cancers. J Oral Maxillofac Surg. 1999;57:294–9.
36. Noske A, Lindenberg JL, Darb-Esfahani S, Weichert W, Buckendahl AC, Roske A, Sehouli J, Dietel M, Denkert C. Activation of mTOR in a subgroup of ovarian carcinomas: correlation with p-eIF-4E and prognosis. Oncol rep. 2008;20:1409–17.
37. Peffley DM, Sharma C, Hentosh P, Buechler RD. Perillyl alcohol and genistein differentially regulate PKB/Akt and 4E-BP1 phosphorylation as well as eIF4E/eIF4G interactions in human tumor cells. Arch Biochem Biophys. 2007;465:266–73.
38. Louis DN, Ohgaki H, Wiestler OD, Cavenee WK, Burger PC, Jouvet A, Scheithauer BW, Kleihues P. The 2007 WHO classification of tumours of the central nervous system. Acta Neuropathol. 2007;114(2):97–109.
39. Stupp R, Mason WP, van den Bent MJ, Weller M, Fisher B, Taphoorn MJ, Belanger K, Brandes AA, Marosi C, Bogdahn U, Curschmann J, Janzer RC, Ludwin SK, Gorlia T, Allgeier A, Lacombe D, Cairncross JG, Eisenhauer E, Mirimanoff RO. Radiotherapy plus concomitant and adjuvant temozolomide for glioblastoma. N Engl J Med. 2005;352(10):987–96.
40. McLendon RE, Halperin EC. Is the long-term survival of patients with intracranial glioblastoma multiforme overstated? cancer. 2003;98(8):1745–8.

41. Wang Y, Jiang T. Understanding high grade glioma: molecular mechanism, therapy and comprehensive management. Cancer Lett. 2013;331(2):139–46.
42. Ohgaki H, Kleihues P. Genetic profile of astrocytic and oligodendroglial gliomas. Brain tumor Pathol. 2011;28(3):177–83.
43. Riddick G, Fine HA. Integration and analysis of genome-scale data from gliomas. Nat Rev Neurol. 2011;7(8):439–50.
44. Ferlay J, Shin H, Bray F, Forman D, Mathers C, Parkin D. Estimates of worldwide burden of cancer in 2008: GLOBOCAN 2008. Int J Cancer. 2010;127:2893–917.
45. Duzgun Z, Eroglu Z, Biray Avci C. Role of mTOR in glioblastoma. Gene. 2016;575(2 Pt 1):187–90.
46. Li X, Wu C, Chen N, Gu H, Yen A, Cao L, Wang E, Wang L. PI3K/Akt/mTOR signaling pathway and targeted therapy for glioblastoma. Oncotarget. 2016;7(22):33440–50.
47. Hegi ME, Diserens AC, Bady P, Kamoshima Y, Kouwenhoven MC, Delorenzi M, Lambiv WL, Hamou MF, Matter MS, Koch A, Heppner FL, Yonekawa Y, Merlo A, Frei K, Mariani L, Hofer S. Pathway analysis of glioblastoma tissue after preoperative treatment with the EGFR tyrosine kinase inhibitor gefitinib—a phase II trial. Mol Cancer Ther. 2011;10(6):1102–12.
48. Van den Bent MJ, Brandes AA, Rampling R, Kouwenhoven MC, Kros JM, Carpentier AF, Clement PM, Frenay M, Campone M, Baurain JF, Armand JP, Taphoorn MJ, Tosoni A, Kletzl H, Klughammer B, Lacombe D, Gorlia T. Randomized phase II trial of erlotinib versus temozolomide or carmustine in recurrent glioblastoma: EORTC brain tumor group study 26034. J Clin Oncol. 2009;27(8):1268–74.
49. Slipczuk L, Bekinschtein P, Katche C, Cammarota M, Izquierdo I, Medina JH. BDNF activates mTOR to regulate GluR1 expression required for memory formation. PLoS one. 2009;4(6):e6007.
50. Tang SJ, Reis G, Kang H, Gingras AC, Sonenberg N, Schuman EM. A rapamycin-sensitive signaling pathway contributes to long-term synaptic plasticity in the hippocampus. Proc Natl Acad Sci U S A. 2002;99(1):467–72.
51. Joy AM, Beaudry CE, Tran NL, Ponce FA, Holz DR, Demuth T, Berens ME. Migrating glioma cells activate the PI3-K pathway and display decreased susceptibility to apoptosis. J Cell Sci. 2003;116(Pt 21):4409–17.
52. Faivre S, Kroemer G, Raymond E. Current development of mTOR inhibitors as anticancer agents. Nat Rev Drug Discov. 2006;5(8):671–88.
53. Sami A, Karsy M. Targeting the PI3K/AKT/mTOR signaling pathway in glioblastoma: novel therapeutic agents and advances in understanding. Tumour Biol. 2013;34(4):1991–2002.
54. Fedrigo CA, Grivicich I, Schunemann DP, Chemale IM, dos Santos D, Jacovas T, Boschetti PS, Jotz GP, Braga Filho A, da Rocha AB. Radioresistance of human glioma spheroids and expression of HSP70, p53 and EGFr. Radiat Oncol. 2011;6:156.
55. Wen PY, Lee EQ, Reardon DA, Ligon KL, Alfred Yung WK. Current clinical development of PI3K pathway inhibitors in glioblastoma. Neuro-oncology. 2012;14(7):819–29.
56. Bendell JC, Rodon J, Burris HA, de Jonge M, Verweij J, Birle D, Demanse D, de Buck SS, Ru QC, Peters M, Goldbrunner M, Baselga J. Phase I, dose-escalation study of BKM120, an oral pan-Class I PI3K inhibitor, in patients with advanced solid tumors. J Clin Oncol. 2012;30(3):282–90.
57. Pitz MW, Eisenhauer EA, MacNeil MV, Thiessen B, Easaw JC, Macdonald DR, Eisenstat DD, Kakumanu AS, Salim M, Chalchal H, Squire J, Tsao MS, Kamel-Reid S, Banerji S, Tu D, Powers J, Hausman DF, Mason WP. Phase II study of PX-866 in recurrent glioblastoma. Neuro-oncology. 2015;17(9):1270–4.
58. Kondapaka SB, Singh SS, Dasmahapatra GP, Sausville EA, Roy KK. Perifosine, a novel alkylphospholipid, inhibits protein kinase B activation. Mol Cancer Ther. 2003;2(11):1093–103.
59. Pitter KL, Galban CJ, Galban S, Tehrani OS, Li F, Charles N, Bradbury MS, Becher OJ, Chenevert TL, Rehemtulla A, Ross BD, Holland EC, Hambardzumyan D. Perifosine and CCI 779 co-operate to induce cell death and decrease proliferation in PTEN-intact and PTEN-deficient PDGF-driven murine glioblastoma. PLoS One. 2011;6(1):e14545.
60. Momota H, Nerio E, Holland EC. Perifosine inhibits multiple signaling pathways in glial progenitors and cooperates with temozolomide to arrest cell proliferation in gliomas in vivo. Cancer res. 2005;65(16):7429–35.

61. Luchman HA, Stechishin OD, Nguyen SA, Lun XQ, Cairncross JG, Weiss S. Dual mTORC1/2 blockade inhibits glioblastoma brain tumor initiating cells in vitro and in vivo and synergizes with temozolomide to increase orthotopic xenograft survival. Clin Cancer Res. 2014;20(22):5756–67.
62. Fan QW, Knight ZA, Goldenberg DD, Yu W, Mostov KE, Stokoe D, Shokat KM, Weiss WA. A dual PI3 kinase/mTOR inhibitor reveals emergent efficacy in glioma. Cancer Cell. 2006;9(5):341–9.
63. Huang TT, Sarkaria SM, Cloughesy TF, Mischel PS. Targeted therapy for malignant glioma patients: lessons learned and the road ahead. Neurotherapeutics. 2009;6(3):500–12.
64. Brown PD, Krishnan S, Sarkaria JN, Wu W, Jaeckle KA, Uhm JH, Geoffroy FJ, Arusell R, Kitange G, Jenkins RB, Kugler JW, Morton RF, Rowland KM Jr, Mischel P, Yong WH, Scheithauer BW, Schiff D, Giannini C, Buckner JC. Phase I/II trial of erlotinib and temozolomide with radiation therapy in the treatment of newly diagnosed glioblastoma multiforme: North Central Cancer Treatment Group Study N0177. J Clin Oncol. 2008;26(34):5603–9.
65. Prados MD, Chang SM, Butowski N, DeBoer R, Parvataneni R, Carliner H, Kabuubi P, Ayers-Ringler J, Rabbitt J, Page M, Fedoroff A, Sneed PK, Berger MS, McDermott MW, Parsa AT, Vandenberg S, James CD, Lamborn KR, Stokoe D, Haas-Kogan DA. Phase II study of erlotinib plus temozolomide during and after radiation therapy in patients with newly diagnosed glioblastoma multiforme or gliosarcoma. J Clin Oncol. 2009;27(4):579–84.
66. Edwards M, Chadda S, Zhao Z, Barber B, Sykes D. A systematic review of treatment guidelines for metastatic colorectal cancer. Assoc Coloproctol Great Britain Ireland. 2010;14:e31–47.
67. Engstrom PF, Benson AB III, Chen YJ, Choti MA, Dilawari RA, Enke CA, Fakih MG, Fuchs C, Kiel K, Knol JA, et al. Colon cancer clinical practice guidelines in oncology. J Natl Compr Cancer Netw. 2005;3:468–91.
68. Akhtar R, Chandel S, Sarotra P, Medhi B, et al. Current status of pharmacological treatment of colorectal cancer. World J Gastrointest Oncol. 2014;6:177–83.
69. Benson AB III, Bekaii-Saab T, Chan E, Chen YJ, Choti MA, Cooper HS, Engstrom PF, Enzinger PC, Fakih MG, Fenton MJ, et al. Localized colon cancer, version 3.2013: featured updates of the NCCN guidelines. J Natl Compr Cancer Netw. 2013;11:519–28.
70. Cheng Y, Yang H, Chen G, Zhang Z. Molecularly targeted drugs for metastatic colorectal cancer. Drug Des Devel Ther. 2013;7:1315–22.
71. Van Loon K, Venook AP. Adjuvant treatment of colon cancer; what is next? Curr Opin Oncol. 2011;23:403–9.
72. Bhalla A, Zulfiqar M, Weindel M, Shidham V. Molecular diagnostics in colorectal carcinoma. Clin Lab Med. 2013;33:835–59.
73. Shao J, Evers B, Sheng H, Cells E. Roles of phosphatidylinositol 3′-kinase and mammalian target of rapamycin/p70 ribosomal protein S6 kinase in K-Ras-mediated transformation of intestinal epithelial. Cancer res. 2004;64:229–35.
74. Weinstein J, Collisson E, Mills G, Mills SK, Ozenberger B, Ellrott K, Shmulevich I, Sander C, Stuart J. The cancer genome atlas Pan Cancer analysis project. Nat Genet. 2012;45:1113–20.
75. Parsons DW, Wang TL, Samuels Y, Bardelli A, Cummins JM, DeLong L, Sillman N, Ptak J, Szabo S, Willson JK, et al. Colorectal cancer: mutations in a signalling pathway. Nature. 2005;436:792.
76. Kim LC, Cook RS, Chen J. mTORC1 and mTORC2 in cancer and the tumor microenvironment. Oncogene. 2016; doi:10.1038/onc.2016.363.
77. Mayer IA, Arteaga CL. The PI3K/AKT pathway as a target for cancer treatment. Annu Rev Med. 2016;67:11–28.
78. Shaw RJ, Cantley LC. Ras, PI(3)K and mTOR signalling controls tumour cell growth. Nature. 2006;441(7092):424–30.
79. Albert L, Karsy M, Murali R, Jhanwar-Uniyal M. Inhibition of mTOR activates the MAPK pathway in glioblastoma multiforme. Cancer genomics proteomics. 2009;6(5):255–61.
80. Carracedo A, Ma L, Teruya-Feldstein J, Rojo F, Salmena L, Alimonti A, Egia A, Sasaki AT, Thomas G, Kozma SC, Papa A, Nardella C, Cantley LC, Baselga J, Pandolfi PP. Inhibition of mTORC1 leads to MAPK pathway activation through a PI3K-dependent feedback loop in human cancer. J Clin invest. 2008;118(9):3065–74.

NF-κB and Its Implication in Liver Health and Cancer Development

6

Kira Bettermann

Contents

Abstract

Nuclear factor kappa-light-chain-enhancer of activated B-cells (NF-κB) belongs to one of the best described and most intensively studied transcription factors in biochemistry in the last 30 years. The NF-κB signaling cascade exists in two variants, the canonical and noncanonical pathway, and its transcription factors

K. Bettermann
Department of Translational Inflammation Research, Medical Faculty, Otto-von-Guericke University, Pfälzer Platz, Building 28, 39106 Magdeburg, Germany
e-mail: kira.bettermann@med.ovgu.de

J. Haybaeck (ed.), *Mechanisms of Molecular Carcinogenesis – Volume 1*,
DOI 10.1007/978-3-319-53659-0_6

are key regulators of several biochemical processes like immune responses, inflammation, survival, and cellular development and growth.

Examination of various transgenic mouse models targeting NF-κB itself or signaling members discovered the implication of NF-κB in chronic inflammatory diseases and cancer development in different organs as in the skin, intestine, and liver.

In this review the focus lies on the central organ of metabolic and inflammatory processes: the liver. It seems that NF-κB is pivotal for the homeostasis in the different hepatic cell types concerning hepatic failure, fibrosis, and HCC progression. NF-κB has the ability to be a potential target in the attempt to circumvent or medicate liver fibrosis and HCC.

6.1 Introduction

Nuclear factor kappa-light-chain-enhancer of activated B-cells (NF-κB) is one of the best and most intensive studied transcription factors in the field of biomedicine. It was discovered and first described 30 years ago by David Baltimore and colleagues [1]. Since that time many studies revealed the outstanding meaning of NF-κB in the development of inflammatory diseases such as arthritis and psoriasis, inflammatory bowel diseases, asthma, and neurodegenerative heart diseases and its contribution to cancer development [2–4]. It is expressed in most mammalian cell types and tissues and controls the transcription of genes involved in immune responses, cell survival, proliferation, and differentiation [5].

Next to *in vitro* approaches, examination of several transgenic mouse models with different NF-κB targets gave the opportunity to raise our understanding of the complex mechanisms behind inflammation-driven diseases *in vivo*. Moreover, the work with conditional murine knockout models in different organs uncovered the central role of NF-κB in mediating innate immune responses and cytokine expression in order to react on pathological outcomes of inflammation like chronical skin inflammation (e.g. psoriasis) and hepatocarcinogenesis [6–10]. The severe affection of organs or tissues during disbalanced NF-κB activation gains importance and constitutes a therapeutic challenge. Further investigations will support the development of clinical trials targeting certain molecules of the NF-κB signaling cascade.

6.2 Members of the NF-κB/Rel Family

The NF-κB signaling pathway is evolutionarily highly conserved and is, besides mammalians, also found in the fruit fly *Drosophila melanogaster*, cnidarians, porifera, viruses, and mollusks [11–14]. *Caenorhabditis elegans* and yeast are the main exceptions [15, 16].

Mammalian NF-κB itself is composed of different types of dimers, appearing as homo- or heterodimers. The single compounds of these dimers are p50 (*NFKB1*), p52 (*NFKB2*), c-Rel (*REL*), p65/RelA (*RELA*), and RelB (*RELB*). All five transcription factors are characterized by an N-terminal Rel homology domain (RHD-NTD) which is needed to mediate DNA binding, homo- and heterodimerization, and nuclear translocation. In the nucleus NF-κB dimers bind at κB sites inside enhancer/promoter regions of target genes where they control transcription by recruiting coactivators and corepressors [17–19]. Furthermore, only p65, RelB, and c-Rel comprise a C-terminal transactivation domain (TAD) which is required for gene transcription. The other two members, p50 and p52, lacking TAD and are processed from the precursor proteins p105 (p50) and p100 (p52) (Fig. 6.1). Both are not directly involved in gene transcription except in combination with p65, RelB, c-Rel, or other proteins which are able to recruit coactivators. After successfully entering into the nucleus, it binds at the following consensus sequence 5′-GGGRNYYYCC-3′ (R, purine; Y, pyrimidine; N, any nucleotide) of DNA κB sites [19]. The TAD and RHD act in each case autonomously and underlay posttranscriptional modifications which might have an influence on NF-κB activation at the level of transcription and/or DNA binding [20].

The most abundant combination of NF-κB dimers are p50/p65 and p50/50, whereas, in contrast, the homodimer p50/p50 can act as a transcriptional repressor [21, 22].

6.3 The Negative Controllers of NF-κB: IκBs

NF-κB dimers are located in the cytoplasm, and translocation from the cytoplasm through the nucleus is regulated by another group of proteins named nuclear factor of kappa light polypeptide gene enhancer in B-cells inhibitor (IκBs) [15]. These proteins are tightly associated with the NF-κB dimers, preventing NF-κB activation by hindering NF-κB translocation through the nuclear membrane. The IκB family includes six members: IκBα (*NFKBIA*), IκBβ (*NFKBIB*), IκBε (*NFKBIE*), IκBγ (*NFKB1*), IκBζ (*NFKBIZ*), and Bcl-3 (*BCL3*) and the NF-κB precursors p100 and p105. All of them share several ankyrin repeat domains (ARD), which are necessary to interact with the RHD of NF-κB [23]. Crystal structure analysis of the IκBα/NF-κB (p50/p65) heterodimer and IκBβ/NF-κB (p50/p50) homodimer allowed a closer look through the binding conditions of each complex and revealed a binding ratio of 1:1. Inside the IκBα/NF-κB complex, the ankyrin repeat six and the C-terminal PEST sequence of IκBα are associated with the p65 RHD-NTD, impeding binding to the DNA κB site (Fig. 6.1). Additionally, p65 undergoes such a conformational change, which strongly supports the linkage to IκBα, holding NF-κB in its inactive state [24–26].

Every IκB member has its own favorite NF-κB dimer. IκBα/β/ε binds to NF-κB dimers, which contain a minimum of one p65 or c-Rel subunit. p100 and p102 are connected to all NF-κB subunits. IκBζ and Bcl-3 have a preference for p50 and p52 homodimers [27, 28].

Here, IκBα is investigated at best and is a central regulatory factor in the canonical NF-κB signaling pathway as described in the next chapter.

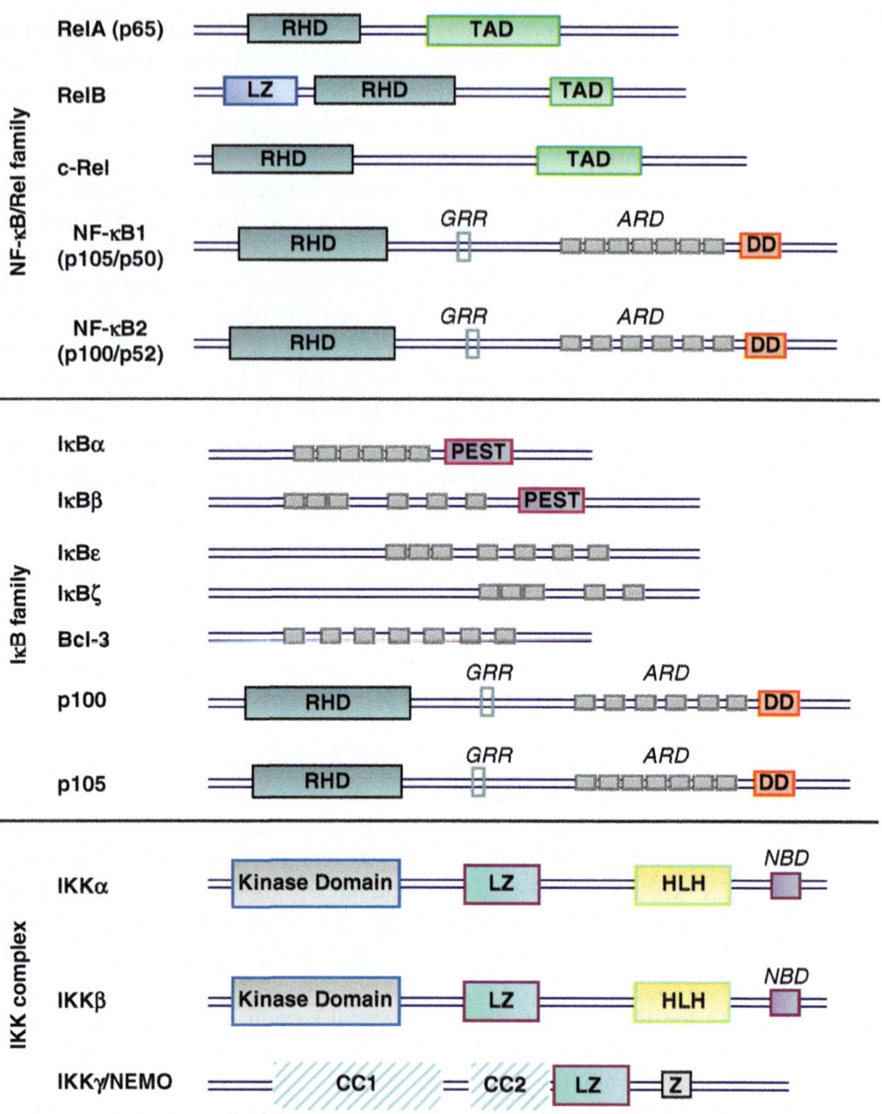

Fig. 6.1 Schematic diagram of the domain structures of each individual NF-κB, IκB, and IKK complex protein family members. All three family groups have a typical domain structure as the Rel homology domain (RHD) for the NF-κB family members, the ankyrin repeat domains (*ARD*) for the IκB family members, and the leucine-zipper (*LZ*) motif for the IKK complex members. On the basis of their function, p100 and p105 are also associated with the NF-κB and IκB family. *CC* coiled coil, *DD* death domain, *GRR* glycine-rich region, *HLH* helix-loop-helix, *NBD* NEMO-binding domain, *PEST* proline-, glutamic acid-, serine-, and threonine-rich region, *TAD* transactivation domain. Adapted from Oeckinghaus et al.: The NF-κB Family of Transcription Factors and its Regulation, Cold Spring Harb Perspect Biol. 2009, 1(4): 1–14

6.4 The Canonical NF-κB Pathway

Various cellular stress stimuli and certain endogenous and exogenous ligands lead to NF-κB activation. The stimulus decides if the canonical or noncanonical NF-κB pathway is activated. Both cascades lead to NF-κB nuclear translocation but are regulated by different checkpoints within the cascade. In the last few years, extensive studies have been performed to characterize both variants of NF-κB activation and to identify important key players. It was shown that the canonical pathway is primarily activated during physiological stress conditions like inflammation, exposure to bacterial products like lipopolysaccharide (LPS), and oxidative stress [29, 30]. The canonical NF-κB signaling cascade is generally activated by the following receptors: tumor necrosis factor receptor (TNFR), interleukin 1 receptor (IL-1R), Toll-like receptor (TLR), B-cell receptor (BCR), and T-cell receptor (TCR) [31]. Receptor ligation leads to the recruitment of certain adapter proteins to TNF receptor-associated factor (TRAF) and receptor-interacting protein kinase 1 (RIPK1). RIPK1 is connected to the TGFβ-activated kinase 1 (*TAK1*)-binding protein (TAB2-TAB3-TAK1) complex and the NF-κB essential modulator (NEMO) via ubiquitin chains, bringing TAK1 into close vicinity to NEMO (IKKγ), the regulatory subunit of the IκB kinase (IKK) complex. This complex is composed of two more members, the catalytic subunits IKKα (IKK1) and IKKβ (IKK2), and represents the crucial step in NF-κB nuclear translocation by controlling proteasomal degradation of IκBα. More precisely, phosphorylation of TAK1 at Thr178 and Thr184 permits a direct phosphorylation of IKKβ inside its activation loop at Ser177 and Ser181 leading straightly to IκBα phosphorylation at Ser32 and Ser36. Phosphorylation of IκBα follows activation of the IKK complex, the second essential regulatory step in NF-κB activation, because IκBα undergoes K48-linked polyubiquitination by the $SCF^{\beta TrCP}$ ubiquitin ligase complex, which induces its fast degradation by the 26S proteasome. Finally, IκBα degradation exposes the nuclear localization site (NLS) of NF-κB, which is needed for nuclear access, DNA binding, and transcription of target genes (Fig. 6.2) [32–37].

6.5 The Noncanonical or Alternative NF-κB Pathway

Next to the extensively studied canonical pathway, an alternative NF-κB activation exists. The so-called noncanonical NF-κB pathway activates NF-κB not by degradation of IκBα/β/ε and p105, but via processing the inactive p100/RelB NF-κB heterodimer through the active p52/RelB heterodimer [38, 39]. This pathway seems to be crucial in lymphoid organogenesis, B-cell maturation and survival, dendritic cell activation, and osteoclastogenesis. The major ligands which initiate these variants of NF-κB activation are the lymphotoxin β receptor (LTβR), B-cell-activating factor belonging to the TNF family receptor (BAFF-R), CD40, and receptor activator of

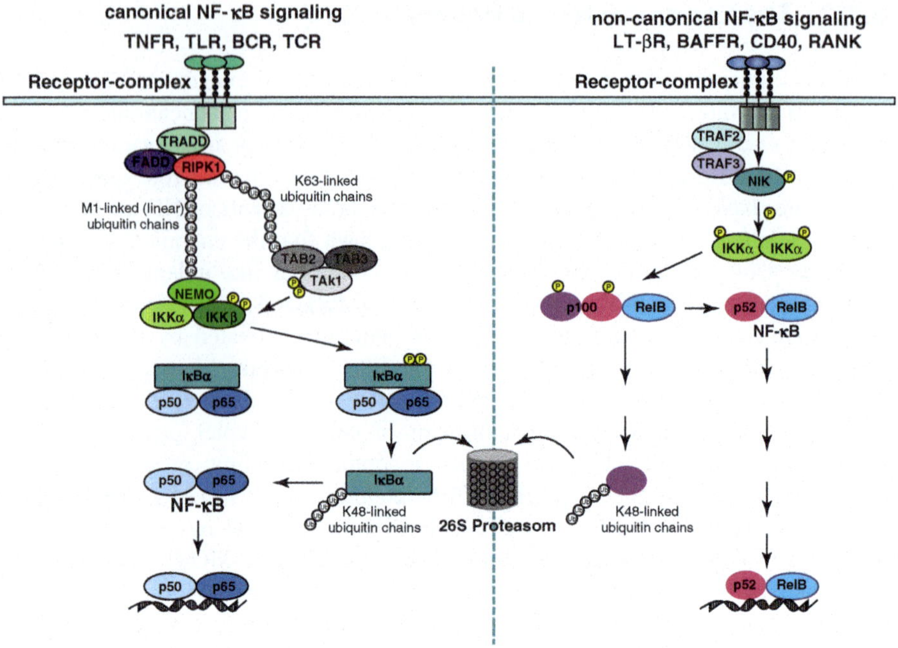

Fig. 6.2 The canonical and noncanonical NF-κB signaling cascade. After activation of the respective receptors, which are able to induce the canonical NF-κB pathway, the TAB-TAK1 complex and NEMO, the regulatory subunit of the IKK complex, get ubiquitinated by RIPK1. These ubiquitin chains bring both complexes into closer vicinity to each other, whereby TAK1 phosphorylates IKKβ, one of the catalytic subunits of the IKK complex. The activated IKK complex phosphorylates two serine residues of the NF-κB inhibitor IκBα, thereby initiating its proteasomal degradation. After IκBα degradation, NF-κB translocates through the nucleus to induce gene expression. The noncanonical NF-κB signaling cascade is controlled at the upper part by NIK. NIK mediates phosphorylation leading to the activation of the homodimer IKKα, another catalytic subunit of the IKK complex. IKKα cleaves the inactive precursor p100 into the active NF-κB subunit p52. In combination with RelB, it forms a heterodimeric NF-κB molecule which enters the nucleus to induce target gene expression. *BCR* B-cell receptor, *BAFF-R* B-cell-activating factor receptor, *FADD* Fas-associated death domain, *IκBα* inhibitor of NF-κB, *IKKα/β* IκB kinase α/β, *LT-βR* lymphotoxin-β receptor, *NEMO* NF-κB essential modulator, *NIK* NF-κB-inducing kinase, *RANK* receptor activator of NF-κB, *RIPK1* receptor-interacting protein 1, *TAB2/3* TGFβ-activated kinase 1 (*TAK1*)-binding protein, *TAK1* TGFβ-activated kinase 1, *TCR* T-cell receptor, *TLR* Toll-like receptor, *TRADD* TNF receptor-associated death domain, *TRAF2/3* TNF receptor-associated factor 2/3, *TNFR* tumor necrosis factor receptor. Adapted from Luedde et al.: The role of NF-κB in hepatic disease models, Translational Research in Chronic Liver Disease, Falk Workshop, Shaker Verlag Aachen 2009: 57–89

NF-κB (RANK). Moreover, uncontrolled stimulation of the noncanonical NF-κB signaling cascade leads to severe diseases such as B-cell lymphomas, ulcerative colitis, and rheumatoid arthritis [40].

Processing of p100 depends on the NF-κB-inducing kinase (NIK), IKKα homodimers, and βTrCP, a subunit of the SCF$^{\beta TrCP}$ ubiquitin ligase. NIK mediates phosphorylation of p100 at Ser866 and Ser870 in its NIK-responsive domain (NRD) and IKKα activation. Activated IKKα phosphorylates p100 at Ser99, Ser108, Ser115, Ser123, and Ser872 which is needed for the recruitment of βTrCP. The SCF^{βTrCP} ubiquitin ligase complex facilitates ubiquitination of p100 and thereby its 26S proteasomal degradation to p52. Interestingly, canonical and noncanonical pathways exhibit some similarities regarding regulatory mechanisms as shown by equivalent phosphorylation sites of p100 and IκBα or ubiquitin-mediated 26S proteasomal degradation of NF-κB inhibitors (Fig. 6.2) [41–43]. In contrast to the canonical pathway, the noncanonical cascade is characterized by a slow and persistent signaling and protein synthesis [44, 45].

Next to ubiquitin, SUMOylation is another regulatory mechanism to shape NF-κB signaling in both pathways. SUMOylation induces posttranslational modifications like ubiquitination and phosphorylation and influences protein-protein interaction and gene transcription [46]. It was shown in different studies that the interplay of SUMO, SUMO proteases, and NF-κB signaling members like NEMO, IκBα, or p100 represent another important level of signal transduction [47, 48].

6.6 NF-κB-Associated Human Diseases and Genetic Mouse Models

Due to the immense effort taken in examining NF-κB and its biological function, the crucial role of NF-κB in cellular homeostasis was uncovered. Several human diseases like psoriasis, colitis ulcerosa, Crohn's disease, rheumatoid arthritis, or cancer development are a result of dysregulated NF-κB activation [2–4].

For a better understanding of the mechanisms behind NF-κB dysregulation *in vivo*, numerous mouse models were developed and examined. These comprise constitutive, tissue-specific conditional knockouts using the cre/loxP technology, gene knock-ins, and reporter mice of single or more IKK subunits or in combination with other members of the NF-κB signaling cascade [6–8, 49–53].

Caused by the tremendous information on the different genetic mouse models and their implication in broadening our understanding of NF-κB-associated human diseases, the next paragraph will focus particularly on one organ which has been extensively studied over the last few years and is a prime example for the importance of NF-κB homeostasis: the liver.

6.7 NF-κB and Its Critical Role for Liver Homeostasis

The liver is the biggest organ in the human body with a weight of 1.2–1.8 kg and makes approximately 2–3% of the whole body weight. The liver is of great importance because it is responsible for different metabolic processes like synthesis of vitally essential proteins (e.g. albumin, blood coagulation factors, hormones), utilization of food residues (e.g. conversion of glucose to amylum), detoxification of intermediate catabolic metabolites (e.g. from medicine), and bile production. Additionally, it is a storage organ for important macromolecules such as hormones or amylum. Besides the wide metabolic functions, the liver has the capability of reacting immunologically as well [54]. During all these biochemical and immunological processes, the different cell types forming the liver are faced with degradation products which might be harmful, such as oxygen radicals, or are attacked by bacterial components like lipopolysaccharides (LPS) or viruses, such as hepatitis viruses. NF-κB activation protects the cells against apoptosis and supports pro-inflammatory responses. Hepatocytes are the dominant hepatic cell type and stress factors, such as cytokines like TNFα or IL-1, and initiate the NF-κB signaling cascade to protect them against cell death [31].

Despite the bad reputation of inflammatory processes, it plays in the liver a central role for wound healing induced by injury processes, such as alcohol abuse or medication intake, which bear a high risk for liver fibrosis and cirrhosis progression and, finally, hepatocellular carcinoma (HCC) development. However, also viral infections such as hepatitis B virus (HBV) and HCV are potent inducers of liver fibrosis and HCC development. During fibrosis, the liver tissue undergoes a perennial process of inflammation, apoptotic and necroptotic events, and compensatory renewal. These chronical processes cause severe side effects, such as the production of highly reactive molecules like reactive oxygen species (ROS), chromosomal aberrations, and possibly malignant alterations of proliferating hepatocytes. Other diseases and reasons which can also trigger HCC formation are nonalcoholic steatohepatitis (NASH), obesity, diabetes, aflatoxin-contaminated nutrients, exposure to toxic compounds like vinyl chloride, and genetic predisposition such as hemochromatosis [55, 56].

6.8 A Deeper View on TNF and NF-κB

Only two possibilities remain if a cell is faced with strong stress: survival or death. TNFα and IL-1 are classical stress inducers and belong to the TNF superfamily. Next to these prominent members, Fas (Apo-1) and TRAIL are also well known and are currently intensively studied proteins. These signaling cascades simultaneously induce expression of pro-survival and proapoptotic proteins and, according to the stimulation strength, determine if the cell dies or survives [57–59].

The balance between life and death is absolutely essential to keep the liver in a healthy status. Apoptosis and necroptosis are different ways of cellular death but are strictly coordinated processes with common molecular characteristics, but with different cellular fragmentation processes into several small pieces [60].

TNFα acts on different biological processes facilitated by TNF-R1 and TNF-R2. Soluble TNFα initiates TNF-R1 signaling, whereas activation of TNF-R2 signaling needs binding of membrane-bound TNFα [61]. Activation of the TNF-R1 by TNFα leads to its trimerization and initiates recruitment of different adaptor proteins building the membrane-bound TNFR-complex I. The complex I comprises TRADD, RIPK1, cIAP1/2, and TRAF2/5. All seven TRAF family members have a C-terminal coiled coil domain which supports protein-protein interaction. Only TRAF2-7 exhibits an N-terminal RING domain which transfers K63-linked ubiquitin to target proteins, acting as E3 ligases. Nevertheless, it is not clarified till now if TRAF proteins mainly act as E3 ligases or as adaptors [20, 62]. It was shown that TRAFs are involved in both, canonical and noncanonical NF-κB pathway, and those TRAFs are also needed for activation of other signaling cascades, therefore acting as a distributor platform for several pathways. RIPK1 is, differently to the TRAFs, exclusively engaged in the canonical NF-κB pathway. RIPK1 and TRAF2/5 seem to interact with each other by TRAF2/5-mediated K63-linked polyubiquitination of RIPK1 (Lys377) [63]. Alternatively, it is also discussed that cIAP1/2 is responsible for RIPK1 polyubiquitination and that TRAF2 only recruits them to the receptor complex. Despite intensive research the exact function of TRAF2/5 and cIAP1/2 concerning RIPK1 ubiquitination could not have been solved satisfactorily to date [64]. Besides K63-linked ubiquitination, RIPK1 undergoes a second, linear Met1-linked ubiquitination simultaneously. This is mediated by the linear ubiquitin chain assembly complex (LUBAC), another E3 ligase complex [65, 66]. RIPK1 is linked with the linear chain to NEMO and with the K63-linked chain to the TAB2/3-TAK1 complex and brings the IKK complex and TAK1 in closer vicinity to each other, leading to the phosphorylation of IKKβ and finally NF-κB activation as described above.

6.9 NF-κB and Its Function in Hepatogenesis

A deeper understanding of NF-κB importance and its components was achieved by using genetically modified mouse models. These models include a setup of diverse genetic approaches such as constitutive knockout models, dominant-negative expression or overexpression of single or double IKK subunits or IκB proteins, tissue-specific conditional knockouts by using cre/loxP recombination, reporter systems, and gene knock-ins [50].

Knockout mice deficient in *Rela* (*p65*) die in the uterus between embryonic days 15 (E15) and E16 because of hepatocyte apoptosis. The primary cause of this event is the failure of a TNFα-mediated IκBα induction and granulocyte/macrophage colony-stimulating factor (GM-CSF) as shown in murine embryonic fibroblasts (MEFs), revealing that RelA has a protective function against TNFα [67, 68]. These findings were approved by generating a double knockout of *Tnf* and *Rela*, which led to a normal embryonic development and a full rescue from lethality [69]. However, another study reported similar results, but these double knockout mice died 10 days after birth from acute hepatitis and neutrophil infiltration [70]. Both studies indicate that RelA and TNF-R1 are not essential for liver development in mice, but seem to sensitize these animals to infections leading to death within a very short time frame. Further investigations of other NF-κB subunits showed that the genetic loss of both *c-Rel* (*Rel*) and *Rela* causes liver failure as well [71]. Moreover, genetic ablation of both transcription factors triggers impaired maturation of B cells, T cells, and macrophages, denoting important roles in controlling genes relevant for immune responses. Genetic manipulation of murine livers by using adenoviral technique allowed transcription of an IκB superrepressor (Ad5IkappaB) which abolished NF-κB linkage to the respective DNA binding sites. After partial hepatectomy the adenoviral infected livers displayed increased apoptosis rates of hepatocytes, proving the importance of NF-κB [72]. For the remaining NF-κB subunits RelB, NF-κB1, and NF-κB2, they were shown to play important roles in the differentiation and proper function of hemopoietic cells [73].

Next to the studies on NF-κB subunits, some studies carried out to define the functional impact of the single IKK members. Constitutive deletion of *Ikkβ* (*Ikk2*$^{-/-}$) in mice leads to embryonic lethality at day E12.5 as a cause of enhanced liver damage. Cell culture experiments of primarily isolated MEFs from these animals revealed impaired NF-κB activation in response to TNFα and IL-1. These results revealed a central function for IKKβ in controlling liver development and NF-κB activity, whereas loss of IKKβ cannot be fully compensated by IKKα (IKK1) [74–76]. Ablation of the catalytic subunit *Nemo* (*Ikkγ*$^{-/-}$) has a similar phenotypical effect as the embryos die between day E12.5 and E13.0 from massive apoptotic liver failure, and experiments with isolated *Ikkγ*$^{-/-}$ MEFs also showed disturbed NF-κB activity after treatment with TNFα, IL-1, LPS, and Poly(I:C), leading to high susceptibility to apoptosis [77]. As stated above, IKKα does not seem to have an impact on liver development and NF-κB induction triggered by pro-inflammatory substances. *Ikkα*$^{-/-}$ mice die perinatally and develop severe skeletal and skin-related defects during embryogenesis [78, 79].

6.10 NF-κB and Its Pivotal Role for Liver Integrity

Generation of tissue-specific knockouts by using the cre/loxP recombination system gives the opportunity of a much more detailed view on NF-κB in particular with regard to protection against cytokine-induced hepatitis. Another great advantage of the cre/loxP technology is the time-dependent loss of target genes determined by activation of the respective promoter during later embryonic developmental stages or postnatally, circumventing embryonic lethality effects. This was demonstrated by a conditional cre-driven hepatocyte knockout of *Rela/p65*. These mice are viable, and isolated primary hepatocytes treated with TNFα were highly sensitive to apoptosis with concurrently enhanced c-Jun N-terminal kinase (JNK) expression and degradation of the anti-apoptotic protein cellular FLICE inhibitory protein long (c-FLIP$_L$) [80]. Investigation of hepatocyte-specific *Ikkβ* deletion in adult mice revealed an unexpected slight sensitivity in response to TNFα or LPS administration contrary to the murine embryonical state [52, 80, 81]. However, treatment with concanavalin A (ConA) promotes severe liver failure in adult *Ikkβ*-deficient animals, which is mainly supported by increased activation of JNK, a key mediator of ConA-induced liver failure. Next to JNK activation, ConA is also a potent activator of T cells, indicating that the anti-apoptotic function of IKKβ is the prevention of T cell-mediated cell death associated with decreased JNK activity [81].

Metabolic diseases represent another potential risk factor in mediating inflammatory processes like type 2 diabetes and obesity. Liver-specific *Ikkβ* ablation abolished insulin sensitivity, while these animals showed insulin resistance in muscle and fat induced by a high-fat diet and upon aging [82].

Double knockout of *Ikkα* and *Ikkβ* (IKKα/β^{LPC-KO}) in liver parenchymal cells (hepatocytes and cholangiocytes) supported increased susceptibility of hepatocytes to LPS *in vivo*, which was not detected in the single knockout conditions, claiming for a more redundant function for both IKKs in canonical NF-κB activation. Of note, simultaneous deficiency of IKKα and IKKβ or the combined ablation of *Ikkα* and *Nemo*, but not *Nemo* alone, led to spontaneous development of cholangitis with disturbed portal bile ducts accompanied by severe jaundice, revealing the importance of both catalytic NF-κB subunits in controlling liver immunology and bile duct integrity [83]. Additional pivotal discoveries concerning the physiological impact of the IKK complex were achieved by examining adult mouse livers lacking the regulatory subunit NEMO. These livers are highly sensitive against TNF- and LPS-mediated inflammation and subsequently cell death *in vivo* and *in vitro* [8, 52].

Further studies regarding genetic ablation of other NF-κB signaling members alone or in combination with different IKKs revealed a great impact on cellular homeostasis as well. Hepatocyte-specific deletion of the mitogen-activated kinase kinase kinase (MAP3K) *TGF-β-activated kinase 1* (TAK1^{LPC-KO}; TAK1$^{\Delta hep}$) leads to a comparable phenotype as seen for IKKα/β^{LPC-KO} with considerable cholangitis, early HCC development, and lethal jaundice at younger age [6, 51]. Deletion of death receptor-associated adaptor proteins like Fas-associated protein with death domain (*Fadd*) or *Tnfr1* in combination with *Nemo* (NEMO/FADD^{LPC-KO}) or *Tak1* (TAK1/TNFR1$^{\Delta hep}$) showed strongly reduced signs of inflammation, fibrosis, and cell death, raising evidence for a pro-apoptotic trigger driving these phenotypes [8, 51]. A look more downstream of the death receptor pathways, regarding casapse-8, highlighted a rescue of Caspase8/NEMO$^{\Delta hep}$ mice from steatosis and HCC development, but these animals developed a severe spontaneous phenotype of liver necrosis, cholestasis, and biliary lesions, most likely caused by a FasR-induced RIPK1-RIPK3-mediated necroptosis [84]. As stated above, the hepatoprotective function of NF-κB is also influenced by a sustained reduced expression level of JNK upon TNFα administration [85–87]. JNK belongs to the MAPK family and is activated via TRAF2, RIPK1, and MKK4/7. JNK is a major mediator of cell death, triggered not only by pro-inflammatory cytokines such as TNFα and IL-1β but also by cellular stressors such as UV radiation, osmotic, oxidative, hypoxic, and genotoxic events [88, 89]. The connective bridge between NF-κB and JNK is TAK1, which is able to phosphorylate IKKβ and MKK4/7, depending on stimulus strength. Next to JNK, TAK1 is also able to phosphorylate MKK3/6, which are activators of p38, another MAPK involved in cell proliferation, differentiation, and cell death [90, 91]. Studies with p38α/IKK2^{LPC-KO} mice showed increased hepatocyte sensitivity against TNFα and LPS administration *in vivo*, which was not the case for the single knockout conditions. Moreover, significantly increased JNK expression levels could be detected, but were not strong enough to induce liver failure [92]. The results revealed impressively that the NF-κB signaling cascade is not an isolated pathway and is embedded in a broad network of stress-related signaling pathways which are tightly regulated in protecting the liver against harmful events.

6.11 NF-κB and Its Implication for HCC Development

Chronic inflammatory liver diseases belong to the main preconditions for generation and progression of liver cirrhosis and subsequently HCC development, affecting 80–90% of patients with liver cirrhosis [55]. The formation of cancer is generally characterized by a disbalance between cell death and survival/proliferation [93]. In most HCCs, NF-κB is constantly active, which drives a continuous burst of pro-inflammatory and anti-apoptotic signals [94, 95]. In the last two decades, murine genetic studies had a great impact on the understanding of molecular mechanisms driving HCC development [6–9, 96]. It has been shown that tumor formation is a process in which different liver cell types are involved and where NF-κB activation is time-dependent altered, particularly at early and late stages of cancer development.

One of the first studies in this direction was done on *Multidrug resistance protein 2* (*Mdr2*$^{-/-}$) mice (human homologue *MDR3*), lacking a permeability (P)-glycoprotein which is located in the bile canalicular membrane of hepatocytes with a function as a phospholipid export pump. Disruption of the functionality of the pump causes a spontaneous phenotype described by cholangitis with a dysfunctional biliary delivery, ending in HCC development at 4–6 months of age [96]. In this phenotype, enhanced NF-κB activity leads to a higher TNFα expression, and impairment of NF-κB activity with a hepatocyte-specific inducible IκB superrepressor transgene negatively affected tumor progression and hepatocyte cell death at later stages of tumor development, whereas NF-κB blockage at the initial stage of tumor formation has no inhibitory effect [97]. Besides this tumor-promoting effect of NF-κB, other studies argue for a tumor-suppressive function. Hepatocyte-specific knockout of *Nemo* promotes hepatitis in these animals at 2 months of age and spontaneous HCC development 12 months after birth, triggered by cytokines and a constant low intrinsic dosage of LPS coming from the commensal gut bacteria [8]. Further studies with different combined hepatocyte-specific knockouts of *Nemo* and *Tnfr1*, *Trail-r*, or *Fas* or quadruple knockout showed no rescue or improvement of this phenotype. Still the combination of NEMO/FADD$^{LPC\text{-}KO}$ caused a much milder progress of liver failure and inflammation and abolished HCC development. Examination of liver sections from 1-year-old NEMO$^{LPC\text{-}KO}$/Tnf$^{-/-}$ mice with developed liver tumors revealed that TNF is not important for the development of spontaneous cell death, hepatitis, and HCC in NEMO$^{LPC\text{-}KO}$ livers. Neither depletion of natural killer cells nor an intercrossing with *Rag-1* ablated mice (NEMO$^{LPC\text{-}KO}$/Rag1$^{-/-}$) prevented liver damage [49]. All these results demonstrated that the spontaneously developed hepatocyte-specific NEMO phenotype is neither the result of death receptor-mediated signaling cascades nor an immune response triggered once, and additional studies are needed to identify the mechanism behind. Interestingly, double knockout of *Nemo* and *Tak1* in hepatocytes rescued the massive phenotype detected in the *Tak1* single knockout condition (as stated above), claiming for a NF-κB-independent tumor formation in the TAK1$^{LPC\text{-}KO}$ livers [6, 8]. Generation of knock-in mice endogenously expressing catalytically inactive RIPK1 D138N (*Ripk1D138N/D138N*) are alive after birth unlike mice conditionally lacking *Ripk1*. Moreover, these mice are protected against TNFα treatment and poly (I:C)-induced necroptosis *in vitro* and TNFα administration *in vivo*, indicating that the kinase activity of RIPK1 is not a prerequisite for cell survival but is crucial for TNFα-induced necroptosis [53]. Moreover, this result is supported by the finding that hepatocyte death and HCC development in $NEMO^{LPC\text{-}KO}$ mice is triggered by RIPK1's kinase activity, independent of NF-κB activity and RIPK1's scaffolding function. A complete NF-κB blockage induced by hepatocyte-specific single or combined knockout of *Rela*, *c-rel*, or *Relb* did not affect the liver, whereas constitutively active IKKβ prevented hepatocarcinogenesis in $NEMO^{LPC\text{-}KO}$ animals. These results revealed a NEMO protective function against HCC development. Hepatocyte-specific ablation of RIPK1 activated a TRADD-related apoptosis and HCC development, showing two different functions of RIPK1 [98].

Hepatocarcinogenesis chemically induced by a single injection of diethylnitrosamine (DEN) in 15-day-old $IKK\beta^{\Delta hep}$ mice causes in 2-month-old mice massive liver tumor development which is not seen in untreated $IKK\beta^{\Delta hep}$ livers [9, 81]. Compared to the untreated mice, DEN treatment supports enhanced ROS production correlating with increased JNK expression levels, hepatocyte death, and compensatory hepatocyte proliferation, which is similar to the results of the ConA-treated $IKK\beta^{\Delta hep}$ mice [81]. Moreover, experiments with $IKK\beta^{-/-}$ fibroblasts showed that antioxidants like manganese superoxide dismutase (MnSOD) or MAPK phosphatases are needed to avoid ROS-mediated sustained JNK activity [99]. Besides ROS, nitric oxide (NO•) is another kind of agent radically synthesized by inducible nitric oxide synthase (iNOS), which is able to induce chronic inflammation and might influence tumor formation by controlling cell proliferation, angiogenesis, survival, medical resistance, and DNA repair [100–102]. Examination of *iNos* knockout mice showed significantly reduced NF-κB activities, and a higher concentration of iNOS is related to tumor proliferation, genomic instability microvascularization, and worse diagnosis for HCC patients. Treatment with iNOS inhibitors, like aminoguanidine, has a negative effect on HCC progression and NF-κB activity and a positive influence on apoptosis *in vivo* and *in vitro*. Moreover, interruption of the NF-κB cascade with sulfasalazine or siRNA led to decreased iNOS expression in different HCC cell lines [103].

Next to the components of the canonical NF-κB pathway, HCC development could also be triggered via the LTβR, one of the activators of the noncanonical NF-κB pathway. Enhanced levels of LTα, LTβ, and LTβR were detected in human HBV-/HCV-infected livers and in HCC. Examination of the transgenic mouse models of LTα and LTβ showed that hepatocyte-specific overexpression promotes fibrosis and HCC development. Ablation of IKKβ, specifically in hepatocytes, rescued mice from HCC development [7].

These results indicate a carcinogenesis promoting function of NF-κB, which is strongly supported by pro-inflammatory mediators.

6.12 NF-κBs Contribution on Liver Fibrosis and Cirrhosis

Liver fibrosis is a disease that develops as a consequence of recurrent "out-of-control" wound healing processes. Persistent removal of damaged and/or inflamed tissue with compensatory renewal leads over time to composition of fibrillar collagen in the affected areas, ending up in scarring liver tissue and loss of metabolic functional areas. Hepatic stellate cells (HSCs) are mainly involved in this process, whereas after certain stimuli this cell population is activated and undergoes a transformation to hepatic myofibroblasts which are built by decomposition of extracellular matrix proteins the collagen scars in the tissue. HBV and HCV infections, autoimmune hepatitis, alcohol abuse, NASH, and cholestasis trigger liver fibrosis and finally cirrhosis. For a long period, it was assumed that activated HSCs/hepatic myofibroblasts are the main source of liver fibrosis, but it seems to be a multifaceted process in which different signaling pathways and other liver cell types such as hepatocytes and Kupffer cells might be of importance [104]. Examination of intercrossed mice with a constitutively active human IKKβ (CAIKK2) allele in postnatal livers, controlled by a tetracycline promoter system, revealed modest liver injury, infiltration of immune cells, enhanced hepatocyte proliferation, and spontaneous liver fibrosis progression. Deeper analysis detected significantly enhanced levels of chemokines and certain chemokine receptors, while interruption of CAIKK2 expression led to declined expression levels in hepatocytes. Moreover, disruption of CAIKK2 expression for few weeks also reduced HSC activation but without significant improvement of fibrosis reduction. Only macrophage removal with liposomal clodronate positively affected liver fibrosis development caused by a reduced NF-κB activation. This indicates that next to transformed HSCs, recruitment of pro-inflammatory immune cells such as macrophages, mediated by prolonged hepatocellular NF-κB activation, are involved in promoting liver fibrosis [105]. Previous studies detected in activated HSCs enhanced levels of NF-κB, correlating not only with increased expression of pro-inflammatory and adhesion molecules, but also with upregulation of anti-apoptotic proteins like TRAF1/2, cIAP1/2, Bcl-X_L, and GADD45β [106]. Moreover, it seems that the CD95/Fas pathway is responsible for HSC death, whereas TGF-β and TNFα induce pro-survival signaling. Disruption of NF-κB activity by overexpression of an IκB superrepressor led to decreased expression levels of anti-apoptotic proteins like Bcl-X_L and enhanced proapoptotic proteins as caspase-3 during TGF-β and TNFα stimulation [107, 108]. Another study revealed a constitutive phosphorylation of p65 at Ser536 mediated by IKKβ and the autocrine renin-angiotensin system in human hepatic myofibroblasts. IKKβ-mediated phosphorylation of p65 allows nuclear translocation, whereas angiotensin II promotes myofibroblast survival in an autocrine and paracrine way [109]. Besides increased NF-κB activation, enhanced p-JNK levels could be detected in human- and murine-activated HSCs, indicating that JNK has an important function in liver fibrosis progression [110].

In addition to the death receptor family, two members of the Toll-like receptor (TLR) family, TLR4 and TLR9, were uncovered to be involved in mediating NF-κB activation during liver fibrosis progression as their activation could be a consequence of gut bacterial components [111, 112].

6.13 NF-κB in HVB/HCV Infections

Approximately 3% of the total world population have been infected with HCV, whereas more than 170 million people are suffering from chronic hepatitis, cirrhosis, and HCC. The annual rate of HCC development is strongly increased in cirrhotic livers [113]. The HCV core, the envelop protein E2, and HCV subgenomic replicons have been shown to enhance p38 and extracellular signal-regulated kinase (ERK) phosphorylation and initiate NF-κB [114, 115]. Moreover, it seems that oxidative stress (ROS) and TGF-β1 are important mediators of fibrogenesis in HCV infection. Silencing of p38, JNK, ERK1/2, and NF-κB (p65) in hepatocytes *in vitro* revealed their implication in enhanced TGF-β1 expression [116]. Experiments using different inhibitors for ROS (diphenyliodonium [DPI]), JNK (SP600125), IRE1 (Irestatin 9389), and NF-κB (6-amino-4-(4-phenoxyphenylethylamino [AQ]), have blocked significantly HCV-induced NF-κB and TGF-β1-mediated SMAD signaling. Silencing of JNK and IRE1 using siRNA inhibited efficiently ER stress, ROS, NF-κB, and TGF-β1 activity [117]. Moreover, it was shown *in vitro* that the hepatitis B virus X (HBx) protein, in addition to HCV infection, maintains enhanced NF-κB and AP-1 activation which might indicate a supportive role of HBx in HCC development [118]. Patients suffering from HBV have a much higher potential to develop HCC confirmed by studies which detected in nearly all examined HCCs of HBV patients chromosomally integrated viral DNA. Next to the HBx protein, the HBV surface antigen PreS2 is able to activate NF-κB and AP-1 via PKC-mediated induction of the c-Raf-1/MAPK signaling pathway, which further leads to increased hepatocyte proliferation, indicating a tumor-promoting function for PreS2 [119].

Stimulation of HCV core protein infected HeLa and HuH-7 cells with TNFα or LTα1/β2 mediates increased or sustained IκBβ degradation, whereas degradation of IκBα was only detected in LTα1/β2-stimulated HeLa cells. Besides cytokine treatment, higher levels of NF-κB activity were also detected in untreated HeLa and HuH-7 cells only harboring the HCV core protein. This finding shows that the HCV core protein is able to positively alter NF-κB initiation and, in combination with cytokine treatment, markedly increased this effect, which might have a direct influence on a stable and continuous NF-κB activation in HCV-infected cells [120].

After viral infection, the immune system reacts with the production of different NF-κB-related cytokines, especially IFNs to defend the infection [121–123]. NF-κB and IFN regulatory factor-3 (IRF-3) initiate alone or together the antiviral Janus kinases/signal transducer and activator (JAK-STAT) signaling cascade [124]. Despite the protective host defense response to eliminate viral infection accompanied by removing and restoring damaged tissue, continuous NF-κB activation also has negative side effects, such as activation of quiescent HSCs or proliferation of hepatocytes with oncogenic mutations, leading to liver fibrogenesis and HCC development. Therefore, therapeutics are needed which keep virus-induced oxidative stress or mediators of fibrosis at a minimum [125, 126].

6.14 NF-κB and Its Meaning for Obesity

Overweight and metabolic diseases have strongly increased all over the world, ranging from childhood to adult stage. The WHO estimated for 2015 that around 2.3 billion adults would be overweighted and approximately 700 million people would be obese [127]. Obesity is a gate opener not only for several diseases, such as insulin resistance, type 2 diabetes, cardiovascular disease, and atherosclerosis, but also for dementia, airway disease, and cancer [128, 129]. Genetic examination of murine and human adipose tissue, as well as murine hepatic tissue, uncovered strongly increased activation of an inflammatory and immune response gene network triggered by a multifaceted genetic loci and environmental issues [130–132]. However, the molecular mechanism of how obesity influences macrophage response is not yet fully understood. Currently, different theories are trying to explain how macrophages and obesity are connected to each other, e.g. by TLR4 induction through a high content of saturated fats, by inflammation of the central nervous system, or by the commensal gut microbiota [133–136].

Adipocytes are the cellular components of the adipose tissue and next to lipid storage they are also able to generate and release pro-inflammatory cytokines and adipokines which attract monocytes and T cells to infiltrate the adipose tissue, where monocytes differentiate to M1 macrophages. It is discussed that one major source triggering this process is metabolic stress caused by overnutrition, which leads to high levels of non-metabolized free fatty acids and ER stress inducing inflammatory responses. However, also removal of apoptotic adipocytes is another potent inducer of macrophage recruitment [128]. Besides macrophage recruitment, activation of TLRs is considered to initiate NF-κB-mediated generation of pro-inflammatory molecules. The NF-κB target IKKε was detected to be needed for high-fat diet (HFD)-induced obesity. Depletion of IKKε revealed a positive influence on high-fat diet fed mice, because these animals were protected from insulin resistance and hepatic steatosis, and they did not show any sign of chronic inflammation in liver or adipose tissue or induction of inflammatory pathways [137]. Similar results were obtained by HFD-treated mice with IKKβ deficiency in hepatocytes (*Ikbkb*$^{\Delta}$*hep*) or in myeloid (*Ikbkb*$^{\Delta}$*mye*) cells. These animals showed liver-specific insulin sensitivity, but revealed insulin resistance in muscle and adipose tissue, which was also examined in obese or aged mice. The *Ikbkb*$^{\Delta}$*mye* mice had preserved total insulin sensitivity with protection against insulin resistance [82]. Constitutively expressed IKKβ in hepatocytes caused increased NF-κB activity comparable to values detected in HFD-treated or obese mice. These animals also developed type 2 diabetes (T2D) with characteristic features such as hyperglycemia, severe hepatic insulin resistance and mild systemic insulin resistance, also affecting muscle tissue. Liver-specific expression of the IκBα superrepressor rescued this phenotype [138]. These results indicate a potential NF-κB function in obesity-related inflammation and T2D locally in the liver and myeloid cells, which seem to be crucial for improving systemic insulin resistance. Sustained excessive food intake also leads to NF-κB initiation via different pro-inflammatory pathways as in adipocytes and macrophages; TLR4 was discovered to be activated by free fatty acids. Ablation of TLR4 protected these mice from insulin resistance in muscle tissue, disturbed glucose metabolism, and inflammatory signaling in liver and adipose tissue [139]. Additional studies showed that loss of function mutation in TLR4 also hindered the development of insulin resistance in adipocytes and in diet-initiated obesity [135, 140, 141]. Next to TLR4, TLR2 seems to have a comparable role in mediating NF-κB activation during HFD-induced obesity and the formation of insulin resistance [142, 143]. Metabolic stress signals attract monocytes to remove apoptotic pancreatic β-cells, thereby secreting TNFα, IL-6, and IL-1 to support β-cell dying. Increased TNFα levels in blood and peripheral tissues were detected in insulin-resistant rodents as well, and HFD promotes enhanced secretion of TNFα, IL-6, and IL-1 in hepatocytes and adipocytes. Inactivation of TNFα in different genetic rodent models mediated a significant peripheral glucose uptake during insulin secretion. Moreover, hepatocyte deficiency of IL-6 receptor (IL-6Rα$^{L\text{-}KO}$) impaired obesity-related insulin resistance and glucose tolerance. Depletion of IL-6Rα induced a massive inflammatory response indicated by enhanced levels of TNFα, IL-6, and IL-10 and IκBα phosphorylation. Glucose tolerance in these animals was reinstated by TNFα neutralization or Kupffer cell deficiency [144–146]. All these results imply a critical role for NF-κB in obesity-mediated insulin resistance and T2D progression and that a controlled release of inflammatory inducers is necessary for a healthy hepatic metabolism.

6.15 NF-κB and Its Function in Hepatic Ischemia/Reperfusion (I/R) Injury

During and after liver transplantation or hepatic resection, ischemia/reperfusion (I/R) is a main reason for liver failure caused by hypoxia or anoxia if oxygen supply and tissue pH are reinstated after clamping of the hepatic blood flow. The pathobiochemical mechanisms behind I/R are versatile and affect all liver cell types [147, 148].

Several studies were done to clarify if I/R induces apoptosis or necroptosis [149–154]. During I/R, hepatocytes express damage-associated molecular patterns (DAMPs)/pathogen-associated molecular patterns (PAMPs) on their surface, and macrophages, Kupffer cells, and dendritic cells bind these DAMPs/PAMPS by TLRs, thereby initiating the immune system. Adaptive and innate immune cells are recruited to the affected tissue sections where they secrete additional inflammatory mediators as TNFα, IL-1, IL-12, and INFγ which fuel these processes [148, 155]. Hepatic adenoviral overexpression of an IκB superrepressor in rats impeded NF-κB activation by decreased TNFα expression, preventing negative effects of hepatic I/R [156]. Furthermore, it was shown that heat shock preconditioning of rat livers significantly impaired I/R-induced NF-κB activation and expression of several inflammatory mediators [157]. Conditional ablation of *Ikkβ* in murine hepatocytes led to impaired liver necrosis and inflammation during I/R compared to the wild-type condition. Also administration of the chemical IKKβ inhibitor AS602868 failed to induce I/R-mediated liver failure without induction of pro-apoptotic side effects, making this inhibitor potentially useful for therapy [52]. A20 is another critical NF-κB target gene activated during inflammatory processes in hepatocytes and is involved in blocking apoptosis, but is also part of a negative feedback loop to control NF-κB activation [158–160]. Murine hepatic recombinant adenoviral overexpression of A20 revealed a significantly increased survival rate after I/R in these animals, indicated by considerably reduced bilirubin and transaminase levels, reduced hemorrhagic necrosis and steatosis, and enhanced hepatocyte proliferation. Moreover, A20 induced the release of peroxisome proliferator-activated receptor alpha (PPARα), an important controller of lipid homeostasis and oxidative damage, which prevented oxidative induced necrosis [161].

These results revealed that interference of IKKβ activation by pharmacological treatment is a putative instrument to protect livers from I/R injury. A20, another crucial NF-κB target, seems to be an additional promising candidate which might have the capacity as a therapeutic target which efficiently blocks I/R [86, 161]. In contrast, mice given recombinant receptor activator of NF-κB ligand (RANKL) before or during hepatic I/R showed enhanced NF-κB activation accompanied by less liver damage [162].

6.16 NF-κB as Therapeutic Target

Liver diseases are characterized by multifaceted biochemical processes, whereas the balance between apoptotic and survival signaling pathways is of great importance for liver homeostasis. Impaired homeostasis leads to chronic inflammation and compensatory proliferation ending up in liver cirrhosis and HCC development. Tremendous work was done in these fields, and particularly the use of genetic modified mouse models showed that NF-κB is one of the most important key players in preserving liver integrity. However, the function of NF-κB in the development of liver diseases is not black and white nor is it equal for all liver cell types. This makes it extremely challenging to design cell type and disease-specific drugs which only affect the cells of interest and not interfering with NF-κB activity or other signaling cascades in healthy cells.

Several biological and chemical compounds have been tested to modify NF-κB activity as steroids, selective estrogen receptor modulators, antioxidants, proteasome inhibitors, and IKK inhibitors [126].

The administration of hormones is critical to assess because of the increased risk of negative side effect induction in the liver or other organs or their prospective impact on the respective gender [126, 163]. Phytochemicals from tea or other plants have a great antioxidant potential which seems to have a significant positive influence on liver health, but further investigations are needed for refining substance purification and antioxidant composition [164–167]. The development of chemical compounds, such as proteasome inhibitors like Bortezomib, mediates reduced NF-κB activation. Treatment of human hepatoma cells with MG132, another protease inhibitor, resulted in apoptosis induction but affected the β-catenin pathway. Currently, the use of proteasome inhibitors as a therapeutic target needs more scientific research because of the diversity of bad side effects induced next to decreased NF-κB activation [168–171]. Regarding specific NF-κB blockage, the IKK complex member IKKβ seems to be a putative target. It has been shown that aspirin and sulfasalazine are able to impair the catalytic activity of the IKK complex. Furthermore, the development of a new class of chemical compounds, referred to as "small molecules," is able to impede NF-κB activation by binding to the ATP-binding pocket of IKKβ, thereby inducing conformational changes. Despite the higher binding specificity of small molecules for their targets and effectiveness compared to other compounds, a potential risk remains for the development of side effects, such as induction of inflammatory responses [126, 172, 173].

In addition to biological and chemical compounds, therapeutics on the DNA/RNA level have also been developed. Different studies were done with small interfering RNAs (siRNAs), microRNAs (miRNAs), and antisense oligodeoxynucleotides (ODNs), placing emphasis on liver fibrosis. Several studies have obtained promising results, but more genetic and clinical trial studies are needed to gain a better mechanistic understanding and to improve application effectiveness of small RNA and DNA products to enhance therapeutic success in the treatment of liver diseases [174].

In summary, further scientific and clinical studies are needed to develop effective therapeutics that are fine-tuned for the control of hepatic NF-κB activation to prevent progression of liver diseases, but without inducing harmful side effects.

References

1. Sen R, Baltimore D. Multiple nuclear factors interact with the immunoglobulin enhancer sequences. Cell. 1986;46:705–16. doi:10.1016/0092-8674(86)90346-6.
2. Barnes PJ, Karin M. Nuclear factor-κB—a pivotal transcription factor in chronic inflammatory diseases. N Engl J Med. 1997;336:1066–71. doi:10.1056/NEJM199704103361506.
3. Camandola S, Mattson MP. NF-kappa B as a therapeutic target in neurodegenerative diseases. Expert Opin Ther Targets. 2007;11:123–32. doi:10.1517/14728222.11.2.123.
4. MacLellan WR, Schneider MD. Death by design programmed cell death in cardiovascular biology and disease. Circ Res. 1997;81:137–44. doi:10.1161/01.RES.81.2.137.
5. Grivennikov SI, Greten FR, Karin M. Immunity, inflammation, and cancer. Cell. 2010;140:883–99. doi:10.1016/j.cell.2010.01.025.
6. Bettermann K, Vucur M, Haybaeck J, Koppe C, Janssen J, Heymann F, Weber A, Weiskirchen R, Liedtke C, Gassler N, Müller M, de Vos R, Wolf MJ, Boege Y, Seleznik GM, Zeller N, Erny D, Fuchs T, Zoller S, Cairo S, Buendia M-A, Prinz M, Akira S, Tacke F, Heikenwalder M, Trautwein C, Luedde T. TAK1 suppresses a NEMO-dependent but NF-κB-independent pathway to liver cancer. Cancer Cell. 2010;17:481–96. doi:10.1016/j.ccr.2010.03.021.
7. Haybaeck J, Zeller N, Wolf MJ, Weber A, Wagner U, Kurrer MO, Bremer J, Iezzi G, Graf R, Clavien P-A, Thimme R, Blum H, Nedospasov SA, Zatloukal K, Ramzan M, Ciesek S, Pietschmann T, Marche PN, Karin M, Kopf M, Browning JL, Aguzzi A, Heikenwalder M. A lymphotoxin-driven pathway to hepatocellular carcinoma. Cancer Cell. 2009;16:295–308. doi:10.1016/j.ccr.2009.08.021.
8. Luedde T, Beraza N, Kotsikoris V, van Loo G, Nenci A, De Vos R, Roskams T, Trautwein C, Pasparakis M. Deletion of NEMO/IKKgamma in liver parenchymal cells causes steatohepatitis and hepatocellular carcinoma. Cancer Cell. 2007;11:119–32. doi:10.1016/j.ccr.2006.12.016.

9. Maeda S, Kamata H, Luo J-L, Leffert H, Karin M. IKKbeta couples hepatocyte death to cytokine-driven compensatory proliferation that promotes chemical hepatocarcinogenesis. Cell. 2005;121:977–90. doi:10.1016/j.cell.2005.04.014.
10. Pasparakis M, Haase I, Nestle FO. Mechanisms regulating skin immunity and inflammation. Nat Rev Immunol. 2014;14:289–301. doi:10.1038/nri3646.
11. Augustin R, Fraune S, Bosch TCG. How Hydra senses and destroys microbes. Semin Immunol. 2010;22:54–8. doi:10.1016/j.smim.2009.11.002.
12. Hemmrich G, Miller DJ, Bosch TCG. The evolution of immunity: a low-life perspective. Trends Immunol. 2007;28:449–54. doi:10.1016/j.it.2007.08.003.
13. Lange C, Hemmrich G, Klostermeier UC, López-Quintero JA, Miller DJ, Rahn T, Weiss Y, Bosch TCG, Rosenstiel P. Defining the origins of the NOD-like receptor system at the base of animal evolution. Mol Biol Evol. 2011;28:1687–702. doi:10.1093/molbev/msq349.
14. Srivastava M, Simakov O, Chapman J, Fahey B, Gauthier MEA, Mitros T, Richards GS, Conaco C, Dacre M, Hellsten U, Larroux C, Putnam NH, Stanke M, Adamska M, Darling A, Degnan SM, Oakley TH, Plachetzki DC, Zhai Y, Adamski M, Calcino A, Cummins SF, Goodstein DM, Harris C, Jackson DJ, Leys SP, Shu S, Woodcroft BJ, Vervoort M, Kosik KS, Manning G, Degnan BM, Rokhsar DS. The Amphimedon queenslandica genome and the evolution of animal complexity. Nature. 2010;466:720–6. doi:10.1038/nature09201.
15. Gilmore TD. Introduction to NF-κB: players, pathways, perspectives. Oncogene. 2006;25:6680–4. doi:10.1038/sj.onc.1209954.
16. Irazoqui JE, Urbach JM, Ausubel FM. Evolution of host innate defence: insights from C. elegans and primitive invertebrates. Nat Rev Immunol. 2010;10:47–58. doi:10.1038/nri2689.
17. Baeuerle PA, Henkel T. Function and activation of NF-kappa B in the immune system. Annu Rev Immunol. 1994;12:141–79. doi:10.1146/annurev.iy.12.040194.001041.
18. Blank V, Kourilsky P, Israël A. NF-κB and related proteins: rel/dorsal homologies meet ankyrin-like repeats. Trends Biochem Sci. 1992;17:135–40. doi:10.1016/0968-0004(92)90321-Y.
19. Chen FE, Ghosh G. Regulation of DNA binding by Rel/NF-kappaB transcription factors: structural views. Oncogene. 1999;18:6845–52. doi:10.1038/sj.onc.1203224.
20. Hayden MS, Ghosh S. Shared principles in NF-κB signaling. Cell. 2008;132:344–62. doi:10.1016/j.cell.2008.01.020.
21. Phelps CB, Sengchanthalangsy LL, Malek S, Ghosh G. Mechanism of κB DNA binding by Rel/NF-κB dimers. J Biol Chem. 2000;275:24392–9. doi:10.1074/jbc.M003784200.
22. Saha A, Hammond CE, Trojanowska M, Smolka AJ. Helicobacter pylori-induced H,K-ATPase alpha-subunit gene repression is mediated by NF-kappaB p50 homodimer promoter binding. Am J Physiol Gastrointest Liver Physiol. 2008;294:G795–807. doi:10.1152/ajpgi.00431.2007.
23. Inoue J, Kerr LD, Rashid D, Davis N, Bose HR, Verma IM. Direct association of pp40/I kappa B beta with rel/NF-kappa B transcription factors: role of ankyrin repeats in the inhibition of DNA binding activity. Proc Natl Acad Sci U S A. 1992;89:4333–7.
24. Huxford T, Huang D-B, Malek S, Ghosh G. The crystal structure of the IκBα/NF-κB complex reveals mechanisms of NF-κB inactivation. Cell. 1998;95:759–70. doi:10.1016/S0092-8674(00)81699-2.
25. Jacobs MD, Harrison SC. Structure of an IkappaBalpha/NF-kappaB complex. Cell. 1998;95:749–58.
26. Malek S, Huang D-B, Huxford T, Ghosh S, Ghosh G. X-ray crystal structure of an IκBβ·NF-κB p65 homodimer complex. J Biol Chem. 2003;278:23094–100. doi:10.1074/jbc.M301022200.
27. Huxford T, Ghosh G. A structural guide to proteins of the NF-κB signaling module. Cold Spring Harb Perspect Biol. 2009;1:a000075. doi:10.1101/cshperspect.a000075.
28. Michel F. Crystal structure of the ankyrin repeat domain of Bcl-3: a unique member of the IkappaB protein family. EMBO J. 2001;20:6180–90. doi:10.1093/emboj/20.22.6180.
29. Ben-Neriah Y, Karin M. Inflammation meets cancer, with NF-κB as the matchmaker. Nat Immunol. 2011;12:715–23. doi:10.1038/ni.2060.
30. Morgan MJ, Liu Z. Crosstalk of reactive oxygen species and NF-κB signaling. Cell Res. 2011;21:103–15. doi:10.1038/cr.2010.178.

31. Oeckinghaus A, Hayden MS, Ghosh S. Crosstalk in NF-κB signaling pathways. Nat Immunol. 2011;12:695–708. doi:10.1038/ni.2065.
32. Delhase M, Hayakawa M, Chen Y, Karin M. Positive and negative regulation of IkappaB kinase activity through IKKbeta subunit phosphorylation. Science. 1999;284:309–13.
33. DiDonato JA, Hayakawa M, Rothwarf DM, Zandi E, Karin M. A cytokine-responsive IκB kinase that activates the transcription factor NF-κB. Nature. 1997;388:548–54. doi:10.1038/41493.
34. Kanayama A, Seth RB, Sun L, Ea C-K, Hong M, Shaito A, Chiu Y-H, Deng L, Chen ZJ. TAB2 and TAB3 activate the NF-κB pathway through binding to polyubiquitin chains. Mol Cell. 2004;15:535–48. doi:10.1016/j.molcel.2004.08.008.
35. Roh YS, Song J, Seki E. TAK1 regulates hepatic cell survival and carcinogenesis. J Gastroenterol. 2014;49:185–94. doi:10.1007/s00535-013-0931-x.
36. Wang C, Deng L, Hong M, Akkaraju GR, Inoue J, Chen ZJ. TAK1 is a ubiquitin-dependent kinase of MKK and IKK. Nature. 2001;412:346–51. doi:10.1038/35085597.
37. Yu Y, Ge N, Xie M, Sun W, Burlingame S, Pass AK, Nuchtern JG, Zhang D, Fu S, Schneider MD, Fan J, Yang J. Phosphorylation of Thr-178 and Thr-184 in the TAK1 T-loop is required for interleukin (IL)-1-mediated optimal NFκB and AP-1 activation as well as IL-6 gene expression. J Biol Chem. 2008;283:24497–505. doi:10.1074/jbc.M802825200.
38. Coope HJ, Atkinson PGP, Huhse B, Belich M, Janzen J, Holman MJ, Klaus GGB, Johnston LH, Ley SC. CD40 regulates the processing of NF-κB2 p100 to p52. EMBO J. 2002;21:5375–85. doi:10.1093/emboj/cdf542.
39. Derudder E, Dejardin E, Pritchard LL, Green DR, Korner M, Baud V. RelB/p50 dimers are differentially regulated by tumor necrosis factor-alpha and lymphotoxin-beta receptor activation: critical roles for p100. J Biol Chem. 2003;278:23278–84. doi:10.1074/jbc.M300106200.
40. Dejardin E. The alternative NF-κB pathway from biochemistry to biology: pitfalls and promises for future drug development. Biochem Pharmacol. 2006;72:1161–79. doi:10.1016/j.bcp.2006.08.007.
41. Fong A, Sun S-C. Genetic evidence for the essential role of β-transducin repeat-containing protein in the inducible processing of NF-κB2/p100. J Biol Chem. 2002;277:22111–4.
42. Shih VF-S, Tsui R, Caldwell A, Hoffmann A. A single NFκB system for both canonical and non-canonical signaling. Cell Res. 2011;21:86–102. doi:10.1038/cr.2010.161.
43. Xiao G, Harhaj EW, Sun SC. NF-kappaB-inducing kinase regulates the processing of NF-kappaB2 p100. Mol Cell. 2001;7:401–9.
44. Liang C, Zhang M, Sun S-C. beta-TrCP binding and processing of NF-kappaB2/p100 involve its phosphorylation at serines 866 and 870. Cell Signal. 2006;18:1309–17. doi:10.1016/j.cellsig.2005.10.011.
45. Sun S-C. Non-canonical NF-κB signaling pathway. Cell Res. 2011;21:71–85. doi:10.1038/cr.2010.177.
46. Bettermann K, Benesch M, Weis S, Haybaeck J. SUMOylation in carcinogenesis. Cancer Lett. 2012;316:113–25. doi:10.1016/j.canlet.2011.10.036.
47. Gao C, Huang W, Kanasaki K, Xu Y, Gao C, Huang W, Kanasaki K, Xu Y. The role of ubiquitination and sumoylation in diabetic nephropathy, the role of ubiquitination and sumoylation in diabetic nephropathy. BioMed Res Int. 2014;2014:e160692. doi:10.1155/2014/160692.
48. Mabb AM, Miyamoto S. SUMO and NF-kappaB ties. Cell Mol Life Sci. 2007;64:1979–96. doi:10.1007/s00018-007-7005-2.
49. Ehlken H, Krishna-Subramanian S, Ochoa-Callejero L, Kondylis V, Nadi NE, Straub BK, Schirmacher P, Walczak H, Kollias G, Pasparakis M. Death receptor-independent FADD signalling triggers hepatitis and hepatocellular carcinoma in mice with liver parenchymal cell-specific NEMO knockout. Cell Death Differ. 2014;21:1721–32. doi:10.1038/cdd.2014.83.
50. Gerondakis S, Grumont R, Gugasyan R, Wong L, Isomura I, Ho W, Banerjee A. Unravelling the complexities of the NF-κB signalling pathway using mouse knockout and transgenic models. Oncogene. 2006;25:6781–99. doi:10.1038/sj.onc.1209944.
51. Inokuchi S, Aoyama T, Miura K, Österreicher CH, Kodama Y, Miyai K, Akira S, Brenner DA, Seki E. Disruption of TAK1 in hepatocytes causes hepatic injury, inflammation, fibrosis,

and carcinogenesis. Proc Natl Acad Sci U S A. 2010;107:844–9. doi:10.1073/pnas. 0909781107.
52. Luedde T, Assmus U, Wüstefeld T, Meyer zu Vilsendorf A, Roskams T, Schmidt-Supprian M, Rajewsky K, Brenner DA, Manns MP, Pasparakis M, Trautwein C. Deletion of IKK2 in hepatocytes does not sensitize these cells to TNF-induced apoptosis but protects from ischemia/reperfusion injury. J Clin Invest. 2005;115:849–59. doi:10.1172/JCI200523493.
53. Polykratis A, Hermance N, Zelic M, Roderick J, Kim C, Van T-M, Lee TH, Chan FKM, Pasparakis M, Kelliher MA. RIPK1 kinase inactive mice are viable and protected from TNF-induced necroptosis in vivo. J Immunol. 2014;193(4):1539–43. doi:10.4049/jimmunol.1400590.
54. Martini FH, Timmons MJ, Tallitsch RB. Human anatomy. 8th ed. Boston, MA: Pearson; 2014.
55. El-Serag HB. Hepatocellular carcinoma. N Engl J Med. 2011;365:1118–27. doi:10.1056/NEJMra1001683.
56. El-Serag HB, Rudolph KL. Hepatocellular carcinoma: epidemiology and molecular carcinogenesis. Gastroenterology. 2007;132:2557–76. doi:10.1053/j.gastro.2007.04.061.
57. Lavrik IN, Krammer PH. Regulation of CD95/Fas signaling at the DISC. Cell Death Differ. 2012;19:36–41. doi:10.1038/cdd.2011.155.
58. Lemke J, von Karstedt S, Zinngrebe J, Walczak H. Getting TRAIL back on track for cancer therapy. Cell Death Differ. 2014;21:1350–64. doi:10.1038/cdd.2014.81.
59. Schleich K, Warnken U, Fricker N, Öztürk S, Richter P, Kammerer K, Schnölzer M, Krammer PH, Lavrik IN. Stoichiometry of the CD95 death-inducing signaling complex: experimental and modeling evidence for a death effector domain chain model. Mol Cell. 2012;47:306–19. doi:10.1016/j.molcel.2012.05.006.
60. Nikoletopoulou V, Markaki M, Palikaras K, Tavernarakis N. Crosstalk between apoptosis, necrosis and autophagy. Biochim Biophys Acta. 2013;1833:3448–59. doi:10.1016/j.bbamcr.2013.06.001.
61. Wajant H, Pfizenmaier K, Scheurich P. Tumor necrosis factor signaling. Cell Death Differ. 2003;10:45–65. doi:10.1038/sj.cdd.4401189.
62. Xie P. TRAF molecules in cell signaling and in human diseases. J Mol Signal. 2013;8:7. doi:10.1186/1750-2187-8-7.
63. Israël A. The IKK complex, a central regulator of NF-κB activation. Cold Spring Harb Perspect Biol. 2010;2:a000158. doi:10.1101/cshperspect.a000158.
64. Wertz IE, Dixit VM. Signaling to NF-κB: regulation by ubiquitination. Cold Spring Harb Perspect Biol. 2010;2:a003350. doi:10.1101/cshperspect.a003350.
65. Haas TL, Emmerich CH, Gerlach B, Schmukle AC, Cordier SM, Rieser E, Feltham R, Vince J, Warnken U, Wenger T, Koschny R, Komander D, Silke J, Walczak H. Recruitment of the linear ubiquitin chain assembly complex stabilizes the TNF-R1 signaling complex and is required for TNF-mediated gene induction. Mol Cell. 2009;36:831–44. doi:10.1016/j.molcel.2009.10.013.
66. Walczak H, Iwai K, Dikic I. Generation and physiological roles of linear ubiquitin chains. BMC Biol. 2012;10:23. doi:10.1186/1741-7007-10-23.
67. Beg AA, Baltimore D. An essential role for NF-kappaB in preventing TNF-alpha-induced cell death. Science. 1996;274:782–4.
68. Beg AA, Sha WC, Bronson RT, Ghosh S, Baltimore D. Embryonic lethality and liver degeneration in mice lacking the RelA component of NF-kappa B. Nature. 1995;376:167–70. doi:10.1038/376167a0.
69. Doi TS, Marino MW, Takahashi T, Yoshida T, Sakakura T, Old LJ, Obata Y. Absence of tumor necrosis factor rescues RelA-deficient mice from embryonic lethality. Proc Natl Acad Sci U S A. 1999;96:2994–9.
70. Rosenfeld ME, Prichard L, Shiojiri N, Fausto N. Prevention of hepatic apoptosis and embryonic lethality in RelA/TNFR-1 double knockout mice. Am J Pathol. 2000;156:997–1007.
71. Grossmann M, Metcalf D, Merryfull J, Beg A, Baltimore D, Gerondakis S. The combined absence of the transcription factors Rel and RelA leads to multiple hemopoietic cell defects. Proc Natl Acad Sci U S A. 1999;96:11848–53.

72. Iimuro Y, Nishiura T, Hellerbrand C, Behrns KE, Schoonhoven R, Grisham JW, Brenner DA. NFkappaB prevents apoptosis and liver dysfunction during liver regeneration. J Clin Invest. 1998;101:802–11. doi:10.1172/JCI483.
73. Gerondakis S, Grumont R, Rourke I, Grossmann M. The regulation and roles of Rel/NF-κB transcription factors during lymphocyte activation. Curr Opin Immunol. 1998;10:353–9. doi:10.1016/S0952-7915(98)80175-1.
74. Li Q, Van Antwerp D, Mercurio F, Lee KF, Verma IM. Severe liver degeneration in mice lacking the IkappaB kinase 2 gene. Science. 1999a;284:321–5.
75. Li ZW, Chu W, Hu Y, Delhase M, Deerinck T, Ellisman M, Johnson R, Karin M. The IKKbeta subunit of IkappaB kinase (IKK) is essential for nuclear factor kappaB activation and prevention of apoptosis. J Exp Med. 1999b;189:1839–45.
76. Tanaka M, Fuentes ME, Yamaguchi K, Durnin MH, Dalrymple SA, Hardy KL, Goeddel DV. Embryonic lethality, liver degeneration, and impaired NF-kappa B activation in IKK-beta-deficient mice. Immunity. 1999;10:421–9.
77. Rudolph D, Yeh W-C, Wakeham A, Rudolph B, Nallainathan D, Potter J, Elia AJ, Mak TW. Severe liver degeneration and lack of NF-κB activation in NEMO/IKKγ-deficient mice. Genes Dev. 2000;14:854–62.
78. Hu Y, Baud V, Delhase M, Zhang P, Deerinck T, Ellisman M, Johnson R, Karin M. Abnormal morphogenesis but intact IKK activation in mice lacking the IKKalpha subunit of IkappaB kinase. Science. 1999;284:316–20.
79. Takeda K, Takeuchi O, Tsujimura T, Itami S, Adachi O, Kawai T, Sanjo H, Yoshikawa K, Terada N, Akira S. Limb and skin abnormalities in mice lacking IKKalpha. Science. 1999;284:313–6.
80. Geisler F, Algül H, Paxian S, Schmid RM. Genetic inactivation of RelA/p65 sensitizes adult mouse hepatocytes to TNF-induced apoptosis in vivo and in vitro. Gastroenterology. 2007;132:2489–503. doi:10.1053/j.gastro.2007.03.033.
81. Maeda S, Chang L, Li Z-W, Luo J-L, Leffert H, Karin M. IKKbeta is required for prevention of apoptosis mediated by cell-bound but not by circulating TNFalpha. Immunity. 2003;19:725–37.
82. Arkan MC, Hevener AL, Greten FR, Maeda S, Li Z-W, Long JM, Wynshaw-Boris A, Poli G, Olefsky J, Karin M. IKK-beta links inflammation to obesity-induced insulin resistance. Nat Med. 2005;11:191–8. doi:10.1038/nm1185.
83. Luedde T, Heinrichsdorff J, de Lorenzi R, De Vos R, Roskams T, Pasparakis M. IKK1 and IKK2 cooperate to maintain bile duct integrity in the liver. Proc Natl Acad Sci U S A. 2008;105:9733–8. doi:10.1073/pnas.0800198105.
84. Liedtke C, Bangen J-M, Freimuth J, Beraza N, Lambertz D, Cubero FJ, Hatting M, Karlmark KR, Streetz KL, Krombach GA, Tacke F, Gassler N, Riethmacher D, Trautwein C. Loss of caspase-8 protects mice against inflammation-related hepatocarcinogenesis but induces non-apoptotic liver injury. Gastroenterology. 2011;141:2176–87. doi:10.1053/j.gastro.2011.08.037.
85. Papa S, Bubici C, Zazzeroni F, Franzoso G. Mechanisms of liver disease: the crosstalk between the NF-κB and JNK pathways. Biol Chem. 2009;390:965–76. doi:10.1515/BC.2009.111.
86. Schwabe RF, Brenner DA. Mechanisms of liver injury. I. TNF-alpha-induced liver injury: role of IKK, JNK, and ROS pathways. Am J Physiol Gastrointest Liver Physiol. 2006;290:G583–9. doi:10.1152/ajpgi.00422.2005.
87. Wullaert A, Heyninck K, Beyaert R. Mechanisms of crosstalk between TNF-induced NF-κB and JNK activation in hepatocytes. Biochem Pharmacol. 2006;72:1090–101. doi:10.1016/j.bcp.2006.07.003.
88. Davis RJ. Signal transduction by the JNK group of MAP kinases. Cell. 2000;103:239–52.
89. Karin M, Gallagher E. From JNK to pay dirt: jun kinases, their biochemistry, physiology and clinical importance. IUBMB Life. 2005;57:283–95. doi:10.1080/15216540500097111.
90. Nebreda AR, Porras A. p38 MAP kinases: beyond the stress response. Trends Biochem Sci. 2000;25:257–60. doi:10.1016/S0968-0004(00)01595-4.

91. Schieven GL. The biology of p38 kinase: a central role in inflammation. Curr Top Med Chem. 2005;5:921–8. doi:10.2174/1568026054985902.
92. Heinrichsdorff J, Luedde T, Perdiguero E, Nebreda AR, Pasparakis M. p38 alpha MAPK inhibits JNK activation and collaborates with IkappaB kinase 2 to prevent endotoxin-induced liver failure. EMBO Rep. 2008;9:1048–54. doi:10.1038/embor.2008.149.
93. Hanahan D, Weinberg RA. Hallmarks of cancer: the next generation. Cell. 2011;144:646–74. doi:10.1016/j.cell.2011.02.013.
94. Li W, Tan D, Zenali MJ, Brown RE. Constitutive activation of nuclear factor-kappa B (NF-kB) signaling pathway in fibrolamellar hepatocellular carcinoma. Int J Clin Exp Pathol. 2010;3:238–43.
95. Tai DI, Tsai SL, Chang YH, Huang SN, Chen TC, Chang KS, Liaw YF. Constitutive activation of nuclear factor kappaB in hepatocellular carcinoma. Cancer. 2000;89:2274–81.
96. Mauad TH, van Nieuwkerk CM, Dingemans KP, Smit JJ, Schinkel AH, Notenboom RG, van den Bergh Weerman MA, Verkruisen RP, Groen AK, Oude Elferink RP. Mice with homozygous disruption of the mdr2 P-glycoprotein gene. A novel animal model for studies of nonsuppurative inflammatory cholangitis and hepatocarcinogenesis. Am J Pathol. 1994;145:1237–45.
97. Pikarsky E, Porat RM, Stein I, Abramovitch R, Amit S, Kasem S, Gutkovich-Pyest E, Urieli-Shoval S, Galun E, Ben-Neriah Y. NF-κB functions as a tumour promoter in inflammation-associated cancer. Nature. 2004;431:461–6. doi:10.1038/nature02924.
98. Kondylis V, Polykratis A, Ehlken H, Ochoa-Callejero L, Straub BK, Krishna-Subramanian S, Van T-M, Curth H-M, Heise N, Weih F, Klein U, Schirmacher P, Kelliher M, Pasparakis M. NEMO prevents steatohepatitis and hepatocellular carcinoma by inhibiting RIPK1 kinase activity-mediated hepatocyte apoptosis. Cancer Cell. 2015;28:582–98. doi:10.1016/j.ccell.2015.10.001.
99. Kamata H, Honda S-I, Maeda S, Chang L, Hirata H, Karin M. Reactive oxygen species promote TNFalpha-induced death and sustained JNK activation by inhibiting MAP kinase phosphatases. Cell. 2005;120:649–61. doi:10.1016/j.cell.2004.12.041.
100. Hussain SP, Harris CC. Inflammation and cancer: an ancient link with novel potentials. Int J Cancer. 2007;121:2373–80. doi:10.1002/ijc.23173.
101. Lasagna N, Fantappiè O, Solazzo M, Morbidelli L, Marchetti S, Cipriani G, Ziche M, Mazzanti R. Hepatocyte growth factor and inducible nitric oxide synthase are involved in multidrug resistance-induced angiogenesis in hepatocellular carcinoma cell lines. Cancer Res. 2006;66:2673–82. doi:10.1158/0008-5472.CAN-05-2290.
102. Ying L, Hofseth AB, Browning DD, Nagarkatti M, Nagarkatti PS, Hofseth LJ. Nitric oxide inactivates the retinoblastoma pathway in chronic inflammation. Cancer Res. 2007;67:9286–93. doi:10.1158/0008-5472.CAN-07-2238.
103. Calvisi DF, Pinna F, Ladu S, Pellegrino R, Muroni MR, Simile MM, Frau M, Tomasi ML, Miglio MRD, Seddaiu MA, Daino L, Sanna V, Feo F, Pascale RM. Aberrant iNOS signaling is under genetic control in rodent liver cancer and potentially prognostic for the human disease. Carcinogenesis. 2008;29:1639–47. doi:10.1093/carcin/bgn155.
104. Seki E, Brenner DA. Recent advancement of molecular mechanisms of liver fibrosis. J Hepatobiliary Pancreat Sci. 2015;22:512–8. doi:10.1002/jhbp.245.
105. Sunami Y, Leithäuser F, Gul S, Fiedler K, Güldiken N, Espenlaub S, Holzmann K-H, Hipp N, Sindrilaru A, Luedde T, Baumann B, Wissel S, Kreppel F, Schneider M, Scharffetter-Kochanek K, Kochanek S, Strnad P, Wirth T. Hepatic activation of IKK/NFκB signaling induces liver fibrosis via macrophage-mediated chronic inflammation. Hepatol. Baltim. Md. 2012;56:1117–28. doi:10.1002/hep.25711.
106. Elsharkawy AM, Wright MC, Hay RT, Arthur MJ, Hughes T, Bahr MJ, Degitz K, Mann DA. Persistent activation of nuclear factor-κB in cultured rat hepatic stellate cells involves the induction of potentially novel rel-like factors and prolonged changes in the expression of IκB family proteins. Hepatology. 1999;30:761–9. doi:10.1002/hep.510300327.
107. Oakley F, Trim N, Constandinou CM, Ye W, Gray AM, Frantz G, Hillan K, Kendall T, Benyon RC, Mann DA, Iredale JP. Hepatocytes express nerve growth factor during liver injury: evidence for paracrine regulation of hepatic stellate cell apoptosis. Am J Pathol. 2003;163:1849–58. doi:10.1016/S0002-9440(10)63544-4.

108. Saile B, Matthes N, El Armouche H, Neubauer K, Ramadori G. The bcl, NFκB and p53/p21WAF1 systems are involved in spontaneous apoptosis and in the anti-apoptotic effect of TGF-β or TNF-α on activated hepatic stellate cells. Eur J Cell Biol. 2001;80:554–61. doi:10.1078/0171-9335-00182.
109. Oakley F, Teoh V, Ching-A-Sue G, Bataller R, Colmenero J, Jonsson JR, Eliopoulos AG, Watson MR, Manas D, Mann DA. Angiotensin II activates IκB kinase phosphorylation of RelA at Ser536 to promote myofibroblast survival and liver fibrosis. Gastroenterology. 2009;136:2334–2344.e1. doi:10.1053/j.gastro.2009.02.081.
110. Kluwe J, Pradere J-P, Gwak G-Y, Mencin A, Minicis SD, Osterreicher CH, Colmenero J, Bataller R, Schwabe RF. Modulation of hepatic fibrosis by c-Jun-N-terminal kinase inhibition. Gastroenterology. 2010;138:347–59. doi:10.1053/j.gastro.2009.09.015.
111. Gäbele E, Mühlbauer M, Dorn C, Weiss TS, Froh M, Schnabl B, Wiest R, Schölmerich J, Obermeier F, Hellerbrand C. Role of TLR9 in hepatic stellate cells and experimental liver fibrosis. Biochem Biophys Res Commun. 2008;376:271–6. doi:10.1016/j.bbrc.2008.08.096.
112. Seki E, De Minicis S, Osterreicher CH, Kluwe J, Osawa Y, Brenner DA, Schwabe RF. TLR4 enhances TGF-beta signaling and hepatic fibrosis. Nat Med. 2007;13:1324–32. doi:10.1038/nm1663.
113. Goossens N, Hoshida Y. Hepatitis C virus-induced hepatocellular carcinoma. Clin Mol Hepatol. 2015;21:105–14. doi:10.3350/cmh.2015.21.2.105.
114. Dolganiuc A, Oak S, Kodys K, Golenbock DT, Finberg RW, Kurt-Jones E, Szabo G. Hepatitis C core and nonstructural 3 proteins trigger toll-like receptor 2-mediated pathways and inflammatory activation. Gastroenterology. 2004;127:1513–24.
115. Zhao L-J, Zhao P, Chen Q-L, Ren H, Pan W, Qi Z-T. Mitogen-activated protein kinase signalling pathways triggered by the hepatitis C virus envelope protein E2: implications for the prevention of infection. Cell Prolif. 2007;40:508–21. doi:10.1111/j.1365-2184.2007.00453.x.
116. Lin W, Tsai W-L, Shao R-X, Wu G, Peng LF, Barlow LL, Chung WJ, Zhang L, Zhao H, Jang J-Y, Chung RT. HCV regulates TGF-β1 production through the generation of reactive oxygen species in an NFκB-dependent manner. Gastroenterology. 2010;138:2509–2518.e1. doi:10.1053/j.gastro.2010.03.008.
117. Chusri P, Kumthip K, Hong J, Zhu C, Duan X, Jilg N, Fusco DN, Brisac C, Schaefer EA, Cai D, Peng LF, Maneekarn N, Lin W, Chung RT. HCV induces transforming growth factor β1 through activation of endoplasmic reticulum stress and the unfolded protein response. Sci Rep. 2016;6:22487. doi:10.1038/srep22487.
118. Kanda T, Yokosuka O, Nagao K, Saisho H. State of hepatitis C viral replication enhances activation of NF-kB- and AP-1-signaling induced by hepatitis B virus X. Cancer Lett. 2006;234:143–8. doi:10.1016/j.canlet.2005.03.030.
119. Hildt E, Hofschneider PH. The PreS2 activators of the hepatitis B virus: activators of tumour promoter pathways. Recent Results Cancer Res. 1998;154:315–29.
120. You LR, Chen CM, Lee YH. Hepatitis C virus core protein enhances NF-kappaB signal pathway triggering by lymphotoxin-beta receptor ligand and tumor necrosis factor alpha. J Virol. 1999;73:1672–81.
121. Bose S, Banerjee AK. Innate immune response against nonsegmented negative strand RNA viruses. J Interferon Cytokine Res. 2003;23:401–12. doi:10.1089/107999003322277810.
122. Hiscott J, Grandvaux N, Sharma S, Tenoever BR, Servant MJ, Lin R. Convergence of the NF-κB and interferon signaling pathways in the regulation of antiviral defense and apoptosis. Ann N Y Acad Sci. 2003;1010:237–48. doi:10.1196/annals.1299.042.
123. Malmgaard L. Induction and regulation of IFNs during viral infections. J Interferon Cytokine Res. 2004;24:439–54. doi:10.1089/1079990041689665.
124. Maher SG, Romero-Weaver AL, Scarzello AJ, Gamero AM. Interferon: cellular executioner or white knight? Curr Med Chem. 2007;14:1279–89.
125. Sasaki R, Kanda T, Nakamura M, Nakamoto S, Haga Y, Wu S, Shirasawa H, Yokosuka O. Possible involvement of hepatitis B virus infection of hepatocytes in the attenuation of apoptosis in hepatic stellate cells. PLoS One. 2016;11:e0146314. doi:10.1371/journal.pone.0146314.

126. Sun B, Karin M. NF-κB signaling, liver disease and hepatoprotective agents. Oncogene. 2008;27:6228–44. doi:10.1038/onc.2008.300.
127. Nguyen DM, El-Serag HB. The epidemiology of obesity. Gastroenterol Clin North Am. 2010;39:1–7. doi:10.1016/j.gtc.2009.12.014.
128. Baker RG, Hayden MS, Ghosh S. NF-κB, inflammation, and metabolic disease. Cell Metab. 2011;13:11–22. doi:10.1016/j.cmet.2010.12.008.
129. Poloz Y, Stambolic V. Obesity and cancer, a case for insulin signaling. Cell Death Dis. 2015;6:e2037. doi:10.1038/cddis.2015.381.
130. Chen Y, Zhu J, Lum PY, Yang X, Pinto S, MacNeil DJ, Zhang C, Lamb J, Edwards S, Sieberts SK, Leonardson A, Castellini LW, Wang S, Champy M-F, Zhang B, Emilsson V, Doss S, Ghazalpour A, Horvath S, Drake TA, Lusis AJ, Schadt EE. Variations in DNA elucidate molecular networks that cause disease. Nature. 2008b;452:429–35. doi:10.1038/nature06757.
131. Emilsson V, Thorleifsson G, Zhang B, Leonardson AS, Zink F, Zhu J, Carlson S, Helgason A, Walters GB, Gunnarsdottir S, Mouy M, Steinthorsdottir V, Eiriksdottir GH, Bjornsdottir G, Reynisdottir I, Gudbjartsson D, Helgadottir A, Jonasdottir A, Jonasdottir A, Styrkarsdottir U, Gretarsdottir S, Magnusson KP, Stefansson H, Fossdal R, Kristjansson K, Gislason HG, Stefansson T, Leifsson BG, Thorsteinsdottir U, Lamb JR, Gulcher JR, Reitman ML, Kong A, Schadt EE, Stefansson K. Genetics of gene expression and its effect on disease. Nature. 2008;452:423–8. doi:10.1038/nature06758.
132. Yang X, Deignan JL, Qi H, Zhu J, Qian S, Zhong J, Torosyan G, Majid S, Falkard B, Kleinhanz RR, Karlsson J, Castellani LW, Mumick S, Wang K, Xie T, Coon M, Zhang C, Estrada-Smith D, Farber CR, Wang SS, van Nas A, Ghazalpour A, Zhang B, Macneil DJ, Lamb JR, Dipple KM, Reitman ML, Mehrabian M, Lum PY, Schadt EE, Lusis AJ, Drake TA. Validation of candidate causal genes for obesity that affect shared metabolic pathways and networks. Nat Genet. 2009;41:415–23. doi:10.1038/ng.325.
133. Ley RE, Turnbaugh PJ, Klein S, Gordon JI. Microbial ecology: human gut microbes associated with obesity. Nature. 2006;444:1022–3. doi:10.1038/4441022a.
134. Nguyen SD, Sok D-E. Effect of 3,4-dihydroxyphenylalanine on Cu(2+)-induced inactivation of HDL-associated paraoxonasel and oxidation of HDL; inactivation of paraoxonasel activity independentofHDLlipidoxidation.FreeRadicRes.2004;38:969–76.doi:10.1080/10715760400000943.
135. Tsukumo DML, Carvalho-Filho MA, Carvalheira JBC, Prada PO, Hirabara SM, Schenka AA, Araújo EP, Vassallo J, Curi R, Velloso LA, Saad MJA. Loss-of-function mutation in toll-like receptor 4 prevents diet-induced obesity and insulin resistance. Diabetes. 2007;56:1986–98. doi:10.2337/db06-1595.
136. Vijay-Kumar M, Aitken JD, Carvalho FA, Cullender TC, Mwangi S, Srinivasan S, Sitaraman SV, Knight R, Ley RE, Gewirtz AT. Metabolic syndrome and altered gut microbiota in mice lacking Toll-like receptor 5. Science. 2010;328:228–31. doi:10.1126/science.1179721.
137. Chiang S-H, Bazuine M, Lumeng CN, Geletka LM, Mowers J, White NM, Ma J-T, Zhou J, Qi N, Westcott D, Delproposto JB, Blackwell TS, Yull FE, Saltiel AR. The protein kinase IKKε regulates energy balance in obese mice. Cell. 2009;138:961–75. doi:10.1016/j.cell.2009.06.046.
138. Cai D, Yuan M, Frantz DF, Melendez PA, Hansen L, Lee J, Shoelson SE. Local and systemic insulin resistance resulting from hepatic activation of IKK-beta and NF-kappaB. Nat Med. 2005;11:183–90. doi:10.1038/nm1166.
139. Shi H, Kokoeva MV, Inouye K, Tzameli I, Yin H, Flier JS. TLR4 links innate immunity and fatty acid–induced insulin resistance. J Clin Invest. 2006;116:3015–25. doi:10.1172/JCI28898.
140. Davis JE, Gabler NK, Walker-Daniels J, Spurlock ME. Tlr-4 deficiency selectively protects against obesity induced by diets high in saturated fat. Obesity. 2008;16:1248–55. doi:10.1038/oby.2008.210.
141. Poggi M, Bastelica D, Gual P, Iglesias MA, Gremeaux T, Knauf C, Peiretti F, Verdier M, Juhan-Vague I, Tanti JF, Burcelin R, Alessi MC. C3H/HeJ mice carrying a toll-like receptor 4 mutation are protected against the development of insulin resistance in white adipose tissue vin response to a high-fat diet. Diabetologia. 2007;50:1267–76. doi:10.1007/s00125-007-0654-8.

142. Davis JE, Braucher DR, Walker-Daniels J, Spurlock ME. Absence of Tlr2 protects against high-fat diet-induced inflammation and results in greater insulin-stimulated glucose transport in cultured adipocytes. J Nutr Biochem. 2011;22:136–41. doi:10.1016/j.jnutbio.2009.12.008.
143. Nguyen MTA, Favelyukis S, Nguyen A-K, Reichart D, Scott PA, Jenn A, Liu-Bryan R, Glass CK, Neels JG, Olefsky JM. A subpopulation of macrophages infiltrates hypertrophic adipose tissue and is activated by free fatty acids via toll-like receptors 2 and 4 and JNK-dependent pathways. J Biol Chem. 2007;282:35279–92. doi:10.1074/jbc.M706762200.
144. Hotamisligil GS, Shargill NS, Spiegelman BM. Adipose expression of tumor necrosis factor-alpha: direct role in obesity-linked insulin resistance. Science. 1993;259:87–91. doi:10.1126/science.7678183.
145. Uysal KT, Wiesbrock SM, Marino MW, Hotamisligil GS. Protection from obesity-induced insulin resistance in mice lacking TNF-α function. Nature. 1997;389:610–4. doi:10.1038/39335.
146. Wunderlich FT, Ströhle P, Könner AC, Gruber S, Tovar S, Brönneke HS, Juntti-Berggren L, Li L-S, van Rooijen N, Libert C, Berggren P-O, Brüning JC. Interleukin-6 signaling in liver-parenchymal cells suppresses hepatic inflammation and improves systemic insulin action. Cell Metab. 2010;12:237–49. doi:10.1016/j.cmet.2010.06.011.
147. Mendes-Braz M, Elias-Miró M, Jiménez-Castro MB, Casillas-Ramírez A, Ramalho FS, Peralta C. The current state of knowledge of hepatic ischemia-reperfusion injury based on its study in experimental models. J Biomed Biotechnol. 2012;2012:298657. doi:10.1155/2012/298657.
148. Peralta C, Jiménez-Castro MB, Gracia-Sancho J. Hepatic ischemia and reperfusion injury: effects on the liver sinusoidal milieu. J Hepatol. 2013;59:1094–106. doi:10.1016/j.jhep.2013.06.017.
149. Gao W, Bentley RC, Madden JF, Clavien P-A. Apoptosis of sinusoidal endothelial cells is a critical mechanism of preservation injury in rat liver transplantation. Hepatology. 1998;27:1652–60. doi:10.1002/hep.510270626.
150. Kohli V, Selzner M, Madden JF, Bentley RC, Clavien PA. Endothelial cell and hepatocyte deaths occur by apoptosis after ischemia-reperfusion injury in the rat liver. Transplantation. 1999;67:1099–105.
151. Linkermann A, Hackl MJ, Kunzendorf U, Walczak H, Krautwald S, Jevnikar AM. Necroptosis in immunity and ischemia-reperfusion injury. Am J Transplant. 2013;13:2797–804. doi:10.1111/ajt.12448.
152. Massip-Salcedo M, Roselló-Catafau J, Prieto J, Avíla MA, Peralta C. The response of the hepatocyte to ischemia. Liver Int. 2007;27:6–16. doi:10.1111/j.1478-3231.2006.01390.x.
153. Rauen U, Kerkweg U, Weisheit D, Petrat F, Sustmann R, de Groot H. Cold-induced apoptosis of hepatocytes: mitochondrial permeability transition triggered by nonmitochondrial chelatable iron. Free Radic Biol Med. 2003;35:1664–78.
154. Theruvath TP, Czerny C, Ramshesh VK, Zhong Z, Chavin KD, Lemasters JJ. C-Jun N-terminal kinase 2 promotes graft injury via the mitochondrial permeability transition after mouse liver transplantation. Am J Transplant. 2008;8:1819–28. doi:10.1111/j.1600-6143.2008.02336.x.
155. Shuh M, Bohorquez H, Loss GE, Cohen AJ. Tumor necrosis factor-α: life and death of hepatocytes during liver ischemia/reperfusion injury. Ochsner J. 2013;13:119–30.
156. Suetsugu H, Iimuro Y, Uehara T, Nishio T, Harada N, Yoshida M, Hatano E, Son G, Fujimoto J, Yamaoka Y. Nuclear factor κB inactivation in the rat liver ameliorates short term total warm ischaemia/reperfusion injury. Gut. 2005;54:835–42. doi:10.1136/gut.2004.043034.
157. Uchinami H, Yamamoto Y, Kume M, Yonezawa K, Ishikawa Y, Taura K, Nakajima A, Hata K, Yamaoka Y. Effect of heat shock preconditioning on NF-κB/I-κB pathway during I/R injury of the rat liver. Am J Physiol Gastrointest Liver Physiol. 2002;282:G962–71. doi:10.1152/ajpgi.00466.2001.
158. Ruland J. Return to homeostasis: downregulation of NF-κB responses. Nat Immunol. 2011;12:709–14. doi:10.1038/ni.2055.
159. Wertz IE, O'Rourke KM, Zhou H, Eby M, Aravind L, Seshagiri S, Wu P, Wiesmann C, Baker R, Boone DL, Ma A, Koonin EV, Dixit VM. De-ubiquitination and ubiquitin ligase domains of A20 downregulate NF-kappaB signalling. Nature. 2004;430:694–9. doi:10.1038/nature02794.

160. Zhang SQ, Kovalenko A, Cantarella G, Wallach D. Recruitment of the IKK signalosome to the p55 TNF receptor: RIP and A20 bind to NEMO (IKKgamma) upon receptor stimulation. Immunity. 2000;12:301–11.
161. Ramsey HE, Da Silva CG, Longo CR, Csizmadia E, Studer P, Patel VI, Damrauer SM, Siracuse JJ, Daniel S, Ferran C. A20 protects mice from lethal liver ischemia reperfusion injury by increasing peroxisome proliferator-activated receptor-α expression. Liver Transpl. 2009;15:1613–21. doi:10.1002/lt.21879.
162. Sakai N, Van Sweringen HL, Schuster R, Blanchard J, Burns JM, Tevar AD, Edwards MJ, Lentsch AB. Receptor activator of nuclear factor-κB ligand (RANKL) protects against hepatic ischemia/reperfusion injury in mice. Hepatol. Baltim. Md. 2012;55:888–97. doi:10.1002/hep.24756.
163. Marinò M, Morabito E, Altea MA, Ambrogini E, Oliveri F, Brunetto MR, Pollina LE, Campani D, Vitti P, Bartalena L, Pincheral A, Marcocci C. Autoimmune hepatitis during intravenous glucocorticoid pulse therapy for Graves' ophthalmopathy treated successfully with glucocorticoids themselves. J Endocrinol Invest. 2005;28:280–4.
164. Bae M-K, Kim S-H, Jeong J-W, Lee YM, Kim H-S, Kim S-R, Yun I, Bae S-K, Kim K-W. Curcumin inhibits hypoxia-induced angiogenesis via down-regulation of HIF-1. Oncol Rep. 2006;15:1557–62.
165. Chen D, Milacic V, Chen MS, Wan SB, Lam WH, Huo C, Landis-Piwowar KR, Cui QC, Wali A, Chan TH, Dou QP. Tea polyphenols, their biological effects and potential molecular targets. Histol Histopathol. 2008a;23:487–96.
166. Toledo LP, Ong TP, Pinho ALG, Jordão A, Vanucchi H, Moreno FS. Inhibitory effects of lutein and lycopene on placental glutathione S-transferase-positive preneoplastic lesions and DNA strand breakage induced in Wistar rats by the resistant hepatocyte model of hepatocarcinogenesis. Nutr Cancer. 2003;47:62–9. doi:10.1207/s15327914nc4701_8.
167. Umemura T, Kai S, Hasegawa R, Kanki K, Kitamura Y, Nishikawa A, Hirose M. Prevention of dual promoting effects of pentachlorophenol, an environmental pollutant, on diethylnitrosamine-induced hepato- and cholangiocarcinogenesis in mice by green tea infusion. Carcinogenesis. 2003;24:1105–9. doi:10.1093/carcin/bgg053.
168. Anan A, Baskin-Bey ES, Bronk SF, Werneburg NW, Shah VH, Gores GJ. Proteasome inhibition induces hepatic stellate cell apoptosis. Hepatology. 2006;43:335–44. doi:10.1002/hep.21036.
169. Cervello M, Giannitrapani L, La Rosa M, Notarbartolo M, Labbozzetta M, Poma P, Montalto G, D'Alessandro N. Induction of apoptosis by the proteasome inhibitor MG132 in human HCC cells: possible correlation with specific caspase-dependent cleavage of beta-catenin and inhibition of beta-catenin-mediated transactivation. Int J Mol Med. 2004;13:741–8.
170. Hegewisch-Becker S, Sterneck M, Schubert U, Rogiers X, Guerciolini R, Pierce JE, Hossfeld DK. Phase I/II trial of bortezomib in patients with unresectable hepatocellular carcinoma (HCC). J Clin Oncol. 2004;22:335S.
171. Joshi-Barve S, Barve SS, Butt W, Klein J, McClain CJ. Inhibition of proteasome function leads to NF-kappaB-independent IL-8 expression in human hepatocytes. Hepatol. Baltim. Md. 2003;38:1178–87. doi:10.1053/jhep.2003.50470.
172. Greten FR, Arkan MC, Bollrath J, Hsu L-C, Goode J, Miething C, Göktuna SI, Neuenhahn M, Fierer J, Paxian S, Van Rooijen N, Xu Y, O'Cain T, Jaffee BB, Busch DH, Duyster J, Schmid RM, Eckmann L, Karin M. NF-kappaB is a negative regulator of IL-1beta secretion as revealed by genetic and pharmacological inhibition of IKKbeta. Cell. 2007;130:918–31. doi:10.1016/j.cell.2007.07.009.
173. Liang M-C, Bardhan S, Li C, Pace EA, Porco JA, Gilmore TD. Jesterone dimer, a synthetic derivative of the fungal metabolite jesterone, blocks activation of transcription factor nuclear factor kappaB by inhibiting the inhibitor of kappaB kinase. Mol Pharmacol. 2003;64:123–31. doi:10.1124/mol.64.1.123.
174. Kim K-H, Park K-K. Small RNA- and DNA-based gene therapy for the treatment of liver cirrhosis, where we are? World J Gastroenterol. 2014;20:14696–705. doi:10.3748/wjg.v20.i40.14696.

Molecular Carcinogenesis of Glial Brain Tumors

7

Peter Strasser and Serge Weis

Contents

P. Strasser
Laboratory of Molecular Biology, Department of Laboratory Medicine, Christian Doppler University Hospital, Paracelsus Medical University, Salzburg, Austria

S. Weis (✉)
Department of Neuropathology, Neuromed Campus, Kepler University Hospital, School of Medicine, Johannes Kepler University, Linz, Austria
e-mail: serge.weis@gkepleruniklinikum.at

J. Haybaeck (ed.), *Mechanisms of Molecular Carcinogenesis – Volume 1*,
DOI 10.1007/978-3-319-53659-0_7

Abstract

Glial tumors represent the most common primary central nervous system tumors. They are classified as astrocytomas, oligodendrogliomas, and oligoastrocytomas. They may occur as benign tumors (WHO grade II) or anaplastic tumors (WHO grade III). The most malignant astrocytoma is called glioblastoma and represents WHO grade IV.

With the recent publication of the revised fourth edition of the WHO classification of CNS tumors, the molecular characterization of the tumors becomes mandatory. Elementary investigations include the determination of the mutation status of *IDH1/2* and co-deletion of *1p/19q*. Additional parameters include mutations in the *ATRX* gene and the *TERT* promoter. A variety of genetic alterations have been described. Specific focus was laid on epigenetic changes, i.e., altered methylation patterns. Studies related to gene or microRNA (miR) expression in brain tumors are still scarce. Brain tumors pose a challenging task for the clinician and require further broad-minded molecular investigations at various levels.

7.1 Introduction

Glial tumors are the most frequently encountered brain tumors. They are tumors of neuroepithelial tissue. Based on the cell of origin, they are classified as astroglial, oligodendroglial, and oligo-astroglial tumors. Other neuroepithelial tumors include ependymal and choroid plexus tumors. Neuronal and mixed neuronal-glial tumors, tumors of the pineal region, and embryonal tumors also belong to the group of tumors of neuroepithelial tissue.

For each tumor entity, predicting the biological behavior by means of histological grading was introduced by the WHO. Thus, (a) grade I tumors have low proliferative potential and the possibility of cure following surgery; (b) grade II tumors are infiltrative in nature and, despite low level of proliferation, often recur. They tend to progress to higher grades; (c) grade III tumors are lesions with histological evidence of malignancy, including nuclear atypia and brisk mitotic activity; and (d) grade IV tumors are cytologically malignant, mitotically active, necrosis-prone neoplasms associated with rapid pre- and postoperative disease evolution and fatal outcome.

According to the Central Brain Tumor Registry of the United States (CBTRUS), the distribution (in %) of primary brain and CNS tumors by histology and by

Table 7.1 Distribution of glial tumor types

	Distribution (in %) of primary brain and CNS tumors		Distribution (in %) of primary brain and CNS gliomas	
Tumor	Overall	Malignant	Tumor subtype (WHO grade)	
Astrocytoma	5.7	17.4	Pilocytic astrocytoma (I)	5.2
			Diffuse astrocytoma (II)	8.6
			Anaplastic astrocytoma (III)	6.1
Glioblastoma	15.1	45.6	Glioblastoma (IV)	55.1
Oligodendroglioma	1.6	4.9	Oligodendroglioma (II)	5.7
Oligoastrocytoma	9.0	2.7	Oligoastrocytoma (II and III)	3.2

histology subtypes is shown in Table 7.1. The overall distribution is based on an analysis of 356,858 tumors, and the distribution of the histological tumor subtypes is based on an analysis of 97,910 tumors [1–3].

7.2 Morphology

7.2.1 Glioblastoma (GBM) (WHO Grade IV)

Glioblastoma or glioblastoma multiforme (GBM) is a highly malignant neuroectodermal tumor composed of densely packed, anaplastic, and highly dedifferentiated tumor cells making the histogenetic typing difficult.

Macroscopically, GBM are tumors of large size which can involve several lobes. GBMs can spread to the contralateral hemisphere through the corpus callosum displaying a symmetrical tumor growth into both hemispheres, i.e., butterfly glioma. The tumor is usually not sharply demarcated, presenting with a broad and diffuse zone of infiltration. The cut surface characteristically shows a varicolored appearance ranging from gray, brown, white, and yellow to dark red. GBMs have a firm consistency. Necroses are present.

At the microscopic level, a high diversity of cell forms is encountered in glioblastoma which encompass anaplastic cells displaying astrocytic features; neoplastic oligodendroglia; high density of small, poorly differentiated cells; marked polymorphism of tumor cells including multinucleated giant cells; areas showing cells with astrocytic, oligodendroglial, and rarely ependymal differentiation; atypical mitoses; vascular endothelial cell proliferation; typical tumor necroses, i.e., palisading with cells arranged side by side in rows and their processes directed toward a central area of necrosis; and large areas of necroses. Based on the predominant features, the following types can be recognized: small cell GBM, classical type, GBM with oligodendroglioma component, and GBM with PNET-like islands.

7.2.2 Astrocytoma (WHO Grade II)

Astrocytoma is a diffusely infiltrating tumor that typically affects young adults and is characterized by a high degree of cellular differentiation and slow growth; the tumor occurs throughout the CNS but is preferentially located supratentorially and has an intrinsic tendency for malignant progression to anaplastic astrocytoma and, ultimately, glioblastoma.

Macroscopically, the borders of the gray or yellow-whitish tumor are blurred. Areas of the tumor tissue might be firm or softened or granular or cystic.

At the microscopic level, astrocytoma is characterized by well-differentiated neoplastic astrocytes in a loosely structured tumor matrix. The neoplastic astrocytes have round to oval nuclei with intermediate-sized masses of chromatin, a distinct nucleolus, and no stainable cytoplasm. There is an absence of mitotic activity, necroses, and microvascular proliferation.

7.2.3 Anaplastic Astrocytoma (AA) (WHO Grade III)

Anaplastic astrocytoma is a diffusely infiltrating, malignant astrocytoma that primarily affects adults, preferentially located in the cerebral hemispheres, which is histologically characterized by nuclear atypia, increased cellularity, and significant proliferative activity. The tumor may arise from diffuse astrocytoma WHO grade II or de novo, i.e., without evidence of a less malignant precursor lesion, and has an inherent tendency to undergo progression.

Macroscopically, AA infiltrates the surrounding brain, accompanied by tissue destruction. It has areas of granularity, opacity, and soft consistency.

At the microscopic level, anaplastic astrocytoma is characterized by an increased cellularity as compared to astrocytoma WHO grade II and distinct nuclear atypia characterized by increased nuclear size, shape, chromatin coarsening, dispersion, and nucleolar prominence. Mitoses might be present as well as multinucleated tumor cells and abnormal mitoses. Microvascular proliferation and necroses are absent.

7.2.4 Pilocytic Astrocytoma (PA) (WHO Grade I)

Pilocytic astrocytoma is a relatively circumscribed, slowly growing, often cystic astrocytoma occurring in children and young adults, histologically characterized by a biphasic pattern with varying proportions of compacted bipolar cells associated with Rosenthal fibers and loose-texture multipolar cells associated with microcysts and eosinophilic granular bodies/hyaline droplets.

Macroscopically, pilocytic astrocytoma is a soft, gray discrete mass with intratumoral and paratumoral cyst formation. Calcifications and hemosiderin deposits might be encountered.

At the microscopic level, PA is characterized by low to moderate cellularity. Heterogeneity of histologic features consists of a biphasic growth pattern of tumor cells with loosely textured multipolar cells (protoplasmic astrocytes), microcysts, and granular bodies/hyaline droplets.

Rosenthal fibers are intracytoplasmic corkscrew-shaped, eosinophilic, and hyaline glial fibrillary acidic protein (GFAP)-positive masses (fibers) which are present in tumor cells. Eosinophilic granular bodies (EGB) are eosinophilic, PAS-positive globular aggregates within astrocytic processes. PAs are highly vascularized tumors. Regressive changes include hyalinized, ectatic vessels (DD cavernous angioma), previous hemorrhage (hemosiderin), calcification, and lymphocytic infiltrates.

7.2.5 Oligodendroglioma (WHO Grade II)

Oligodendroglioma is a diffusely infiltrating, well-differentiated glioma of adults, typically located in the cerebral hemispheres, composed of neoplastic cells morphologically resembling oligodendroglia and often harboring deletions of chromosomal arms 1p and 19q.

Macroscopically, it is a well-defined mass of soft to gelatinous consistency (mucoid degeneration) and grayish-pink color. Calcifications are frequently encountered.

At the microscopic level, oligodendroglioma consists of monomorphic tumor cells with round nuclei, perinuclear halo only seen on paraffin sections (honeycomb appearance), and moderate cellularity.

7.2.6 Anaplastic Oligodendroglioma (WHO Grade III)

Anaplastic oligodendroglioma is an oligodendroglioma with focal or diffuse histological features of malignancy and a less favorable prognosis.

Macroscopically, anaplastic oligodendroglioma presents as a well-defined mass of soft to gelatinous consistency (mucoid degeneration) and grayish-pink color. Calcifications are frequent.

At the microscopic level, anaplastic oligodendroglioma is characterized by diffusely infiltrating tumor cells reminiscent of oligodendrocytes with round hyperchromatic nuclei, perinuclear halo, and scant cellular processes. Tumor cells display marked cellular and nuclear pleomorphism. Multinucleated giant cells might be present. The dense network of branching capillaries (chicken-wire pattern) is still preserved. Necroses, when present, are not indicative of shorter survival.

7.2.7 Oligoastrocytoma (WHO Grade II)

Oligoastrocytoma is a diffusely infiltrating glioma composed of a conspicuous mixture of two distinct neoplastic cell types morphologically resembling the tumor cells in oligodendrogliomas and diffuse astrocytoma of WHO grade II.

Macroscopically, the tumors have blurred borders, enlarge and distort the invaded structures, and are of gray or yellow-whitish color. Areas of the tumor tissue might be firm or softened or granular or cystic.

At the microscopic level, oligoastrocytoma is a tumor of moderate cellularity characterized by the presence of neoplastic glial cells with astrocytic or oligodendroglial phenotypes. Microcalcifications and microcysts might be present. The tumor lacks necroses and microvascular proliferation.

7.2.8 Anaplastic Oligoastrocytoma (WHO Grade III)

Anaplastic oligoastrocytoma is an oligoastrocytoma with histological features of malignancy, such as increased cellularity, nuclear atypia, pleomorphism, and increased mitotic activity.

Macroscopically, the tumor infiltrates the surrounding brain with tissue destruction. The tumor shows areas of granularity, opacity, and soft consistency.

At the microscopic level, anaplastic oligoastrocytoma is characterized by increased cellularity as compared to oligoastrocytoma WHO grade II; distinct nuclear atypia with increased variations of nuclear size, shape, chromatin coarsening, and dispersion; nucleolar prominence; cellular pleomorphism; and high mitotic activity. Tumors with necroses should be classified as "glioblastoma with oligodendroglial component."

In the revised version of the fourth WHO classification of CNS tumors [4], the above-described tumor entities are labeled as follows:

- Diffuse astrocytoma, IDH mutant
- Diffuse astrocytoma, IDH wild type
- Diffuse astrocytoma, NOS (not otherwise specified)
- Anaplastic astrocytoma, IDH mutant
- Anaplastic astrocytoma, IDH wild type
- Anaplastic astrocytoma, NOS
- Glioblastoma, IDH wild type
- Oligodendroglioma, IDH mutant and 1p/19q-co-deleted
- Oligodendroglioma, NOS
- Anaplastic oligodendroglioma, IDH-mutant and 1p/19q-co-deleted
- Anaplastic oligodendroglioma, NOS
- Oligoastrocytoma, NOS
- Anaplastic oligoastrocytoma, NOS

The NOS (not otherwise specified) category includes both tumors that have not been tested for the genetic parameter(s) and tumors that have been tested but did not show the diagnostic genetic alterations.

7.3 Genetics

In addition to "classical" histopathology, molecular genetic markers have found their way into diagnostic schemes designed for glioma classification and prognosis. Of particular interest are marke rs which allow an unambiguous distinction of tumor subtypes. Ideally, the requirements for such markers are their significant prevalence in some tumor subtypes and their low frequency, or even virtual absence, in others. In routine diagnostics, comprehensive genetic testing includes the synopsis of all analyzed markers to obtain a reliable classification of the investigated tumor(s).

Some molecular markers that have emerged as powerful tools in recent years are *IDH1/IDH2* mutations, *ATRX* mutations, *TERT* promoter mutations, and a chromosomal aberration typically found in oligodendrogliomas, the 1p/19q co-deletion [reviewed in [5]].

7.3.1 *IDH* Mutations

Important genetic traits in glial tumors are recurrent point mutations in the *IDH1* and *IDH2* genes. *IDH1* is located on chromosome 2q33.3 and encodes the $NADP^+$-dependent cytosolic enzyme isocitrate dehydrogenase 1 (IDH1) [6]. The *IDH2* gene at chromosome locus 15q26.1 encodes the $NADP^+$-dependent mitochondrial enzyme isocitrate dehydrogenase 2 (IDH2).

A frequently observed aberration is a missense *IDH1* point mutation, resulting in the transition of arginine to histidine at amino acid position 132 (R132H) which is located in the enzyme's substrate binding site. Wild-type IDH1 is involved in cytosolic NADPH production, but to date, the role of the mutated gene in gliomagenesis is not yet fully understood. It is important to note that the *IDH1* mutation is found at a high frequency in secondary GBM (~80%) but only rarely in primary GBM. Moreover, the IDH1-R132H mutation was also observed in up to 80% of grade II and grade III astrocytomas. The detection of mutated IDH1 (R132H) is therefore routinely used as a specific diagnostic marker in these tumors, and it supports discrimination between primary and secondary GBM.

At a much lower frequency, functional mutations are also reported in the *IDH2* gene; here, the major target is codon 172 which corresponds to an arginine residue in the wild-type enzyme. R172 represents the site analogous to R132 in IDH1. Noteworthy, *IDH1* and *IDH2* mutations appear to occur mutually exclusive in glioma patients.

7.3.2 *ATRX* Mutations

The *ATRX* gene (α-thalassemia/mental retardation syndrome X-linked) is located at Xq21.1 on the long arm of the X chromosome. ATRX is expressed exclusively in the nucleus where it plays an important role in telomere stabilization and chromatin remodeling. Inactivating mutations in the *ATRX* gene result in loss of functional protein which in turn triggers a mechanism known as the ALT (alternative lengthening of telomeres) phenotype. *ATRX* mutations are frequently found in astrocytomas WHO grades II and III as well as in oligoastrocytomas, and with a lower incidence in secondary glioblastomas and oligodendrogliomas [7].

7.3.3 *TERT* Promoter Mutations

The *TERT* gene (telomerase reverse transcriptase) encodes a subunit of the telomerase complex which is crucial for maintaining telomere length and stability. In dividing cells, telomeres become shorter with each division cycle which eventually leads to cellular senescence. In proliferating cells, e.g., during developmental or regenerative processes, telomerase activity counteracts telomere shortening, thus maintaining the cells' replicative potential, a feature that is also characteristic for tumor tissue. Mutations in the *TERT* promoter resulting in increased expression of the gene have been described in a variety of cancer cells. High mutation frequencies

Table 7.2 Synopsis of important genetic markers in astrocytic tumors

Genetic marker	Primary GBM WHO grade IV	Secondary GBM WHO grade IV	Anaplastic astrocytoma WHO grade III	Diffuse astrocytoma WHO grade II
IDH1 mutation	–	+	+	+
ATRX mutation	–	+	+	+
TERT promoter mutation	+	–	–	–

were observed in adult primary glioblastomas (83%) and oligodendrogliomas (78%); lower frequencies were seen in oligoastrocytomas (25%) and astrocytomas WHO grades II and III (~10%) [8].

7.3.4 1p/19q Co-deletion

The most commonly detected genomic aberration in oligodendrogliomas is a heterozygous loss (LOH) of the short arm of chromosome 1 associated with LOH of the long arm of chromosome 19 (1p/19q co-deletion). This genetic anomaly is observed in the vast majority of oligodendrogliomas, with incidences of up to 90% reported for grade II and somewhat lower for grade III tumors (50–70%). Moreover, the 1p/19q co-deletion apparently occurs mutually exclusive of the TP53 mutations and chromosome 17p losses which are more common in astrocytic tumors.

Table 7.2 illustrates the mutual relationship of the presence of *IDH* mutations, *ATRX* mutations, and *TERT* promoter mutations in astrocytic tumors; "+" indicates high prevalence of the mutations; "–" refers to very low frequencies/absence of the mutations.

7.3.5 Glioblastoma (WHO Grade IV)

Genome-wide analyses revealed that basically three major signaling cascades are affected by genetic aberrations in GBM: (a) the TP53 (tumor protein 53) pathway; (b) the RTK (receptor tyrosine kinase)/RAS/PI3K (phosphoinositide 3-kinase) pathway, both involved in the regulation of cellular growth, apoptosis, and proliferation; and (c) the RB1 (retinoblastoma) pathway, controlling the G1 to S phase transition in the cell cycle.

Genes which are mutated in GBM are involved in the regulation of cell signaling, cell proliferation and survival, cell cycle, apoptosis, and NADPH production.

7.3.5.1 Pathway-Related Genes

The compilation outlined below (Table 7.3) contains a selection of pathway-related genes which are commonly altered in GBM, with respect to their properties, chromosomal location, and nature of pathogenic changes.

Table 7.3 Pathway-related genes affected in GBM

Pathway	Gene	Role	Chromosome location	Function	Changes
TP 53 pathway	*TP53* tumor protein p53	Tumor suppressor gene	17p13.1		Mutations, deletions
	MDM2 (E3 ubiquitin protein ligase)	Oncogene	12q14-15	Overexpression inhibits p53 function	Amplification
	CDKN2A: (cyclin-dependent kinase inhibitor 2A)		9p21	Encoding tumor suppressors p14ARF and p16INK4A (involved in RB1 pathway; see below)	Deletions
RTK (receptor tyrosine kinase)/ RAS/PI3K (phosphoinositide 3-kinase)	RTK: *EGFR* (epidermal growth factor receptor)	Oncogenic properties	7p12		Mutations, amplification [gain of chromosome 7, double minutes]
	RTK: *PDGFRA* (platelet-derived growth factor receptor; α-peptide)	Oncogenic properties	4q12		Amplification
	PTEN (phosphatase and tensin homolog)	Tumor suppressor gene	10q23.3	Inhibitor in the PI3 kinase pathway (PI3K activates the oncogenic Akt pathway)	Mutations, deletion
	PIK3CA	Oncogene	3q26.3	Encoding the PI3 kinase catalytic subunit α	Mutations, amplification
	PIK3R1	Oncogene	5q13.1	Encoding the PI3 kinase regulatory subunit 1 (α)	Mutations
	NF1 (neurofibromin 1)	Tumor suppressor gene	17q11.2	Negative regulator of the oncogenic Ras pathway	Mutations, deletions

(continued)

Table 7.3 (continued)

Pathway	Gene	Role	Chromosome location	Function	Changes
RB1 pathway	*RB1* (retinoblastoma 1)	Tumor suppressor gene	13q14.2		Mutations, deletions
	CDKN2A (cyclin-dependent kinase inhibitor 2A)		9p21	Encoding tumor suppressors p16INK4A and p14ARF (involved in TP53 pathway)	Deletions
	CDKN2B (cyclin-dependent kinase inhibitor 2B)		9p21	Encoding tumor suppressor p15INK4B (inhibits CDK4)	Deletions
	CDK4 (cyclin-dependent kinase 4)	Oncogenic properties	12q14		Amplification

7.3.5.2 *IDH1* Mutations

In addition to the pathway-related gene alterations, point mutations in the *IDH1* gene are a prominent feature in GBM. As described above, the IDH1-R132H mutation is detected predominantly in secondary GBM (~80%) but only rarely in primary GBM. Analysis of the IDH1 mutation is therefore routinely used as a supporting parameter to differentiate between GBM subtypes (Table 7.2).

7.3.5.3 *TERT* Promoter Mutations

Unlike the IDH1 mutation, *TERT* promoter mutations are found in the vast majority of adult primary glioblastomas (~80%) (Table 7.2). On the contrary, very low mutation frequencies were observed in secondary GBM [9].

7.3.5.4 *ATRX* Mutations

ATRX mutations were frequently detected in secondary glioblastomas (57%) and rarely in primary glioblastomas (4%) [10], an observation which provides an additional tool for the characterization of GBM subtypes (Table 7.2).

7.3.5.5 Mutations Affecting Oncogenes/Tumor Suppressor Genes

Somatic mutations frequently found to cause activation of oncogenes and/or inactivation of tumor suppressor genes are listed in Table 7.4 (mutation frequencies compiled from [11, 12]).

7.3.5.6 Chromosomal Aberrations

Apart from somatic mutations, genomic instability resulting in somatic copy number alterations (SCNAs) is a major determining factor in GBM. The most important chromosomal abnormalities are depicted in Table 7.5 (LOH = loss of heterozygosity).

7.3.5.7 Summary of Genetic Characteristics of Primary and Secondary Glioblastomas

The differentiation between primary and secondary GBM is reflected by distinct pathogenetic patterns in the two tumor subtypes. Major differences that are consistently observed mainly involve key regulatory genes such as *EGFR*, *PTEN*,

Table 7.4 Somatic mutations in oncogenes/tumor suppressor genes

Gene name	Gene symbol	Somatic mutations (% GBM samples)
Tumor protein p53	*TP53*	31–42
Phosphatase and tensin homolog	*PTEN*	24–37
Neurofibromin 1	*NF1*	15–21
Epidermal growth factor receptor	*EGFR*	14–18
Retinoblastoma 1	*RB1*	8–13
Phosphoinositide-3-kinase, regulatory subunit 1 (alpha)	*PIK3R1*	7–10
Phosphoinositide-3-kinase, catalytic, alpha polypeptide	*PIK3CA*	7–10

Table 7.5 Chromosomal aberrations in GBM

Chromosomal abnormality	Genes affected	
	Oncogenes	Tumor suppressor genes
Amplification of chromosome region 3q26	*PIK3CA*	
Amplification of chromosome region 4q12	*PDGFRA*	
Gain of chromosome 7	*EGFR*	
Amplification of chromosome region 7p11	*EGFR*	
Loss of chromosome 9p regions		*CDKN2A, CDKN2B*
Loss of chromosome 10		*PTEN*
LOH in chromosome region 10q		*PTEN*
Amplification of chromosome region 12q14-15	*MDM2, CDK4*	
LOH in chromosome region 13q		*RB1*
LOH in chromosome region 17p		*TP53*
Loss of chromosome region 17q11.2		*NF1*

TP53, and, as described earlier, *IDH1*, *ATRX*, and *TERT* promoter mutations. In the following, the most significant differences are summarized.

Primary GBM is characterized by:

(a) Amplification of the *EGFR* oncogene accompanied by LOH on chromosome 10q where the tumor suppressor gene *PTEN* is located.
(b) Complete loss of chromosome 10.
(c) Mutated *PTEN* in about one third of primary GBMs (not seen in secondary glioblastoma).
(d) *TP53* mutations found at a significantly lower rate in primary than in secondary GBM.
(e) A preferred correlation pattern of *EGFR* amplification and mutations in *PTEN* and *TP53*, i.e., *EGFR* amplification and *PTEN* mutations are associated with low *TP53* mutation frequencies.
(f) Absence of *IDH1* mutations.
(g) A high frequency of *TERT* promoter mutations (~80%).

Secondary GBM is characterized by:

(a) A high frequency of *TP53* mutations, often occurring together with LOH in chromosome 17p.
(b) The occurrence of high *TP53* mutation rates which is largely complemented by a lack of *EGFR* amplification, probably indicating a mutually exclusive relationship.
(c) A chromosomal aberration associated predominantly with secondary GBM is LOH in chromosome 19q (54%), in contrast to primary GBM (6%).

(d) A high frequency of *IDH1* mutations (~80%).
(e) *ATRX* mutations (57%).

A graphical representation of molecular alterations occurring in primary GBM as compared to secondary GBM is illustrated as follows:

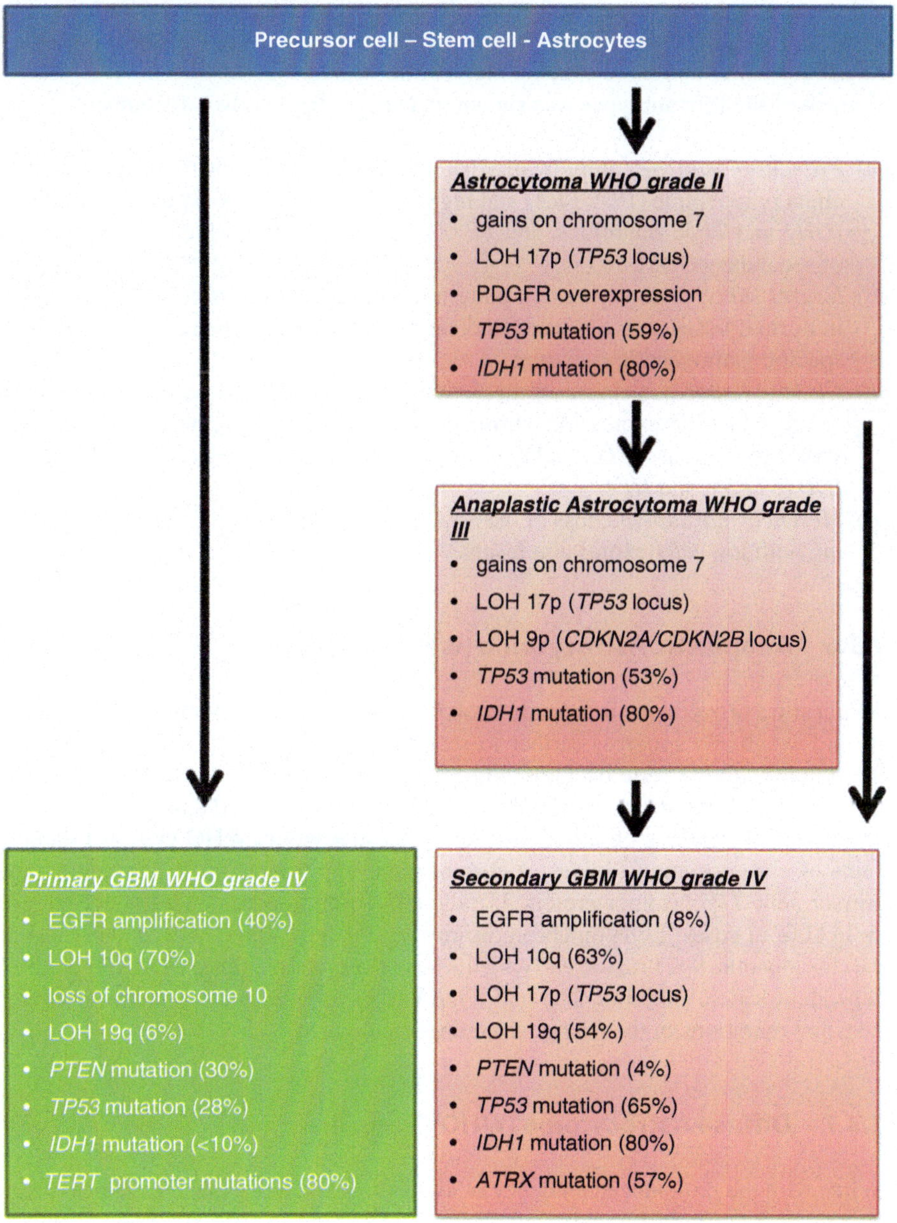

7.3.5.8 Novel Molecular Classification of GBM into Four Distinct Subtypes

Based entirely on genomic and gene expression profiling, a new classification model was recently proposed for glioblastoma. This scheme suggests to replace the currently accepted primary and secondary GBM subtypes with four redefined subgroups: *proneural*, *neural*, *classical*, and *mesenchymal* [13]. The essential intention of this approach was to relate the specific molecular signature of each tumor subgroup to the progenitor cell type from which it may have developed, thus providing a basis for better and more specific therapeutic strategies.

Each of the four subtypes was shown to display characteristic features:

(a) *Proneural subtype*: amplification of platelet-derived growth factor receptor, alpha polypeptide (*PDGFRA*) and high levels of *PDGFRA* expression, frequent *IDH1* and *TP53* mutations, and *TP53* LOH, all features which are reminiscent of secondary GBM.
(b) *Neural subtype*: an expression pattern of neuronal markers very similar to that in normal tissue was observed, indicating that this subtype is not defined by a specific pathogenetic signature.
(c) *Classical subtype*: chromosome 7 amplification (*EGFR*) with corresponding enhanced *EGFR* expression, chromosome 10 loss (*PTEN*), and loss of chromosome 9p regions (*CDKN2A*); no abnormalities were detected in *PDGFRA*, *IDH1*, *TP53*, and *NF1*.
(d) *Mesenchymal subtype*: loss of the *NF1* locus on chromosome 17q11.2, correlating with low *NF1* expression levels.

7.3.6 Anaplastic Astrocytoma (WHO Grade III)

Anaplastic astrocytomas frequently evolve from diffuse astrocytomas (WHO grade II) and eventually progress further to secondary glioblastomas (WHO grade IV). Accordingly, the genetic background of anaplastic astrocytomas includes features that are also found in astrocytic tumors of both WHO grades II and IV.

Genetic aberrations observed in anaplastic astrocytoma WHO grade III include gains of chromosome 7; losses of chromosome 17p; mutations in the tumor suppressor gene *TP53* (tumor protein 53); the *IDH1* point mutation (IDH1-R132H) in up to 80% of cases; LOH on chromosomes 6, 10q, 11p, 19q, and 22q; deletions of the chromosome 9p21 region carrying the tumor suppressor genes *CDKN2A* (cyclin-dependent kinase inhibitor 2A) and *CDKN2B* (cyclin-dependent kinase inhibitor 2B); and mutations in the *RB1* gene (retinoblastoma 1) (~25% of cases).

7.3.7 Diffuse Astrocytoma (WHO Grade II)

In line with the pronounced tendency of astrocytoma WHO grade II tumors to progress to higher grade gliomas (anaplastic astrocytoma, grade III, and secondary glioblastoma, grade IV), some of the genetic traits associated with these malignant tumors are already laid out in diffuse astrocytoma. In particular, gains of chromosome 7 and losses of chromosome 17p which occur at high frequencies should be noted.

Genes affected in astrocytomas WHO grade II include:

(a) *TP53* (tumor suppressor gene, on chromosome 17p13.1, encoding tumor protein 53):

In more than 60% of diffuse astrocytomas, monoallelic deletion (loss of heterozygosity, LOH) of the chromosome 17p region harboring *TP53* has been described; moreover, in the majority of cases, mutations in the remaining *TP53* allele ultimately result in a total lack of the functional gene product.

(b) *PDGFRA* (oncogenic; encoding platelet-derived growth factor receptor, α-peptide):

In WHO grade II astrocytomas, upregulation of PDGFRA expression can be observed, although supporting evidence is based on relatively low sample sizes. In addition, elevated PDGFRA levels are more frequently correlated with higher grade gliomas. Taken together, it therefore remains uncertain to which extent PDGFRA overexpression contributes to tumorigenesis in diffuse astrocytoma.

(c) *IDH1/2* (isocitrate dehydrogenase 1, isocitrate dehydrogenase 2):

The IDH1-R132H mutation was found in up to 80% of grade II gliomas and, at a much lower frequency, functional mutations in the *IDH2* gene affecting codon 172 (R172).

7.3.8 Pilocytic Astrocytoma (WHO Grade I)

Patients suffering from neurofibromatosis type 1 (NF1), a hereditary tumor syndrome, frequently develop pilocytic astrocytoma. About one third of pilocytic astrocytomas are observed in NF1 patients, whereas the sporadic types of this tumor are NF1 independent.

NF1-associated pilocytic astrocytoma: Neurofibromin, encoded by the *NF1* gene on chromosome 17q11.2, functions as a tumor suppressor by inhibiting oncogenic Ras (=*Ra*t *s*arcoma) signaling. In neurofibromatosis type 1 and in NF1-associated pilocytic astrocytoma, *NF1* gene deletions and mutations result in loss of functional neurofibromin.

NF1-independent pilocytic astrocytoma: In sporadic pilocytic astrocytoma, typical genetic aberrations (>60%) are duplications at chromosome region 7q34, affecting the *BRAF* gene (v-raf murine sarcoma viral oncogene homolog B). These duplications create in-frame fusions of *BRAF* with the upstream *KIAA1549* gene. The resulting aberrant fusion proteins contain the BRAF kinase domain and were shown to exhibit constitutive BRAF kinase activity which in turn activates the oncogenic MAPK (mitogen-activated protein kinase) signaling pathway.

Oncogenic BRAF activation not only occurs via gene duplication but may also be the result of mutations occurring around codon 600: (a) a T > A mutation at nucleotide position 1799, creating the replacement of the wild-type valine 600 by a glutamate residue in the protein (referred to as the BRAF^{V600E} mutation), and (b) two different 3bp insertions, both resulting in an additional threonine residue at amino acid position 599. These activating mutations occur at a much lower frequency than the *KIAA1549/BRAF* fusions (~9%).

Similar to the *BRAF* fusions, albeit less common, are fusions on chromosome 3p25 between the *SRGAP3* (SLIT-ROBO Rho GTPase-activating protein 3) gene and the *RAF1* (v-raf-1 murine leukemia viral oncogene homolog 1) gene. RAF1 is a positive regulator of the oncogenic MAPK signaling pathway; in *SRGAP3/RAF1* fusions, the auto-inhibitory region of RAF1 is lost, leading to a constitutive activation of the MAPK pathway.

Somatic mitochondrial mutations, mostly single nucleotide exchanges, were recently reported in pilocytic astrocytoma. Some of the mutations resided in coding regions, causing amino acid alterations. The affected gene products were identified as proteins involved in electron transport/oxidative phosphorylation [14].

7.3.9 Oligodendroglioma (WHO Grade II) and Anaplastic Oligodendroglioma (WHO Grade III)

As described above, about 90% of WHO grade II and 50–70% of WHO grade III oligodendrogliomas exhibit the heterozygous 1p/19q co-deletion. Thus, this chromosomal aberration represents a genetic hallmark in oligodendrogliomas.

Mutations in the *IDH1* and *IDH2* genes are another characteristic feature in oligodendrogliomas. The IDH1-R132H mutation is most frequently observed (>70%). In IDH2, the homologous site (R172) was found to be mutated, however, in only a small fraction of the tumors [6]. In oligodendrogliomas, IDH1/2 mutations appear to be strongly associated with *TERT* promoter mutations and the 1p/19q co-deletion.

Novel genetic anomalies have recently been described in oligodendrogliomas and encompass point mutations in the *CIC* (capicua transcriptional repressor) gene, located at chromosome 19q13.2; its gene product acts as a transcriptional repressor downstream of the receptor tyrosine kinase (RTK) pathway. Interestingly, *CIC* mutations occur in the majority (~70%) of oligodendrogliomas exhibiting the 1p/19q co-deletion plus IDH mutations and point mutations in the *FUBP1* (far upstream element binding protein 1) gene on chromosome 1p31.1; FUBP1 is a transcriptional regulator of the c-*Myc* oncogene. Most of the *FUBP1* mutations (>70%) are found in oligodendrogliomas that also carry *CIC* mutations.

7.3.10 Oligoastrocytoma (WHO Grade II)

Similar to oligodendrogliomas, albeit at a lower frequency, heterozygous chromosome 1p/19q co-deletions have been described in oligoastrocytomas (30–50% of the cases). Notably, 19q deletion without 1p loss is often observed in these tumors.

Several genetic aberrations which are reminiscent of astrocytic tumors can be detected in about 30% of cases, e.g., (a) mutations in the tumor suppressor gene TP53, (b) LOH of chromosome 17p, (c) anomalies of chromosome 10, and (d) amplification of the *EGFR* (epidermal growth factor receptor) gene.

7.3.11 Genetic Markers

Taken together, the determination of the following genetic markers is requested/useful:

- Astrocytomas and/or glioblastomas:
  Mutation status of *IDH1* and *IDH2* (mutated versus wild type), allelic loss of chromosomes 1p and 19q (co-deletion), loss-of-function mutations in the *TP53* and *ATRX* genes. Additional genes include *EGFR*, *PTEN*, *PDGFRA*, *MET*, *PI3K*, chromatin-related genes (*H3F3A*, *HIST1H3B/C*), and *TERT* promoter mutations.
- Oligodendrogliomas and/or oligoastrocytomas:
  Mutation status of *IDH1* and *IDH2* (mutated versus wild type), allelic loss of chromosomes 1p and 19q (co-deletion), and mutations in the *CIC* and *FUBP1* genes as well as in the *TERT* promoter region.

7.4 Epigenetics

7.4.1 DNA Methylation

7.4.1.1 Glioblastoma (WHO Grade IV)

DNA hypermethylation of promoter regions is a frequently observed mechanism in glioblastomas by which tumor suppressor genes (such as *TP53*, *PTEN*, or *CDKN2A*) are silenced.

In recent years, the *MGMT* gene, encoding the DNA repair enzyme O^6-methylguanine-DNA methyltransferase, has received particular attention. Guanine alkylated at its O^6 position represents a mutagenic DNA lesion that is normally repaired by the MGMT enzymatic activity. It transfers the methyl group from the nucleobase to an active cysteinyl residue in its own sequence. Thus, by removing methyl groups from mutagenic O^6-methylguanine residues, the enzyme contributes to genome integrity.

In glioblastoma chemotherapy, alkylating drugs like the widely used temozolomide (TMZ) are employed to introduce DNA damage in tumor cells with the intention to trigger apoptosis and cell death. Active MGMT counteracts this mechanism, thus conferring resistance to the treatment. However, hypermethylation of the *MGMT* promoter, abolishing the transcription of the gene, was demonstrated in a high percentage of GBMs, i.e., in up to 75% of secondary and approximately 35% of primary glioblastomas. It is obvious that patients with glioblastoma lacking

MGMT expression show better responsiveness to TMZ chemotherapy which also has been implicated with improved prognosis.

7.4.1.2 Diffuse Astrocytoma (WHO Grade II)

So far, the best studied epigenetic feature associated with grade II astrocytomas is promoter hypermethylation of two tumor suppressor genes, *ARF* and *MGMT*:

The *ARF* gene codes for the tumor suppressor protein p14ARF which acts as a supporting factor in the TP53 pathway.

The *MGMT* gene encodes the DNA repair enzyme O^6-methylguanine-DNA methyltransferase.

Promoter hypermethylation results in reduced expression of the tumor suppressor genes and, as a result, in diminished protein function.

7.4.1.3 Oligodendroglioma (WHO Grade II)

Several genes were shown to be affected by promoter hypermethylation, resulting in reduced expression levels: (a) the *MGMT* gene, encoding the DNA repair enzyme O^6-methylguanine-DNA methyltransferase, (b) *CDKN2A* (cyclin-dependent kinase inhibitor 2A) and *CDKN2B* (cyclin-dependent kinase inhibitor 2B), (c) *RB1* (retinoblastoma 1), (d) *DAPK1* (death-associated protein kinase 1), and *ESR1* (estrogen receptor 1).

7.4.2 MicroRNAs

7.4.2.1 Glioblastoma (WHO Grade IV)

In GBM, most of the microRNAs (miRNAs) surveyed were shown to be overexpressed, and some have been functionally studied, e.g., miR-10b, miR-17, miR-21, miR-93, miR-221, and miR-222. The smaller cluster of downregulated miRNAs includes, among others, miR-7, miR-34a, miR-128, and miR-137. For a detailed update on miRNAs and their specific targets in glioblastoma, the reader is referred to a systematic review [15].

7.4.2.2 Pilocytic Astrocytoma (WHO Grade I)

To date, epigenomic investigations in pilocytic astrocytoma are still sparse. One study on microRNA (miR) expression [16] revealed overexpression of miR-29a, miR-34a, miR-138, miR-299–5p, and miR-432 and underexpression of miR-93, miR-106b, miR-129, miR-135a, and miR-135b.

7.5 Gene Expression

7.5.1 Glioblastoma (WHO Grade IV)

Microarray-based analyses of differentially expressed genes have lately been used to characterize GBM subtypes correlating with patients' response to treatment and prognosis of survival. Recently, an intensified search for a "consensus" expression profile from several independent GBM data sets produced a limited number of statistically robust marker panels [17–19]. Each proposed panel contains only a handful of marker genes which were validated for their potential to support classification of GBM subtypes for predicting clinical outcome.

In summary, the marker genes included in the panels described above are *POLD2*, *CYCS*, *MYC*, *AKR1C3*, *YME1L1*, *ANXA7*, and *PDCD4* [17]; *EDNRB*, *CHAF1B*, *PDLIM4*, and *HJURP* [19]; and *AQP1*, *CHI3L1*, *EMP3*, *GPNMB*, *IGFBP2*, *LGALS3*, *OLIG2*, *PDPN*, and *RTN1* [18]. Genes associated with a better prognosis of survival were *EDNRB*, *OLIG2*, and *RTN1*. Genes associated with worse prognosis include *CHAF1B*, *PDLIM4*, *HJURP*, *AQP1*, *CHI3L1*, *EMP3*, *GPNMB*, *IGFBP2*, *LGALS3*, and *PDPN*.

7.5.2 Diffuse Astrocytoma (WHO Grade II)

Several genes differentially expressed in astrocytomas grade II, as compared to controls, were identified in recent years. Upregulated expression was described for *CD9*; *CSPG2*, also known as *VCAN* (versican); *NTF3* (neurotrophin 3); *EGFR* (epidermal growth factor receptor); *PDGFRA* (platelet-derived growth factor receptor, alpha polypeptide); and *TIMP3* (TIMP metallopeptidase inhibitor 3). Downregulated expression was noted for *TYRO3* (TYRO3 protein tyrosine kinase).

7.5.3 Oligodendroglioma (WHO Grade II)

Commonly observed in oligodendrogliomas are elevated expression levels of PDGFA (platelet-derived growth factor, alpha polypeptide), PDGFB (platelet-derived growth factor, beta polypeptide), PDGFRA (platelet-derived growth factor receptor, alpha polypeptide), and PDGFRB (platelet-derived growth factor receptor, beta polypeptide).

7.6 Pathogenesis

The pathogenesis of glial tumors is not well understood. Most of the tumors might arise de novo, i.e., astrocytomas (WHO grade II), anaplastic astrocytoma (WHO grade III), glioblastoma (WHO grade IV), pilocytic astrocytoma (WHO grade I), oligodendroglioma (WHO grade II), anaplastic oligodendroglioma (WHO grade III), oligoastrocytoma (WHO grade II), and anaplastic oligoastrocytoma (WHO grade III).

Malignant progression from WHO grade II to WHO grade III occurs in anaplastic astrocytoma, oligodendroglioma, and oligoastrocytoma.

Pilocytic astrocytoma (WHO grade I) derived from piloid cells, i.e., cells similar to those found around chronic lesions of the hypothalamus, cerebellum, spinal cord, or glial stromal cells of the pineal gland. Oligodendroglioma (WHO grade II) arises from oligodendrocytes or other glial precursor cells.

Glioblastoma (WHO grade IV) arises via de novo genesis of highly anaplastic tumor cells (primary glioblastoma) or from anaplastic areas which develop rapidly within astrocytomas, oligodendrogliomas, or ependymomas and overgrow the primary less anaplastic areas (secondary glioblastoma). This pathogenic mechanism becomes obvious when the clinical history is of long duration with several surgical interventions. With progression of the disease, the grade of anaplasia of the tumors removed is also progressing.

7.7 Further Reading

For more detailed information on the different tumor entities, the following articles are suggested: astrocytoma [20, 21], anaplastic astrocytoma [22, 23], glioblastoma [11–13, 24–28], pilocytic astrocytoma [29, 30], oligodendroglioma [20, 31–33], anaplastic oligodendroglioma WHO grade III [20, 31–33], and oligoastrocytoma [20, 32].

References

1. Ostrom QT, Gittleman H, Farah P, Ondracek A, Chen Y, Wolinsky Y, Stroup NE, Kruchko C, Barnholtz-Sloan JS. CBTRUS statistical report: primary brain and central nervous system tumors diagnosed in the United States in 2006-2010. Neuro Oncol. 2013;15(Suppl 2):ii1–56. doi:10.1093/neuonc/not1151.
2. Ostrom QT, Gittleman H, Fulop J, Liu M, Blanda R, Kromer C, Wolinsky Y, Kruchko C, Barnholtz-Sloan JS. CBTRUS statistical report: primary brain and central nervous system tumors diagnosed in the United States in 2008-2012. Neuro Oncol. 2015;17(Suppl 4):iv1–iv62. doi:10.1093/neuonc/nov1189.
3. Ostrom QT, Gittleman H, Liao P, Rouse C, Chen Y, Dowling J, Wolinsky Y, Kruchko C, Barnholtz-Sloan J. CBTRUS statistical report: primary brain and central nervous system tumors diagnosed in the United States in 2007-2011. Neuro Oncol. 2014;16(Suppl 4):iv1–63. doi:10.1093/neuonc/nou1223.
4. Louis DN, Ohgaki H, Wiestler OD, Cavenee WK, Ellison DW, Figarella-Branger D, Perry A, Reifenberger G, Von Deimling A. WHO classification of tumours of the central nervous system. Revised 4th ed. Lyon: WHO/IARC; 2016.
5. Rodriguez FJ, Vizcaino MA, Lin MT. Recent advances on the molecular pathology of glial neoplasms in children and adults. J Mol Diagn. 2016;18(5):620–34. doi:10.1016/j.jmoldx.2016.1005.1005.
6. Yan H, Parsons DW, Jin G, McLendon R, Rasheed BA, Yuan W, Kos I, et al. IDH1 and IDH2 mutations in gliomas. N Engl J Med. 2009;360(8):765–73. doi:10.1056/NEJMoa0808710.
7. Watson LA, Goldberg H, Berube NG. Emerging roles of ATRX in cancer. Epigenomics. 2015;7(8):1365–78. doi:10.2217/epi.1315.1382.
8. Killela PJ, Reitman ZJ, Jiao Y, Bettegowda C, Agrawal N, Diaz Jr LA, Friedman AH, et al. TERT promoter mutations occur frequently in gliomas and a subset of tumors derived from cells with low rates of self-renewal. Proc Natl Acad Sci U S A. 2013;110(15):6021–6. doi:10.1073/pnas.1303607110.
9. Koelsche C, Sahm F, Capper D, Reuss D, Sturm D, Jones DT, Kool M, et al. Distribution of TERT promoter mutations in pediatric and adult tumors of the nervous system. Acta Neuropathol. 2013;126(6):907–15. doi:10.1007/s00401-00013-01195-00405.

10. Jiao Y, Killela PJ, Reitman ZJ, Rasheed AB, Heaphy CM, de Wilde RF, Rodriguez FJ, et al. Frequent ATRX, CIC, FUBP1 and IDH1 mutations refine the classification of malignant gliomas. Oncotarget. 2012;3(7):709–22.
11. Bleeker FE, Molenaar RJ, Leenstra S. Recent advances in the molecular understanding of glioblastoma. J Neurooncol. 2012;108(1):11–27. doi:10.1007/s11060-11011-10793-11060.
12. Dunn GP, Rinne ML, Wykosky J, Genovese G, Quayle SN, Dunn IF, Agarwalla PK, et al. Emerging insights into the molecular and cellular basis of glioblastoma. Genes Dev. 2012;26(8):756–84. doi:10.1101/gad.187922.187112.
13. Verhaak RG, Hoadley KA, Purdom E, Wang V, Qi Y, Wilkerson MD, Miller CR, et al. An integrated genomic analysis identifies clinically relevant subtypes of glioblastoma characterized by abnormalities in PDGFRA, IDH1, EGFR and NF1. Cancer Cell. 2010;17(1):98.
14. Lueth M, Wronski L, Giese A, Kirschner-Schwabe R, Pietsch T, von Deimling A, Henze G, Kurtz A, Driever PH. Somatic mitochondrial mutations in pilocytic astrocytoma. Cancer Genet Cytogenet. 2009;192(1):30–5. doi:10.1016/j.cancergencyto.2009.1003.1002.
15. Moller HG, Rasmussen AP, Andersen HH, Johnsen KB, Henriksen M, Duroux M. A systematic review of microRNA in glioblastoma multiforme: micro-modulators in the mesenchymal mode of migration and invasion. Mol Neurobiol. 2013;47(1):131–44. doi:10.1007/s12035-12012-18349-12037.
16. Birks DK, Barton VN, Donson AM, Handler MH, Vibhakar R, Foreman NK. Survey of MicroRNA expression in pediatric brain tumors. Pediatr Blood Cancer. 2011;56(2):211–6. doi:10.1002/pbc.22723.
17. Bredel M, Scholtens DM, Harsh GR, Bredel C, Chandler JP, Renfrow JJ, Yadav AK, et al. A network model of a cooperative genetic landscape in brain tumors. JAMA. 2009;302(3):261–75. doi:10.1001/jama.2009.1997.
18. Colman H, Zhang L, Sulman EP, McDonald JM, Shooshtari NL, Rivera A, Popoff S, et al. A multigene predictor of outcome in glioblastoma. Neuro Oncol. 2010;12(1):49–57. doi:10.1093/neuonc/nop1007.
19. de Tayrac M, Aubry M, Saikali S, Etcheverry A, Surbled C, Guenot F, Galibert MD, et al. A 4-gene signature associated with clinical outcome in high-grade gliomas. Clin Cancer Res. 2011;17(2):317–27. doi:10.1158/1078-0432.CCR-1110-1126.
20. Marko NF, Weil RJ. The molecular biology of WHO Grade II gliomas. Neurosurg Focus. 2013;34(2):E1. doi:10.3171/2012.3112.FOCUS12283.
21. Ohgaki H, Kleihues P. Genetic profile of astrocytic and oligodendroglial gliomas. Brain Tumor Pathol. 2011;28(3):177–83. doi:10.1007/s10014-10011-10029-10011.
22. Riemenschneider MJ, Reifenberger G. Astrocytic tumors. In: von Deimling A, editor. Gliomas, Recent results cancer res, vol. 171. Berlin: Springer; 2009. p. 3–24.
23. Sarkar C, Jain A, Suri V. Current concepts in the pathology and genetics of gliomas. Indian J Cancer. 2009;46(2):108 19.
24. Gupta K, Salunke P. Molecular markers of glioma: an update on recent progress and perspectives. J Cancer Res Clin Oncol. 2012;138(12):1971–81. doi:10.1007/s00432-00012-01323-y.
25. Martinez R. Beyond genetics in glioma pathways: the ever-increasing crosstalk between epigenomic and genomic events. J Signal Transduct. 2012;2012:519807.
26. Ohgaki H, Kleihues P. Genetic pathways to primary and secondary glioblastoma. Am J Pathol. 2007;170(5):1445–53.
27. Purow BW, Schiff D. Glioblastoma genetics: in rapid flux. Discov Med. 2010;9(45):125–31.
28. Westermark B. Glioblastoma--a moving target. Ups J Med Sci. 2012;117(2):251–6. doi:10.3109/03009734.03002012.03676574.
29. Marko NF, Weil RJ. The molecular biology of WHO grade I astrocytomas. Neuro Oncol. 2012;14(12):1424–31. doi:10.1093/neuonc/nos1257.
30. Pfister S, Witt O. Pediatric gliomas. In: von Deimling A, editor. Gliomas, Recent results cancer res, vol. 171. Berlin: Springer; 2009. p. 67–81.

31. Alentorn A, Sanson M, Idbaih A. Oligodendrogliomas: new insights from the genetics and perspectives. Curr Opin Oncol. 2012;24(6):687–93. doi:10.1097/CCO.1090b1013e328357f328354ea.
32. Harris BT, Hattab EM. Molecular pathology of the central nervous system. In: Cheng L, Eble JN, editors. Molecular surgical pathology. New York, NY: Springer Science+Business Media; 2013. p. 357–405.
33. Hartmann C, von Deimling A. Molecular pathology of oligodendroglial tumors. In: von Deimling A, editor. Gliomas, Recent results cancer res, vol. 171. Berlin: Springer; 2009. p. 25–49.

Molecular Carcinogenesis of Non-glial Brain Tumors

8

Serge Weis and Peter Strasser

Contents

Abstract

Besides glial brain tumors, non-glial tumors, i.e., meningiomas and schwannomas, constitute the most frequent tumors encountered in daily routine. Both tumor types are in their majority benign tumors which are cured by total surgical resection. Their malignant forms are rare. The molecular alterations in both tumor categories involve various pathways.

S. Weis (✉)
Department of Neuropathology, Neuromed Campus, Kepler University Hospital, School of Medicine, Johannes Kepler University, Linz, Austria
e-mail: serge.weis@kepleruniklinikum.at

P. Strasser
Laboratory of Molecular Biology, Department of Laboratory Medicine, Christian Doppler University Hospital, Paracelsus Medical University, Salzburg, Austria

J. Haybaeck (ed.), *Mechanisms of Molecular Carcinogenesis – Volume 1*,
DOI 10.1007/978-3-319-53659-0_8

8.1 Introduction

Non-glial tumors are quite frequent and include meningiomas and peripheral nerve sheath tumors. Both tumor types are in their majority benign tumors which are cured by total surgical resection. Their malignant forms are rare.

Meningiomas arise from meningothelial cells. Three WHO grades are distinguished (I–III) for meningioma (I), atypical meningioma (II), and anaplastic meningioma (III). The molecular alterations involve various pathways. Meningiomas affect middle-aged and elderly patients. Female patients outweigh males by a factor of 2:1 to reach 3.5:1 in patients aged 40–49 years. Overall, meningiomas represent 36.4% of brain tumors.

Schwannomas represent tumors of peripheral nervous system including cranial nerves. Nerve sheath tumors make up approximately 8.1% of all tumors affecting the brain and its coverings.

8.2 Meningioma

8.2.1 Morphology

Meningiomas are usually benign tumors arising from meningeal or arachnoithelial cells. They occur frequently and make up 40–45% of tumors seen in the daily clinical practice. According to their biologic behavior, three groups of meningiomas are distinguished: meningioma (with various histological subtypes) (WHO grade I, 85% of meningiomas), atypical meningioma (WHO grade II, 12% of meningiomas), and anaplastic meningioma (WHO grade III, 3% of meningiomas).

Each WHO grade meningioma has various histological subtypes as follows: WHO grade I, meningothelial, transitional, fibroblastic, angiomatous, metaplastic, psammomatous, secretory, microcystic, and lymphoplasmacyte rich; WHO grade II, clear cell and chordoid; WHO grade III, rhabdoid and papillary.

Meningiomas occur most frequently in middle-aged and elderly patients (sixth to seventh decade). The male-to-female ratio is 1:2. Meningiomas are found in intracranial locations (over the cerebral convexities, parasagittal in connection with the falx and venous sinuses, sphenoid ridges, para-/suprasellar region, and olfactory grooves) as well as intraspinally in the orbita and near the optic nerve and rarely intraventricularly, epidurally, or in other organs (e.g., lung, pleura, liver, bone).

8.2.1.1 Meningioma

Meningiomas are meningothelial (arachnoidal) cell neoplasms, typically attached to the inner surface of the dura mater. Macroscopically they are rubber or firm, well demarcated, nodular, or lobulated.

Meningothelial meningioma tumor cells are of uniform size with oval nuclei, delicate chromatin, and resemblance of arachnoid cap cells. Tumor cells form syncytia with undiscernible cell processes. Tumor cells form lobules, separated by thin collagenous septae.

Fibrous (fibroblastic) meningiomas are made up of spindle cells forming parallel, storiform, and interlacing bundles in a collagen-rich matrix. The tumor cells form wide fascicles with varying amounts of intercellular collagen.

Transitional meningioma is characterized by a coexistence of meningothelial and fibrous patterns and lobular or fascicular arrangements, and whorl and psammoma body formation is possible.

Psammomatous meningiomas contain more psammoma bodies than tumor cell mass. Psammoma bodies are irregular calcified masses which may become confluent.

8.2.1.2 Atypical Meningioma

Atypical meningiomas are characterized by increased mitotic activity (four or more mitoses per 10 high-power (40×) fields). Three or more of the following histological features are present: increased cellularity, small cells with a high nuclear-to-cytoplasmic ratio, prominent nucleoli, uninterrupted patternless or sheet-like growth, and foci of "spontaneous" or "geographic" necrosis.

8.2.1.3 Anaplastic Meningioma

Anaplastic meningiomas are characterized by histological features of frank malignancy with malignant cytology resembling carcinoma, melanoma, or high-grade sarcoma and marked elevated mitotic index (20 or more mitoses per 10 high-power (40×) fields).

8.2.2 Pathogenesis

Radiation (low, moderate, and high dose) might lead to atypical or aggressive multifocal highly proliferative meningiomas; sex hormones (overrepresentation of women) and hormonal medications are considered as potential pathogenetic causes for tumor formation.

8.2.3 Genetics

In contrast to glioblastoma, where comprehensive information on the underlying molecular pathology is available, studies on genetic mechanisms involved in meningioma development and progression have been intensified only more recently. Below, some current major findings in this field are summarized.

8.2.3.1 Signaling Pathways

Several signaling pathways were shown to malfunction in meningioma; among those, the hedgehog (Hh) and the Wnt ("wingless") pathways appear to be most frequently affected.

Table 8.1 Hedgehog pathway-related genes affected in meningioma

Gene	Function	Chromosome location
SMO (smoothened)	Oncogenic properties	Chromosome 7q32.3
GLI1, *GLI2* (GLI family zinc finger proteins)	Oncogenic properties	Chromosome regions 12q13.2–12q13.3 and 12q14, respectively
GLIS2 (GLI-similar zinc finger 2)	Oncogenic properties	Chromosome 16p13.3
FOXM1 (forkhead box protein M1)	Oncogenic properties	Chromosome 12p13
IGF2 (insulin-like growth factor 2)	Oncogenic properties	Chromosome 11p15.5
SPP1 (secreted phosphoprotein 1)	Encoding osteopontin, oncogenic properties	Chromosome 4q22.1
PTCH1 (patched)	Tumor suppressor	Chromosome 9q22.3

Hh (Hedgehog) Pathway
The primary physiological role of the Hh signaling cascade is its regulatory function in embryonic development and differentiation. Defects in this pathway are often associated with tumorigenesis. In meningioma, several affected Hh pathway-related genes were identified (Table 8.1):

Wnt (Wingless) Pathway
Aberrant functions of this pathway in meningioma are related to the downregulation of (1) *APC* (adenomatous polyposis coli), tumor suppressor, chromosome region 5q21-q22, and LOH in benign meningioma; (2) *BCR* (breakpoint cluster region), putative tumor suppressor and chromosome 22q11.23; (3) *CDH1* (cadherin 1), encoding E-cadherin, tumor suppressor, and chromosome 16q22.1; and (4) *SFRP1* (secreted frizzled-related protein 1), tumor suppressor and chromosome 8p11.21.

Additional signaling cascades that were found to be affected in meningioma [1] include the RB (retinoblastoma)/TP53 (tumor protein 53) pathways, the PI3K (phosphatidylinositol 3-kinase)/Akt pathway, the MAPK (mitogen-activated protein kinase) pathway, and the Notch signaling pathway.

8.2.3.2 Chromosomal Aberrations and Mutations

The best studied chromosomal abnormality in meningioma is loss of heterozygosity (LOH) of chromosome region 22q12.2 where the tumor suppressor gene *NF2* (neurofibromin 2), encoding the merlin (schwannomin) protein, is located. Additional somatic mutations in the second *NF2* allele lead to complete loss of functional merlin, thus triggering tumorigenesis. *NF2* somatic mutations in meningiomas were reported at frequencies of 30–60% in WHO grade I and 70–80% in WHO grades II/III.

Loss of chromosome 1p is found more often in higher grade than in benign meningiomas, suggesting an association with tumor progression and recurrence. Possible tumor suppressor candidate genes are, e.g., *CDKN2C* (cyclin-dependent kinase inhibitor 2C) at 1p32 encoding protein p18INK4C or *ALPL* (alkaline phosphatase, liver/bone/kidney) at 1p36.12.

Losses of chromosome 9p regions are found predominantly in the more aggressive tumors, with the highest incidence in grade III meningiomas. At region 9p21, two tumor suppressor genes are lost: *CDKN2A* (cyclin-dependent kinase inhibitor 2A) encoding proteins p16INK4A and p14ARF and *CDKN2B* (cyclin-dependent kinase inhibitor 2B) encoding p16INK4B.

Loss of chromosome 14q regions is associated more frequently with increasing tumor grade. Genes that are affected include the tumor suppressors *MEG3* (maternally expressed gene 3) at 14q32 or *NDRG2* (N-Myc downstream-regulated gene 2) at 14q11.2.

Amplification of chromosome 17q23 encompassing the oncogene *RPS6K* (ribosomal protein S6 kinase) was identified in some malignant meningiomas.

Another genetic aberration is LOH at chromosome 18p11.32 which was reported at frequencies between 20% and 70%. This locus contains the *DAL-1* (differentially expressed in adenocarcinoma of the lung-1) gene, encoding tumor suppressor protein 4.1B, which in turn interacts with the tumor suppressor in lung cancer-1 protein (TSLC1).

A germline missense mutation was reported in the tumor suppressor gene *SMARCB1* (SWI/SNF-related matrix-associated actin-dependent regulator of chromatin subfamily B member 1) in members of a single family suffering from schwannomas and multiple meningiomas. Additional losses of the *SMARCB1* wild-type allele were demonstrated in these patients, frequently accompanied by somatic *NF2* mutations and loss of the *NF2* wild-type allele [2].

A short overview of the described chromosomal aberrations in meningioma is summarized in Table 8.2.

Table 8.2 Chromosomal aberrations in meningioma

Chromosomal abnormality	Genes affected	
	Oncogenes	Tumor suppressor genes
LOH in chromosome regions 1p32/1p36		*CDKN2C/ALPL*
LOH in chromosome region 9p		*CDKN2A, CDKN2B*
LOH in chromosome regions 14q11/14q32		*NDRG2/MEG3*
Amplification of chromosome 17q23	*RPS6K*	
LOH in chromosome region 18p11.32		*DAL-1*
LOH in chromosome regions 22q12.2/22q12.3		*NF2/TIMP-3*

8.2.4 Epigenetics

8.2.4.1 DNA Methylation

Promoter hypermethylation as a means of inhibiting gene expression in meningioma has been reported for several tumor suppressor genes, e.g.:

1. *TIMP-3* (tissue inhibitor of metalloproteinase-3), located on chromosome 22q12.3, *TIMP-3* hypermethylation is almost exclusively accompanied by LOH at 22q12.3.
2. *MEG3* (maternally expressed gene 3), chromosome 14q32. Methylation is associated more frequently with LOH at 14q32 in higher grade meningiomas.
3. *NDRG2* (N-Myc downstream-regulated gene 2), chromosome 14q11.2. In meningiomas WHO grade III, downregulated expression is consistently observed.
4. *HOX* (homeobox) genes of the *HOXA* cluster: *HOXA7*, *HOXA9*, and *HOXA10.* Hypermethylation appears to be associated with higher grades of meningioma and recurrence [3, 4].

8.2.4.2 MicroRNAs

Several studies focused on dysregulated microRNA (miRNA) expression in meningiomas:

1. Downregulated miR-200a was reported in a limited number of benign meningiomas and was identified as a tumor suppressor molecule [5].
2. miR-335 overexpression was demonstrated in meningiomas of all grades, with the highest levels found in grade III tumors. The miR-335 molecule exhibits oncogenic properties by decreasing expression of the tumor suppressor protein Rb-1 (retinoblastoma 1) [6].
3. In a more recent study [7], a signature of 14 differentially expressed miRNAs was proposed to discriminate meningeal from normal tissue. Some of those, miR-96-5p, miR-190a (both upregulated), miR-29c-3p, and miR-219-5p (both downregulated), were associated with tumor progression and higher recurrence rates.

8.2.5 Gene Expression

Genes involved in the Hh and Wnt pathways are closely associated with meningioma development and/or progression which is illustrated by distinct expression patterns [8, 9].

For the hedgehog pathway, elevated mRNA levels were determined for pathway-activating genes such as *SMO*, *FOXM1*, *SPP1*, *IGF2*, and the *GLI* family genes, whereas *DHH* (desert hedgehog), *PTH1R* (parathyroid hormone 1 receptor), and the tumor suppressor *PTCH1* appear to be downregulated, the latter, however, only in low-grade meningiomas.

Wnt pathway-related genes frequently underexpressed include, e.g., *CDH1* in atypical and malignant meningioma and *BCR* in all grades of meningioma; *SFRP1* was found to be downregulated in recurrent vs. primary meningiomas and, thus, is considered a putative marker for recurrent meningiomas.

Downregulation of the *DLC1* gene (deleted in liver cancer 1) was observed in benign meningiomas and interpreted as a potential tumor suppressor.

Loss of expression was reported for TSLC1, a tumor suppressor interacting with protein 4.1B; this phenomenon was more pronounced in higher grade meningiomas where it correlated with increased proliferation rates and poorer prognosis.

A comprehensive microarray-based study identified differentially expressed genes (DEGs) in original and recurrent meningiomas. Most DEGs were located on chromosomes 1p, 6q, and 14q, and the majority were found to be underexpressed in recurrent meningiomas. On the other hand, genes of the histone cluster 1, located on chromosome 6p, were overexpressed in recurrent meningiomas [10].

In microcystic meningiomas, a tissue inhibitor of metalloproteinases, TIMP-1 (TIMP metallopeptidase inhibitor 1) interacting with matrix metalloproteinase 9 (MMP-9), was shown to be underexpressed, as compared to a control group. In contrast, high levels of MMP-9 were detected in the tumor tissues; therefore, it was speculated that increased MMP-9/TIMP-1 ratios might be involved in the pathogenesis of microcystic meningioma [11].

Expression of *STAT3* (signal transducer and activator of transcription 3), an oncogene, was found to be upregulated in grade II meningiomas, with concomitant induction of VEGF (vascular endothelial growth factor) expression.

Several oncogenes, upregulated in some meningiomas, include *c-sis* (encodes the platelet-derived growth factor beta polypeptide, PDGFB), the transcription factors *c-myc* and *c-fos*, *bcl-2* (bcl-2 protein, a regulator of apoptosis), or *TP73* (tumor protein p73).

hTERT mRNA, encoding the telomerase catalytic subunit of human telomerase reverse transcriptase, is more frequently detected in grade II and III than in benign meningiomas which correlates well with the high levels of telomerase activity found in the aggressive tumors.

8.3 Tumors of the Peripheral Nervous System

Tumors affecting the peripheral nervous system can be classified into schwannoma, neurofibroma, perineurioma, malignant peripheral nerve sheath tumor (MPNST), miscellaneous benign tumors, non-neurogenic tumors, hemopoietic neoplasms, hyperplastic lesions, hamartoma, and choristoma.

Tumors of the peripheral nervous system present with radicular pain, signs of nerve root or spinal cord compression, and signs of eighth cranial nerve affection. They can be an incidental finding and present as a mass in painless neurofibromas and as multiple masses as hallmark of neurofibromatosis 1 (NF1) associated with pigmented cutaneous macules (café-au-lait spots) and freckling and progressively enlarging mass with or without neurologic symptoms in cases of malignant peripheral nerve sheath tumor.

8.3.1 Morphology

8.3.1.1 Schwannoma

Schwannomas are benign nerve sheath tumors that are typically encapsulated and composed entirely of well-differentiated Schwann cells. They are mainly found in the cerebellopontine angle and at the spinal nerve root.

The following types of schwannoma are histologically distinguished: (1) conventional schwannoma, (2) cellular schwannoma, (3) plexiform schwannoma, and (4) melanotic schwannoma.

Conventional schwannoma is composed of neoplastic Schwann cells with moderate quantities of eosinophilic cytoplasm, without discernible cell borders. Antoni A pattern consists of areas of compact elongated cells with nuclear palisading (Verocay bodies), while Antoni B pattern consists of areas of less cellularity, loose textured cells with indistinct processes, and variable lipidization. Thick-walled and hyalinized vessels are usually encountered.

8.3.1.2 Neurofibroma

Neurofibroma is a well-demarcated intraneural or diffusely infiltrative extraneural tumor consisting of a mixture of cell types, including Schwann cells, perineurial-like cells, and fibroblasts. They occur frequently as sporadic solitary nodules and less frequently as solitary, multiple, or numerous lesions in NF1.

Neurofibroma is composed of Schwann cells with ovoid to thin curved to elongated nuclei and scant cytoplasm and of fibroblasts in a matrix of collagen fibers. Numerous atypical nuclei (atypical neurofibroma) and significantly increased cellularity (cellular neurofibroma) might be encountered.

8.3.1.3 Malignant Peripheral Nerve Sheath Tumor (MPNST)

MPNST is a malignant tumor arising from a peripheral nerve; 50% of malignant peripheral nerve sheath tumors are associated with neurofibromatosis type 1. 50–70% arise from neurofibroma.

The following histological types are distinguished: (a) epithelioid MPNST, (b) MPNST with divergent mesenchymal and/or epithelial differentiation, (c) glandular MPNST, and (d) melanotic MPNST.

The common histological features of MPNST include herringbone or interwoven-fasciculated pattern of cell growth; tightly packed spindle cells with variable quantities of cytoplasm; elongated, waved nuclei; alternating loose and dense cellular areas or diffuse growth pattern; and mitotic activity with >4 mitotic figures per high field.

8.3.2 Pathogenesis

The pathogenesis of schwannomas is unknown.

8.3.3 Genetics

8.3.3.1 Schwannoma

The *NF2* gene encodes the merlin protein (for more details, see Sect. 8.3.4.2). Alterations of the *NF2* gene include inactivating mutations in 60%, small frameshift mutations resulting in truncated protein products, loss of the remaining wild-type allele on chromosome 22q, and loss of merlin expression. Other genetic changes include loss of chromosome 1p, gains of 9q34 and 17q, and loss of the *PRKAR1A* (protein kinase cAMP-dependent type I regulatory subunit alpha) region on 17q.

8.3.3.2 MPNST

Inactivation of both NF1 alleles occurs frequently in MPNSTs of NF1 patients. It has been observed that MPNSTs predominantly arise from plexiform neurofibromas and are associated with the *NF1−/−* genotype. Some of the major molecular biology findings in MPNSTs include 50% of MPNSTs manifest in patients with NF1. Complex numerical and structural karyotypic abnormalities include near-triploid or hypodiploid chromosome numbers, chromosomal losses, and recombinations involving almost all chromosomes. Chromosome 17 shows abnormalities in *NF1* and *TP53* (tumor protein 53) loci. Losses of chromosomes 13, 17, 18, and 22 and gains of chromosomes 2 and 14, as well as *TP53* mutations and altered protein expression, are reported. Inactivation of the p53- and Rb (retinoblastoma) regulatory pathways are reported in 75% of cases.

CDKN2A (cyclin-dependent kinase inhibitor 2A) on chromosome 9p21 shows homozygous deletions in the gene encoding the tumor suppressors p16INK4A and p14ARF in 50% of the cases and inactivates the neighboring gene *CDKN2B* (cyclin-dependent kinase inhibitor 2B; encoding the tumor suppressor p15INK4B).

MicroRNAs (microRNAs) regulating expression of oncogenes or tumor suppressor genes have been shown to participate in the malignant transformation of benign neurofibromas to MPNSTs, a process which is reflected in altered microRNA expression levels in the MPNST cells. Table 8.3 shows a selection of several miR's including some associated protein-coding genes [compiled from [12]].

8.3.4 Neurofibromatoses

The neurofibromatoses comprise a group of hereditary tumor syndromes including the three genetically distinct diseases neurofibromatosis type 1 (NF1), neurofibromatosis type 2 (NF2), and schwannomatosis. Since specific tumor suppressor genes are identified to correlate with each of these syndromes, their molecular pathology is now well established:

The *NF1* (neurofibromin 1) gene, located at chromosome 17q11.2, is affected in NF1 patients; *NF1* encodes the tumor suppressor neurofibromin which acts as a negative regulator of the Ras (Rat sarcoma) protein.

Table 8.3 MicroRNA expression in MPNST

	Activity	miR expression Upregulated (↑) Downregulated (↓)	Associated expression Upregulated (↑) Downregulated (↓)
miR-10b	Oncogenic	↑	↓ Neurofibromin; ↑ TWIST1
miR-21	Oncogenic	↑	↓ PDCD4
miR-29c	Tumor suppressor	↓	↓ COL1A1, COL21A1, COL5A2; ↓ TDG
miR-34a	Tumor suppressor	↓	↓ Tp53
miR-204	Tumor suppressor	↓	↑ HMGA2
miR-214	Oncogenic	↑	↓ PTEN; ↑ TWIST1

COL1A1 collagen, type I, alpha 1; *COL5A2* collagen, type V, alpha 2; *COL21A1* collagen, type XXI, alpha 1; *HMGA2* high mobility group AT-hook 2; *PDCD4* programmed cell death 4 (neoplastic transformation inhibitor); *PTEN* phosphatase and tensin homologue; *TDG* thymine-DNA glycosylase; *TWIST1* twist basic helix-loop-helix transcription factor 1

The *NF2* (neurofibromin 2) gene, encoding the tumor suppressor "merlin" (moesin-ezrin-radixin-like protein), also termed schwannomin, is involved in NF2 and schwannomatosis; *NF2* is located at chromosome 22q12.2.

The *SMARCB1* (SWI-/SNF-related, matrix-associated, actin-dependent regulator of chromatin, subfamily b, member 1) gene is associated with schwannomatosis; *SMARCB1* is located at chromosome 22q11.23.

8.3.4.1 Neurofibromatosis Type 1

Neurofibromatosis type I (NF1) is an autosomal dominant disorder characterized by neurofibromas, multiple café-au-lait spots, axillary and inguinal freckling, optic gliomas, osseous lesions, and iris hamartomas (Lisch nodules), caused by mutations of the *NF1* gene on chromosome 17q11.2. The encoded protein is neurofibromin, a tumor suppressor gene, which plays a role in cell proliferation and differentiation.

The diagnostic criteria include the presence of two or more of the following signs [13]:

1. Six or more café-au-lait macules (1.5 cm or larger in postpubertal individuals, 0.5 cm or larger in prepubertal individuals)
2. Two or more neurofibromas of any type
3. One or more plexiform neurofibromas
4. Freckling of armpits or groin
5. Pilocytic astrocytoma of optic pathway ("optic glioma")
6. Two or more Lisch nodules (iris hamartomas)
7. Dysplasia/absence of the sphenoid bone or dysplasia/thinning of long bone cortex
8. First-degree relative with NF1

Tumors associated with NF1 include neurofibromas, gliomas, sarcomas, and neuroendocrine/neuroectodermal and hematopoietic tumors.

Genetics

The molecular genetic events observed in NF1 occur according to the "two-hit" model: (1) germline mutations inactivate the *NF1* gene on one allele (*NF1+/−*), resulting in decreased cellular levels of functional neurofibromin; (2) in a subsequent step ("second-hit"), somatic *NF1* gene alterations in the other allele occur (loss of heterozygosity, LOH), leading to the complete absence of active neurofibromin protein (*NF1−/−* cells).

Reduced or abolished activity of neurofibromin triggers deregulation and ultimately constitutive activation of the Ras signaling pathway, thereby also affecting members of downstream signaling cascades driving cell proliferation, such as MAPK (mitogen-activated protein kinase), PI3K (phosphatidylinositol 3-kinase), PKB (protein kinase B, also known as Akt), and mTOR (mammalian target of rapamycin).

8.3.4.2 Neurofibromatosis Type 2

NF2 is an autosomal dominant disorder characterized by neoplastic and dysplastic lesions that primarily affect the nervous system; bilateral vestibular schwannomas are the hallmark, with other manifestations including schwannomas of other cranial nerves, spinal and cutaneous schwannomas, intracranial and spinal meningiomas, gliomas, meningioangiomatosis, glial hamartomas, ocular abnormalities, and neuropathies, caused by mutations of the *NF2* gene on chromosome 22q12. The encoded protein is merlin (schwannomin).

Definite NF2 is characterized by bilateral schwannomas or first-degree relative with NF2 and either unilateral vestibular schwannoma at <30 years or any two of the following: meningioma, schwannoma, glioma, and posterior subcapsular lens opacity.

Probable NF2 is characterized by unilateral vestibular schwannoma at <30 years and at least one of the following, meningioma, schwannoma, glioma, and posterior lens opacity; or two or more meningiomas and either unilateral vestibular schwannoma at <30 years; or one of the following, schwannoma, glioma, and posterior lens opacity.

A clinical study on 150 NF2 patients is available in the literature [14].

Genetics

The genetic background underlying neurofibromatosis type 2 (NF2) is linked to changes in the *NF2* (neurofibromin 2) gene which encodes merlin (schwannomin; see below). The aberrations observed in *NF2* include missense mutations, frame shifts resulting in truncated gene products, small base insertions/deletions, and large deletions. The genetic mechanism driving the development of NF2 involves an initial monoallelic inactivating germline mutation in the *NF2* gene, followed by an inactivating somatic *NF2* mutation in the second allele ("two-hit" model). Together, both events lead to a complete loss of functional merlin protein.

Although NF2 and schwannomatosis are recognized as two clinically distinct syndromes, it appears that aberrations in the *NF2* gene are involved in both diseases.

However, inactivating germline *NF2* mutations, typically found in NF2, are lacking in schwannomatosis where only somatic mutations in tumor tissues have been observed.

Merlin is an ERM (ezrin/radixin/moesin) protein acting as a tumor suppressor. It functions as a multi-suppressor from cell membrane to nucleus, a linker between extracellular cues and intracellular signaling pathways that regulate cell motility, proliferation, and survival. Additional roles of merlin include regulation of receptor distribution and signaling at the cell cortex, coordination of receptor signaling and intercellular contact, and promotion of Schwann cell elongation and myelin segment length.

Merlin exerts its growth-suppressive activity through a folded conformation that is tightly controlled via phosphorylation by numerous protein kinases including PAK, PKA, and Akt. It inhibits cell proliferation by modulating the growth activities of its binding partners, including the cell surface glycoprotein CD44, membrane-cytoskeleton linker protein ezrin and PIKE (PI 3-kinase enhancer) GTPase, etc.

Merlin is involved in the regulation of several cellular processes, depending on its subcellular location [15–18]:

1. *Nucleus*: Merlin is able to translocate to the nucleus where it inhibits the pro-proliferative E3 ubiquitin ligase CRL4/DCAF1, thereby suppressing oncogenic gene expression.
2. *Cell membrane*: Active merlin inhibits the activities of RTKs (receptor tyrosine kinases) and integrins, thus participating in the regulation of downstream intracellular signal cascades, PI3K/Akt (phosphoinositide 3 kinase/Akt, or protein kinase B) pathway, Raf/ERK/MAPK (rat fibrosarcoma/extracellular signal-regulated kinase/mitogen-activated protein kinase) pathway, and Rac/PAK (rac GTPase/p21-activated kinase) pathway.

Merlin is a key regulator of the Hippo pathway which governs organ size and cell number by controlling the proliferation/apoptosis ratio. The proliferation-repressive activity of merlin is also partially regulated by S518 phosphorylation. Mutations inactivating merlin function result in increased cellular proliferation and survival. Merlin is degraded through multistep phosphorylation by oncogenic kinases.

8.3.5 Schwannomatosis

Schwannomatosis is a usually sporadic and sometimes autosomal dominant disorder characterized by multiple spinal, cutaneous, and cranial nerve schwannomas, without vestibular schwannomas or other manifestations of NF1 or NF2, associated with inactivation of the *NF2* gene in tumors, but not in the germline.

Definite schwannomatosis is characterized by two or more (pathologically proven) schwannomas and lack of vestibular schwannomas on MRI study at >30 years and no known constitutional *NF2* mutation, or one (pathologically proven) schwannoma and first-degree relative with schwannomatosis.

Probable schwannomatosis is characterized by two or more schwannomas at age <30 years and no evidence of vestibular schwannomas on MRI scan and no known constitutional *NF2* mutation, or two or more schwannomas at age <45 years and no symptoms of cranial nerve VII dysfunction and no known constitutional *NF2* mutation, or radiographic evidence of one schwannoma and first-degree relative with schwannomatosis.

8.3.5.1 Genetic and Molecular Biology Findings

Mutations in the *NF2* gene are frequently detected in schwannomatosis-associated schwannoma tissue but are absent in the germline and in non-tumor tissue.

As the major predisposing factor in schwannomatosis, however, the tumor suppressor gene *SMARCB1* (SWI-/SNF-related, matrix-associated, actin-dependent regulator of chromatin, subfamily b, member 1) has been identified. Germline mutations of the *SMARCB1* gene were reported in 45% of familial cases and 7% of sporadic cases. The mutations include an exon 1 mutation (c.41C>A) and 3′ untranslated region mutation (c.*82C>T).

Mutant SMARCB1 proteins retain the ability to suppress cyclin D1 activity. Further causative genes might be found. The gene is located on chromosome 22q11.2, which is in proximity to the *NF2* gene (22q12.2). In fact, germline *SMARCB1* mutations associated with somatic *NF2* mutations were frequently described in patients developing schwannomas. This observation led to the proposition of a "four-hit" mechanism which is thought to trigger tumorigenesis: (1) the presence of a germline *SMARCB1* mutation on one of the two alleles, (2) and (3) deletion of the chromosome 22 region spanning the second (wild-type) *SMARCB1* allele and one of the *NF2* alleles, and (4) mutation of the second (wild-type) *NF2* allele.

The physiological function of the *SMARCB1* gene product is its involvement in the regulation of gene expression by participating in chromatin remodeling. To date, however, it is unclear by which mechanism(s) the mutated and/or missing proteins promote tumorigenesis.

8.4 Further Reading

For more detailed information on the different tumor entities, see [12, 15–21].

References

1. Choy W, Kim W, Nagasawa D, Stramotas S, Yew A, Gopen Q, Parsa AT, Yang I. The molecular genetics and tumor pathogenesis of meningiomas and the future directions of meningioma treatments. Neurosurg Focus. 2011;30(5):E6. doi:10.3171/2011.3172.FOCUS1116.
2. van den Munckhof P, Christiaans I, Kenter SB, Baas F, Hulsebos TJ. Germline SMARCB1 mutation predisposes to multiple meningiomas and schwannomas with preferential location of cranial meningiomas at the falx cerebri. Neurogenetics. 2012;13(1):1–7. doi:10.1007/s10048-10011-10300-y.
3. Di Vinci A, Brigati C, Casciano I, Banelli B, Borzi L, Forlani A, Ravetti GL, et al. HOXA7, 9, and 10 are methylation targets associated with aggressive behavior in meningiomas. Transl Res. 2012;160(5):355–62. doi:10.1016/j.trsl.2012.1005.1007.

4. Kishida Y, Natsume A, Kondo Y, Takeuchi I, An B, Okamoto Y, Shinjo K, et al. Epigenetic subclassification of meningiomas based on genome-wide DNA methylation analyses. Carcinogenesis. 2012;33(2):436–41. doi:10.1093/carcin/bgr1260.
5. Saydam O, Shen Y, Wurdinger T, Senol O, Boke E, James MF, Tannous BA, et al. Downregulated microRNA-200a in meningiomas promotes tumor growth by reducing E-cadherin and activating the Wnt/beta-catenin signaling pathway. Mol Cell Biol. 2009;29(21):5923–40. doi:10.1128/MCB.00332-00309.
6. Shi L, Jiang D, Sun G, Wan Y, Zhang S, Zeng Y, Pan T, Wang Z. miR-335 promotes cell proliferation by directly targeting Rb1 in meningiomas. J Neurooncol. 2012;110(2):155–62. doi:10.1007/s11060-11012-10951-z.
7. Zhi F, Zhou G, Wang S, Shi Y, Peng Y, Shao N, Guan W, et al. A microRNA expression signature predicts meningioma recurrence. Int J Cancer. 2013;132(1):128–36. doi:10.1002/ijc.27658.
8. Laurendeau I, Ferrer M, Garrido D, D'Haene N, Ciavarelli P, Basso A, Vidaud M, Bieche I, Salmon I, Szijan I. Gene expression profiling of the hedgehog signaling pathway in human meningiomas. Mol Med. 2010;16(7–8):262–70. doi:10.2119/molmed.2010.00005.
9. Pham MH, Zada G, Mosich GM, Chen TC, Giannotta SL, Wang K, Mack WJ. Molecular genetics of meningiomas: a systematic review of the current literature and potential basis for future treatment paradigms. Neurosurg Focus. 2011;30(5):E7. doi:10.3171/2011.3172.FOCUS1117.
10. Perez-Magan E, Rodriguez de Lope A, Ribalta T, Ruano Y, Campos-Martin Y, Perez-Bautista G, Garcia JF, et al. Differential expression profiling analyses identifies downregulation of 1p, 6q, and 14q genes and overexpression of 6p histone cluster 1 genes as markers of recurrence in meningiomas. Neuro Oncol. 2010;12(12):1278–90. doi:10.1093/neuonc/noq1081.
11. Paek SH, Kim DG, Park CK, Phi JH, Kim YY, Im SY, Kim JE, Park SH, Jung HW. The role of matrix metalloproteinases and tissue inhibitors of matrix metalloproteinase in microcystic meningiomas. Oncol Rep. 2006;16(1):49–56.
12. Sedani A, Cooper DN, Upadhyaya M. An emerging role for microRNAs in NF1 tumorigenesis. Hum Genomics. 2012;6:23. doi:10.1186/1479-7364-1186-1123.
13. Gutmann DH, Aylsworth A, Carey JC, Korf B, Marks J, Pyeritz RE, Rubenstein A, Viskochil D. The diagnostic evaluation and multidisciplinary management of neurofibromatosis 1 and neurofibromatosis 2. JAMA. 1997;278(1):51–7.
14. Evans DG, Huson SM, Donnai D, Neary W, Blair V, Newton V, Harris R. A clinical study of type 2 neurofibromatosis. Q J Med. 1992;84(304):603–18.
15. Carroll SL. Molecular mechanisms promoting the pathogenesis of Schwann cell neoplasms. Acta Neuropathol. 2012;123(3):321–48.
16. Plotkin SR, Blakeley JO, Evans DG, Hanemann CO, Hulsebos TJ, Hunter-Schaedle K, Kalpana GV, et al. Update from the 2011 International Schwannomatosis Workshop: from genetics to diagnostic criteria. Am J Med Genet A. 2013;161A(3):405–16. doi:10.1002/ajmg.a.35760.
17. Uhlmann EJ, Plotkin SR. Neurofibromatoses. Adv Exp Med Biol. 2012;724:266–77.
18. Zhou L, Hanemann CO. Merlin, a multi-suppressor from cell membrane to the nucleus. FEBS Lett. 2012;586(10):1403–8.
19. Pasmant E, Vidaud M, Vidaud D, Wolkenstein P. Neurofibromatosis type 1: from genotype to phenotype. J Med Genet. 2012;49(8):483–9. doi:10.1136/jmedgenet-2012-100978.
20. Patil S, Chamberlain RS. Neoplasms associated with germline and somatic NF1 gene mutations. Oncologist. 2012;17(1):101–16. doi:10.1634/theoncologist.2010-0181.
21. Rodriguez FJ, Stratakis CA, Evans DG. Genetic predisposition to peripheral nerve neoplasia: diagnostic criteria and pathogenesis of neurofibromatoses, Carney complex, and related syndromes. Acta Neuropathol. 2012;123(3):349–67. doi:10.1007/s00401-00011-00935-00407.

Molecular Carcinogenesis of Uveal Melanoma

9

Werner Wackernagel, Lisa Tarmann, Gerald Langmann, and Arun D. Singh

Contents

W. Wackernagel (✉) • L. Tarmann • G. Langmann
Department of Ophthalmology, Medical University Graz, Graz, Austria
e-mail: werner.wackernagel@medunigraz.at; lisa.tarmann@medunigraz.at; gerald.langmann@medunigraz.at

A.D. Singh
Department of Ophthalmic Oncology, Cleveland Clinic Foundation, Cole Eye Insitute,
e-mail: singha@ccf.org

J. Haybaeck (ed.), *Mechanisms of Molecular Carcinogenesis – Volume 1*,
DOI 10.1007/978-3-319-53659-0_9

Abstract

Melanocytes are derived from the neural crest and enter the eye during embryogenesis. Uveal melanoma of the eye is a rare but deadly disease. About 50% of patients will eventually develop metastatic disease with an inevitable fatal end. Predisposing factors are race, skin and hair color, and familial tumor predisposition syndromes.

Although clinically uveal melanoma phenotype gives the impression of one disease, genetically uveal melanoma can be classified into at least two subgroups which can be distinguished by DNA-based and mRNA-based technologies. While patients with disomy 3 and gene expression profile class 1 have only a low risk of developing metastatic disease, patients with monosomy 3 and/or class 2 gene expression profile are likely to die from metastases. In addition to prognostic information, genetic testing also provides new insights into molecular pathobiology of uveal melanoma. Mutations in GNAQ, GNA11, and BAP1 have been found to be the crucial steps in tumor development.

Those insights raise the hope for targeted therapies and improved prognoses for uveal melanoma patients in the near future.

9.1 Background

Melanocytes are found ubiquitously in the eye and periocular region. It is our current understanding that during embryogenesis pigment cells derived from the neural crest migrate along the nerve sheaths of the trigeminal branch V1 and reach the eye via the branches of the ciliary nerves. Still being melanoblasts, those cells enter the eye close to the optic disc to be distributed throughout the uvea—the choroid, the ciliary body, and the iris—where they mature and become melanocytes [1]. Disturbance of this process of maturation may lead to proliferation of melanocytic cells. They appear benign, round, or oval in shape if the proliferation started after the cells had reached their final destination. The lesion may affect a large sector of the iris and choroid or even the periocular skin if proliferation began early in posteroanterior migration, at a central branch of the trigeminal nerve (e.g., nevus of Ota) [2, 3]. Additional genetic alterations acquired later in life may cause further growth and increased proliferation rate and finally result in the development of a malignant uveal melanoma. Over the past decade, our understanding of the molecular and genetic mechanisms which deregulate the cell cycle of melanocytes and eventually turn nevi into melanomas has increased significantly. Many researchers and research groups contribute to elucidation of the pathogenesis of uveal melanoma, with the ultimate goal of finding a cure for or even prevent this disease.

In the following chapter, we present the current status of research in the molecular carcinogenesis of uveal melanoma, show how the clinical appearance results from molecular and genetic mechanisms, and demonstrate the clinical impact which might result from those. For readers who are not experts in ophthalmic oncology, we will give a brief general overview of the disease, as far as this is necessary for the understanding.

9.2 Epidemiology of Uveal Melanoma

The incidence of uveal melanoma varies among different races and ethnicities over the world. An analysis of the SEER (Surveillance, Epidemiology, and End Results) program database in the United States covering 36 years, from 1973 to 2008, revealed a mean age-adjusted incidence of 5.1 cases per million per year [4]. However, the majority of cases (97.8%) were observed in the white population. The ratio of black population-Asian population-Hispanic population-non-Hispanic white population is 1:1.2:5:19 (SEER program data, 1992–2000) [5]. Whereas the incidence rate was not significantly different between the registries in the United States, an analysis of the data from the European Cancer Registry-based (EUROCARE) study, which combined the data from 67 individual European cancer registries over the years 1983–1994, showed a dependency on latitude. The incidence decreased from north to south, with eight cases per million per year in Scandinavia and only two cases in the south of Europe (Spain and Southern Italy) [6, 7]. In comparison to cutaneous melanoma, with an incidence of 21.8 cases per 100,000 men and women per year, uveal melanoma appears rare (data from SEER Stat Fact Sheets 2016). Only 5% of all melanomas arise from the eye [8]. However, uveal melanoma is still the most common primary intraocular malignancy in adults [9].

9.3 Predisposing Factors

Predisposing factors can help elucidate the genetic mechanisms underlying carcinogenesis. Epidemiological, familial, clinical, and occupational factors contribute to tumor development. The incidence of uveal melanoma increases up to an age of 70 years, and the incidence peaks at 24.5 cases per million males and 17.8 cases per million females per year in the United States, resulting in a mean age at diagnosis of 60 years [6]. Overall, sex does not seem to be a predisposing factor. Data from the EUROCARE study and from the United States demonstrate that race (see the above), skin and eye color, and the ability to tan are the main predisposing factors [4, 7, 10]. This suggests a possible role of UV radiation in the carcinogenesis of uveal melanoma [11–13]. Though this hypothesis is still under discussion and there is no good evidence available, the preferred occurrence of iris melanoma in the inferior half of the iris and the higher incidence of choroidal melanomas at the posterior pole of the eye could be explained by a higher exposure to UV light in those locations. A cohort effect found in the data from the EUROCARE study, with a higher incidence of uveal melanoma in the birth cohorts from 1910 to 1935, might be interpreted that way, as changes in the profiles of work and occupational UV light exposure and the use of sunglasses could explain the decrease in incidence in the younger cohorts [7].

Familial uveal melanoma: Occurrence of uveal melanoma in more than one family member is a rare event, accounting for only 0.6% of patients [14]. However, several cases have been reported, dating back until 1905 [15]. In the published cases, no pattern of inheritance could be found, and only few individuals were affected in each family [14–16]. Other features of a genetic predisposition, such as early age at diagnosis, bilateral involvement, or phenotypic associations were not present [17]. Therefore, it might be possible that two individual family members might be affected by chance alone, though the likelihood of such an occurrence is small [18]. On the other hand, uveal melanoma has been reported to occur more frequently in patients showing oculodermal melanocytosis, familial atypical mole and melanoma syndrome, neurofibromatosis, Li-Fraumeni syndrome, or germline BAP1 mutations, which all have a hereditary and therefore familial background.

Oculodermal melanocytosis: Patients with oculodermal melanocytosis present with congenital increased pigmentation within the distribution of the first and/or second branch of the trigeminal nerve. The affected organs include not only the periocular skin/eyelid, episclera/sclera, uvea, and orbit but also the meninges and tympanic membrane. Patients with oculodermal melanocytosis have a significantly increased risk for uveal melanoma, with an estimated lifetime risk of 1:400 for developing uveal melanoma [19]. It remained unclear whether the increased risk for uveal melanoma results simply from the increased number of melanocytes in the uvea or from an underlying predisposing condition. The large affected sector of the trigeminal nerve points toward an early event in melanogenesis, altering melanoblasts before final maturation [3]. The frequent occurrence of mutations in GNAQ found in intradermal melanocytic proliferations/blue nevi and uveal melanomas appears to drive melanocytic proliferation and might explain the increased risk of malignant transformation [20].

Familial atypical mole and melanoma syndrome (FAM-M): According to the NIH consensus conference, the FAM-M syndrome is diagnosed in individuals with multiple atypical cutaneous nevi, showing distinct histological features and with cutaneous melanoma in one or more first- or second-degree relatives [21]. Dysplastic or atypical nevi of the skin may also occur isolated and are associated with an increased risk of cutaneous melanoma [22]. However, the FAM-M syndrome is a hereditary syndrome, caused by a mutation of the CDKN2A gene, coding for the INK4a and ARF proteins, which regulate the cell cycle and act as tumor suppressors. Germline mutations in CDKN2A are associated with an increased risk of cutaneous melanoma, glioblastoma, and pancreatic cancer [23, 24]. Several case series suggested a role for CDKN2A and the FAM-M syndrome in the development of uveal nevi and melanoma. Uveal nevi and uveal melanomas seem to occur more frequently in patients with FAM-M syndrome. And the FAM-M syndrome has been observed more frequently in patients with uveal melanoma [25–32]. Despite this clinical evidence, the connection between FAM-M is still not fully understood and unproven, as CDKN2A mutations do not seem to play a role in uveal melanoma development and/or progression [33–36].

Neurofibromatosis-1 is a frequent autosomal dominant inherited genetic disorder affecting about 1 in 3000 individuals [37]. Because of the high prevalence, repeated cases of uveal melanoma in NF 1 patients may be coincidental. A connection has been suggested because half of uveal melanomas show reduced expression of NF1 tumor suppressor gene causing an increased activity of Ras and of MAPK activation [38].

Li-Fraumeni syndrome is an autosomal dominant inherited cancer predisposition syndrome caused by a germline p53 mutation [39].

BAP1 mutation: Germline mutations of BAP1 cause a hereditary tumor predisposition syndrome, and affected individuals frequently develop uveal melanoma, lung adenocarcinoma, mesothelioma, and meningioma [40–42]. The spectrum and number of tumors vary between individuals and the affected kindreds, and several cases of familial uveal melanoma have been associated with BAP1 mutations [43, 44]. The reduced penetrance of the germline mutation, however, can impede the identification of those patients as familial cases. Somatic mutations of BAP1 have been identified in metastasizing uveal melanoma and are thought to be a critical step for development of an aggressive tumor phenotype [45]. In case of a germline BAP1 mutation, loss of the wild-type allele of BAP1 on the remaining allele on chromosome 3 will promote melanoma development. The role of BAP1 mutations and the presumed genetic mechanisms are described later in this chapter.

9.4 Prognosis

Uveal melanoma is generally reported to have an overall chance of 50% for spreading to the liver and cause metastatic disease, which is usually fatal within a few months. However, for counselling patients, this information is not very helpful. It has been shown that for most patients, the prognosis is actually much better or much worse and that we need to obtain an individualized risk, by combining clinical, histological, and genetic risk factors [46].

Clinical factors: Older age and male gender have been associated with reduced survival. However, there remains the possibility that those results are biased by a delay in diagnosis in older patients, a higher general mortality rate in older patients (competing risks), and prolonged survival of younger patients with metastatic disease compared to older ones [47–50].

Histopathological factors: Uveal melanoma is commonly classified on cytomorphology according to a classification scheme proposed by Callender in 1931 and modified by McLean in 1983 [51]. Cells are divided into fusiform spindle cells and the larger polygonal and pleomorphic epithelioid cells. Both cell types may be present in the same tumor, which is then classified as "mixed cell type." Epithelioid cell type has been associated with a higher metastasis rate [52]. Problems with the classification lie in the significant inter- and intraobserver variability and the missing consensus on how many epithelioid cells must be present for a melanoma to be classified as "mixed" or "epithelioid" [53]. Cytological tumor heterogeneity here indicates an underlying genetic heterogeneity. It has been shown by several authors that epithelioid cell type and cytogenetic risk factors (monosomy 3) are correlated [54, 55]. As the presence of even a low number of cells with monosomy 3 already causes deterioration of prognosis, agreement of a cutoff level for a histopathological classification appears critical. Recently, after the mutations in the gene encoding BRCA1-associated protein 1 (BAP1) on chromosome 3p21.1 have been identified as critical factor for the development of metastatic tumor phenotype, it was shown that depletion of BAP1 resulted in less differentiated spindle morphology, again drawing a connection between molecular changes and histopathological findings [56]. In addition to epithelioid cell type, the presence of specific extracellular matrix patterns (closed loops; networks) has been found to be associated with poor prognosis [57–59]. Those patterns are best depicted on PAS-stained sections and were initially thought to represent blood vessels but then identified as fluid-conducting channels lacking endothelial cells (vasculogenic mimicry). As cell type, they also correlate with other risk factors for metastasis, reflecting putative molecular changes in the melanoma cells [60]. In addition, increased microvascular density and the presence of tumor infiltrating macrophages and lymphocytes worsen prognosis [61–63]. Today, histopathological risk factors have lost much of their importance for counselling patients about their likely prognosis, as the identification of genetic risk factors has been proven to be superior predictors. However, they are still in use in multivariate mathematical models for prognostication as the inclusion of numerous predictors enables the model to correct for errors in even genetic results to some extent [46, 60].

Tumor parameters: Tumor dimensions, especially the largest tumor basal diameter (LBD), are strongly correlated with patient survival. While less than 5% of melanomas with a LBD less than 10 mm will be fatal within 5 years from diagnosis, 5-year mortality rises to over 50% in melanomas with a LBD over 15 mm [64, 65]. LBD remains an independent risk factor adding prognostic information even in addition to gene expression profiling (GEP) [66]. However, we do not know whether metastasis happens because the tumor has grown large and has had more opportunity to spread because it has been there for a longer time or whether large tumor size is simply an indicator of a greater growth rate [46]. Tumor height appears to be a less important parameter for patient survival. Tumor location, on the other hand, has long been recognized as an important predictor of metastasis. Ciliary body involvement is associated with reduced survival; however, it correlates with larger tumor size and it did not remain an independent risk factor for metastatic disease when information from GEP was available [53, 66, 67]. Melanomas of the iris, in contrast, have a favorable prognosis and a much lower mortality rate [68, 69]. At the time of diagnosis, iris melanomas are usually smaller than choroidal or ciliary body melanomas, and they show less aggressive histological features [53]. Cytogenetic changes with known unfavorable impact could be demonstrated in iris melanoma as well. Due to the small number of cases, however, the impact on survival has not been analyzed in detail [70].

9.5 Cytogenetics

Since the first report on cytogenetic changes in uveal melanoma more than 30 years ago, cytogenetic testing has gradually evolved from a research tool to a routine clinical test in the management of uveal melanoma [71]. Twenty years ago, Prescher et al. demonstrated the prognostic significance of cytogenetic abnormalities (i.e., monosomy 3), which was shown to be far superior to any clinical or histopathological marker [72]. Since then, the techniques to identify chromosomal anomalies have evolved and replaced chromosome G-banding. Today, fluorescence in situ hybridization (FISH), spectral karyotyping (SKY), (array) comparative genomic hybridization (a-CGH), and multiplex ligation-dependent probe amplification (MLPA) are in routine use for predictive testing and have been evaluated in large patient series [73, 74]. Microsatellite analysis (MSA) and single-nucleotide polymorphism (SNP) arrays offer the additional opportunity to identify loss of heterogeneity (LOH) and detect isodisomy of chromosomes which might be missed by the other techniques. Using SNP data, an attempt at creating an evolutionary tree for uveal melanoma has been published [75].

9.5.1 Chromosome 3

The loss of one copy of chromosome 3 (monosomy 3) is the most frequent and prognostically most important chromosomal aberration in uveal melanoma and can be found in 50–61% of tumors [72, 76–81]. In 1996, a dramatic reduction in the 3-year survival probability from approximately 100% to less than 50% was reported for the first time in patients with monosomy 3 melanoma [72]. Since then, this finding has been confirmed by other authors several times [46, 55, 60, 82]. Only about 5–20% of patients with disomy 3 melanoma are expected to eventually develop metastatic disease [83]. Several possible explanations why metastasis occurs in a disomy 3 tumor have been discussed. First, intratumoral heterogeneity may lead to sampling errors if only a single small biopsy is used for analysis. The risk for misclassification with fine needle aspiration biopsy (FNAB) has been estimated to be less than 1% and however has to be considered [78, 84]. Second, e.g., FISH, CGH, and MLPA cannot detect loss of heterozygosity (LOH) caused by isodisomy and give the impression of metastasis occurring without monosomy 3. Partial deletions of chromosome 3 might also be missed by FISH. However, the prognostic significance of partial deletions is still unclear. Identified deletions affected regions of the short arm (3p11–3p14, 3p25–3p26, 3p25.1–3p25.2), as well as smaller regions on 3q (3q13–3q21 and 3q24–3q26) [85–88]. A critical region of deletion causing metastatic disease has not been identified, until Harbour et al. identified mutations of BAP1 gene located on chromosome 3p21.1 in the majority of metastasizing uveal melanomas by exome sequencing [45]. Loss of chromosome 3 seems to be an early event in development of uveal melanoma and can be found in combination with other chromosomal aberrations [89, 90]. Unmasking of the second allele with mutated BAP1 induces tumor progression and also determines an aggressive tumor phenotype [56]. It is therefore not surprising that the loss of chromosome 3 is associated with other predictors of a poor prognosis, such as increased tumor diameter, ciliary body involvement, epithelioid cell type, and extravascular matrix patterns [55, 72, 91, 92].

9.5.2 Chromosome 8

Aberrations of chromosome 8 may affect both the short and the long arm. Depending on the technique used, gains of 8q can be found in 35–75% and loss of 8p in 15–30%

[76, 93]. Gains of 8q and losses of 8p may occur together by formation of an isochromosome [90, 94, 95].

As for chromosome 3 changes of chromosome 8 are also associated with poor prognosis. Combining information on chromosome 3 and on chromosome 8 status improves the accuracy of prediction of metastatic disease compared to monosomy 3 or chromosome 8 status alone [46, 60]. Monosomy 3 and 8q gains occur together in about 45% of tumors and are associated with large tumor size, ciliary body involvement, and aggressive histology [80, 82]. In addition, chromosome 8 abnormalities are found in virtually all metastases from uveal melanoma, either to the liver or the brain [71, 94, 96]. Because of this and the variable copy number of 8q in one tumor, gains of chromosome 8 are thought to be a secondary event in uveal melanoma development. Improved analysis technique showed 8q amplification to be far more common than previously thought and high-resolution CGH identified frequent partial deletions on chromosome 8 [93]. Gains could be localized to 8q23–8q24 in many cases, and several oncogenes in that region have been evaluated for the potential significance. MYC, located on chromosome 8q24 and coding for a transcription factor, DDEF1 (development and differentiation factor 1), enhancing the motility of uveal melanoma cells, and NBS1 (Nijmegen breakage syndrome 1) were all analyzed for their possible role in promoting metastasis [97–100]. Though frequently overexpressed in uveal melanoma, an association to prognosis could not be established. ENPP2, also located on 8q24 and coding for autotaxin or ectonucleotide pyrophosphatase/phosphodiesterase family member 2 (ENPP2), has been identified by gene expression profiling as a predictor of survival [101]. Autotaxin is an enzyme producing lysophosphatidic acid (LPA), a small motility-enhancing and angiogenic lipid molecule [102]. Its role for tumor progression and prognosis in uveal melanoma still needs to be established.

Losses of chromosome 8p have also been analyzed in detail to identify possible tumor suppressor genes. Deletions could be localized on 8p12–8p22 by array CGH, and silencing of LZTS1 on the retained hemizygous allele was found [103]. This tumor suppressor genes code for the leucine zipper tumor suppressor-1 and inhibit motility and invasion of uveal melanoma in cell cultures. Interestingly, loss of 8p correlates with the prognostically unfavorable class 2B on gene expression profiling, further emphasizing the importance of 8p loss [89, 104].

9.5.3 Chromosome 6

Gains of chromosome 6p are observed in 28–54% of uveal melanomas and losses of 6q in 35–37% [76, 105]. Gains of 6p seem to be associated with a good prognosis. Possible explanations are that 6p gains are preferably found in spindle cell melanomas and rarely occur together with monosomy 3 [90, 106, 107]. Gains of 6p also correlate with gene expression profile subclass 1b, which is associated with a good prognosis. The protective effect of 6p gain might result from an alternative molecular pathway in tumor development dividing UM in two separate subgroups with good (6p gain) or worse (monosomy 3) prognosis. Loss of 6q, on the other hand, has been associated with a metastasizing phenotype. The effect of simultaneous 6p gain and 6q loss caused by chromosome rearrangements is unknown [77, 80].

Some authors have tried to identify tumor suppressor genes and oncogenes on chromosome 6; however, the genetic mechanism underlying chromosome 6 abnormalities remains unclear [108–110].

9.5.4 Chromosome 1

About a quarter of uveal melanomas show deletion on the short arm of chromosome 1, indicating a poorer prognosis. Because deletions of 1p are frequently found in larger tumors and in association with monosomy 3 and changes of chromosome 8, they are thought to develop later in tumor progression [78, 80, 84, 94, 111, 112]. However, 1p loss adds prognostic information independent of chromosome 3 status [112].

Several genetic mechanisms could be affected by 1p deletion, e.g., the NOTCH pathway and TP73 [113].

9.6 Molecular Genetics

9.6.1 MYC

The MYC gene is located on chromosome 8q24, a region frequently amplified in uveal melanoma, and therefore has been supposed to play a role in tumor progression [76, 90, 94, 95]. It codes for a transcription factor regulating the expression of numerous genes, controlling the cell cycle and apoptosis, and modifies the chromatin structure via histone acetyltransferases (HATs) [114, 115]. Many tumors show constitutive overexpression of MYC, which has also been identified in uveal melanoma and proposed as a potential prognostic marker [100, 116]. In cutaneous melanoma, overexpression of MYC seems to be associated with poor prognosis in cutaneous melanoma [117]. In uveal myeloma MYC is also frequently overexpressed but surprisingly associated with improved prognosis [97, 100]. Hence, the role of MYC overexpression in uveal melanoma remains unclear.

9.6.2 TGF-b

TGF-b1 is located on chromosome 19, a region without frequent structural abnormalities in uveal melanoma. It has become of interest, because intraocular melanomas are growing in an immune privileged location, and the immunosuppressive properties inside the eye are mediated by cytokines, such as TGF-b [118]. TGF-b has antiproliferative and anti-apoptotic effects on various cell types (epithelial, endothelial, neuronal, leukocytes) [119]. TGF-b is secreted into the extracellular matrix and stored until its activation by numerous often unspecific stimuli, like as irradiation, reactive oxygen, and proteases such as plasmin and metalloproteinases [120, 121]. Metalloproteinases (MMP) have been of interest as a marker for tumor progression as they are involved in tumor invasion, angiogenesis, and metastasis, and TGF-b enhance and modify these effects [122, 123]. Whereas MMP-9 is predominantly expressed in epithelioid melanomas and associated with poorer prognosis, immunohistochemical staining showed TGF-b positivity in uveal melanoma regardless of cell type, tumor size, or location [118]. TGF might also play a direct role in hematogenous metastasis to the liver, by increasing adhesion of uveal melanoma to the hepatic endothelium [124]. For metastases to grow to detectable size, induction of angiogenesis is a crucial step in tumor progression [125]. By its angiogenic effects, TGF-b could further promote growth of uveal melanoma metastases, and expression of TGF-b receptor endoglin has been found to correlate with metastatic death [126, 127]. A high number of tumor infiltrating leukocytes and macrophages as well as high HLA class I and II expression have also been linked with bad prognosis, as those are preferably found in monosomy 3 melanomas [128]. By downregulation of MHC class I antigen, on the other hand, TGF-b renders uveal melanoma cells more susceptible to cytolysis by natural killer cells [129, 130]. It has been suggested that uveal melanoma might prepare its own microenvironment for growth by secretion of local factors into the vitreous and aqueous humor [131].

9.6.3 Bcl-2

The Bcl family of proto-oncogenes comprises pro-apoptotic proteins, like Bax, Bad, and Bak, and anti-apoptotic proteins, like Bcl-2, Bcl-xL, and Bcl-w [132]. Blc-2 seems to be of specific importance in regulation of tumor cell survival and apoptosis. As the pro-apoptotic members, Bcl-2 resides on the outer membrane of the mitochondria, inhibiting the initiation of the apoptotic cascade by the intrinsic pathway [133, 134]. Bcl-2 is strongly expressed in uveal melanoma, as shown by immunohistochemistry [97, 135–137]. Uveal melanoma is known to be resistant against radiation, and high radiation dose has to be used for the treatment of primary tumor. It is also resistant to standard chemotherapy. This is explained by the anti-apoptotic effects of Bcl-2. Therefore Bcl-2 has become a possible starting point in the search for a targeted therapy [138, 139]. Bcl-2 inhibitors could revert the inhibition of apoptosis and allow for the initiation of apoptosis by pro-apoptotic members of the Bcl family. Nemati et al. showed increased response of uveal melanoma to chemotherapy with fotemustine after the administration of a Bcl-2 inhibitor in animal xenografts [140]. Until recently no Bcl inhibitor with acceptable toxicity was available for use in humans [141, 142]. However, novel Bcl-2 inhibitors with low toxicity have been approved for use in chronic lymphocytic leukemia (CLL) with 17p deletion (affecting p53), and clinical testing in uveal melanoma is expected.

9.6.4 P53

P53 plays an important role as a tumor suppressor, inactivated in more than 50% of all tumors. The gene for p53 is located on chromosome 17p13.1. It is a transcription factor controlling the expression of other downstream genes to regulate cell proliferation and growth. Upregulation of p53 has been described in uveal melanoma after irradiation [143, 144]. This appears plausible as p53 is a main factor in the cellular response to stress and DNA damage. Until DNA repair is completed, the cell is stopped in the G1 or G2 phase of the cell cycle [145]. If the damage is too severe to be repaired, the cell will undergo apoptosis. In contrast to other tumors, which frequently show altered expression or mutations of p53, the signaling pathway upstream of p53 seems to be intact in uveal melanoma [146]. However, there seems to be a functional inhibition of p53 in uveal melanoma. P53 interacts with several downstream molecules in a negative-feedback loop [147]. First, the cyclin-dependent kinase inhibitor CDKN2A stabilizes p53 by degrading Mdm2 through p14(ARF) [148, 149]. Second, it induces the expression of Hdm2/Mdm2 (human/mouse double minute 2). Overexpressed Hdm2/Mdm2 in reverse represses p53 transcriptional activity and also enhances its degradation [150–152]. This mechanism seems to be important in uveal melanoma, and overexpression of Mdm2 has been shown to be of prognostic value [143, 153]. Some authors evaluated the possible therapeutic effect of inhibition of Mdm2 by a small synthetic peptide in animal models; however, this has never been evaluated in clinical trials [141, 154, 155].

9.6.5 Rb

The Rb pathway has been shown to be frequently altered and functionally inhibited in uveal melanoma [143, 156–158]. To arrest cells in the G1 or G1/S phase, the Rb protein has to be kept in a hypophosphorylated state. This is accomplished by the cyclin-dependent kinase inhibitor CDKN2A coding for p16(INK)4a, keeping Rb active. In this hypophosphorylated state, Rb binds E2Fs, thereby repressing their transcriptional activity. Phosphorylation of Rb by cyclin-dependent kinases (CDK) can occur at multiple phosphorylation sites in Rb throughout the cell cycle, which gradually inhibits Rb function and releases E2F [159, 160]. Loss or inactivation of Rb will result in deregulated cell cycle progression and cell proliferation. In uveal melanoma Rb protein is frequently hyperphosphorylated and inactivated because of cyclin D1 overexpression and CDKN2A promotor methylation [143]. Still, most tumors, including retinoblastoma, need additional mutations of other tumor suppressor genes to develop [161].

9.6.6 Ras-Raf-MAPK Pathway (GNAQ, GNA11, CYSLTR2, PLCB4)

Mutations of proteins in the mitogen-activated protein kinase pathway (MAPK) have long been implicated in the development of conjunctival melanomas [162–164]. BRAF mutations are found in more than 65% of cutaneous melanoma, resulting in a constitutively active protein leading to MAPK activation [165]. Such BRAF mutations are rarely observed, on the other hand, in uveal melanoma [166–169]. Activation of the MAPK pathway has been demonstrated to be a frequent event in UM; however, this does not seem to occur through mutations of BRAF [170]. The mechanisms behind the activation remained obscure until the discovery of mutations in GNAQ and GNA11 in uveal melanocytic lesions. A connection between mutations in GNAQ/GNA11 and proliferation of melanocytes was first established in 2004 during the analysis of determinants of skin color in mice [171]. Thereupon those mutations were found in almost 85% of uveal melanomas [20, 172, 173]. GNAQ and GNA11 encode the alpha subunit of heterotrimeric g-proteins (g-alpha-q and g-alpha-11). G-protein-coupled receptors (GPCR) are a large group of transmembrane receptors exerting multiple physiologic functions, such as in the sensory system, vision, olfaction, and taste [174]. Their function is transmitted by g-proteins from the extracellular to intracellular space. In their basal and inactive state, the alpha, beta, and gamma subunits are bound together with GDP. Upon activation through ligand binding to the g-protein-coupled receptor, the heterotrimeric g-protein dissociates and exchanges GDP to GTP [175]. The GTPase activity of the alpha subunit hydrolysis GTP to GDP. Mutations in the alpha subunit of GNAQ and GNA11 keep them in the activated state, leading to constitutive MAP-kinase pathway activation and cell proliferation without further extracellular stimuli. This raised the hope for targeted therapy with MEK inhibitors. In fact, a randomized open-label phase II clinical trial comparing selumetinib versus chemotherapy showed for the first time prolonged progression-free survival in patients with uveal melanoma metastasis [176]. However, this has not been confirmed in later studies [177].

In patients without detectable mutations in GNAQ/GNA11 of the g-protein-coupled receptor CYSLTR2 and PLCB4, the gene encoding for the 1-phosphatidyl inositol-4,5-bisphosphate phosphodiesterase beta-4, a downstream target of GNAQ/GNA11, could be identified, further emphasizing the importance of MAP-kinase pathway activation through this mechanism [178, 179].

9.6.7 BAP1

Enormous efforts have been made to identify the critical mutation on chromosome 3 that promotes metastasis. Harbour et al. identified mutations of BAP1, located on chromosome 3p21.1, in 47% of metastasizing monosomy 3 uveal melanomas [45]. Interestingly, one of their patients had a germline mutation which was uncovered by loss of the second allele (monosomy 3). Soon after germline BAP1 mutations were found to cause a hereditary tumor predisposition syndrome, the spectrum of this is still evolving [40, 41, 180]. The BRCA1-associated protein-1 (BAP1) was first described by Jensen et al. as a potential tumor suppressor gene [181, 182]. BAP1 encodes a deubiquitinating enzyme and acts on BRCA1, histone H2A, host cell factor-1, and O-linked N-acetylglucosamine transferase (OGT) [181]. RNAi-mediated depletion of BAP1 in uveal melanoma cells resulted in loss of differentiation and gain of stemlike properties, similar to GEP class 2 tumors [56]. Though the molecular mechanisms of BAP1 are still not fully understood, a targeted therapy seems possible. Histone deacetylase (HDAC) inhibitors might restore histone H2A function which has been shown to be accompanied by increased melanocytic differentiation [183, 184].

9.6.8 SF3B1 and EIF1AX

Recently, additional somatic mutations have been identified in UM by exon sequencing, specifically occurring in patients with disomy 3 and partial monosomy 3 [185, 186]. SF3B1 is found in about 15–20% of uveal melanomas and associated with a favorable prognosis and prolonged metastasis-free interval compared to uveal melanomas with BAP1 mutations [186–189]. Mutations were mutually exclusive with BAP1 mutations and seem to support the bifurcated model of tumor progression in uveal melanoma [108, 190]. However, within the disomy 3 and partial monosomy 3 tumors, uveal melanomas with SF3B1 mutations had a worse prognosis than those without this mutation [189]. Interestingly, the mutational spectrum in SF3B1 of tumors with and without metastasis was different, and a further subgrouping of tumors according to mutation type might help to identify patients at risk of metastatic disease [186]. Tumors harboring EIF1AX mutations also are associated with a longer disease-free survival and rarely demonstrate metastasis [189]. Both mutations EIF1AX and SF3B1 are associated with good prognostic features, such as disomy 3; spindle cell type, positive BAP1 immunohistochemistry staining; and the absence of closed vascular loops [191].

9.6.9 Gene Expression Profiling (GEP)

While over the past decades methods for detection of cytogenetic changes and mutations have evolved and allowed for more detailed analyses, some other groups chose another approach offering a different view of the tumor and its microenvironment. Gene expression profiling (GEP), based on mRNA signature, has been described as a functional snapshot of the tumor microenvironment, which is perhaps less variable across the tumor and therefore less affected by intratumoral heterogeneity [192]. GEP was studied first in uveal melanoma cell lines, comparing those to normal melanocytes [193]. Then, Tschentscher et al. developed oligonucleotide microarrays to describe and compare the expression profile of monosomy and disomy 3 melanomas [194]. They also classified uveal melanomas according to their gene expression profile.

Soon afterwards, Onken et al. improved this technique and, leaving chromosomal status behind, classified uveal melanomas based on results from GEP, using unsupervised clustering methods to divide uveal melanomas in class 1 melanomas with low risk and class 2 tumors with a high risk of metastasis. Since then, numerous other investigators have compared the accuracy of GEP versus cytogenetic prognostic indicators. Unfortunately, as DNA-based techniques continuously evolved (MLPA, a-CGH, and SNP array), no direct comparison of GEP classification versus latest DNA-based techniques has been made. Onken et al. reduced the number of genes necessary for classification and developed a PCR-based microfluidics platform for routine clinical testing [192, 195].

Gene expression profiling closely correlates with cytogenetic findings in most cases [196, 197]. Class 1 tumors usually show disomy 3 and spindle cell type melanoma, whereas class 2 profile is associated with epithelioid cell type, monosomy 3. A further subclassification is possible into four groups (1A, 1B,

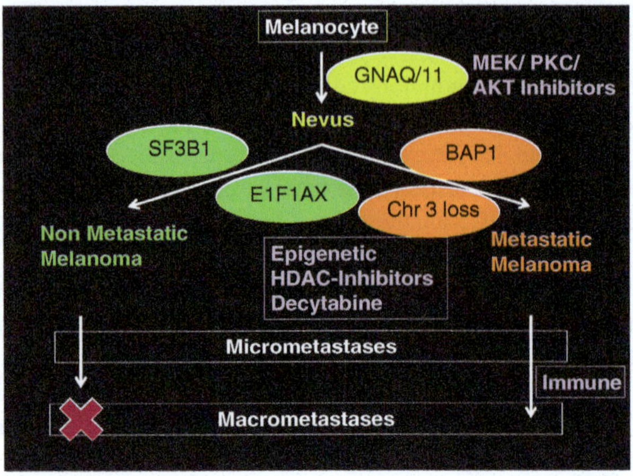

Fig. 9.1 Proposed bifurcated tumor progression model. GNAQ/GNA11 mutations induce growth of uveal nevi, which subsequently evolve into melanoma by acquiring (**a**) mutations in SF3B1 and/or EIF1AX, associated with a rather good prognosis, or (**b**) mutations in BAP1 and loss of the second allele by monosomy 3, which is associated with a high risk for metastasis. A more advanced model of an evolutionary tree for uveal melanoma was published by Nakul Singh et al. [75]

and 2A, 2B). Class 1A predicts the best prognosis, and class 1B is slightly worse, corresponding to disomy 3 melanomas with 6p gain. For class 2 melanomas, additional loss of chromosome 8p corresponds to the subclass 2B with the worst prognosis [89]. Based on this, a modified bifurcated tumor progression model has been proposed which is shown in Fig. 9.1. Some problems, however, remain. First, the significance and predictive value of GEP after radiation has not been evaluated so far and, as in contrast, has been done for DNA-based tests [198–201]. Also biopsy of non-melanoma tissue will reveal class 1 profile. For DNA-based tests, identification of GNAQ/GNA11 mutations can easily be done which will be positive in 85% of cases and ensure that melanocytic tissue has been biopsied.

9.7 Conclusion and Future Prospective

New genetic techniques have provided new and sometimes unsuspected insights into the pathobiology and the molecular carcinogenesis of uveal melanoma. Extended use of next-generation sequencing and proteomics will provide further insights, and targeted therapies will hopefully lead to an improved prognosis for patients with metastatic disease. Still, one should not forget to consider and reconsider again basic facts of melanogenesis in the light of latest findings as explained in the introduction of this chapter. It is still unknown whether choroidal nevi are congenital and slowly grow to clinically detectable size during life or whether they arise from normal melanocytes, which acquire their first transforming event (mutation of GNAQ/GNA11) later in life.

Findings from clinical syndromes like the oculodermal melanocytosis still pose many questions to be answered. Unifying concepts of pigment cell distribution and ocular melanogenesis may answer some questions but even more arise.

References

1. Schwab C, Wackernagel W, Grinninger P, Mayer C, Schwab K, Langmann G, Richtig E, Wedrich A, Hofmann-Wellenhof R, Zalaudek I. A unifying concept of uveal pigment cell distribution and dissemination based on an animal model: insights into ocular melanogenesis. Cells Tissues Organs. 2016;201:232–8.
2. Saida T. Histogenesis of congenital and acquired melanocytic nevi: a unifying concept. Am J Dermatopathol. 2006;28:377–9.
3. Schwab C, Zalaudek I, Mayer C, Riedl R, Wackernagel W, Juch H, Aigner B, Brunasso AM, Langmann G, Richtig E. New insights into oculodermal nevogenesis and proposal for a new iris nevus classification. Br J Ophthalmol. 2015;99:644–9.
4. Singh AD, Turell ME, Topham AK. Uveal melanoma: trends in incidence, treatment, and survival. Ophthalmology. 2011;118:1881–5.
5. Hu DN, Yu GP, Mccormick SA, Schneider S, Finger PT. Population-based incidence of uveal melanoma in various races and ethnic groups. Am J Ophthalmol. 2005;140:612–7.
6. Singh AD, Topham A. Incidence of uveal melanoma in the United States: 1973-1997. Ophthalmology. 2003;110:956–61.
7. Virgili G, Gatta G, Ciccolallo L, Capocaccia R, Biggeri A, Crocetti E, Lutz JM, Paci E, Group EW. Incidence of uveal melanoma in Europe. Ophthalmology. 2007;114:2309–15.
8. Chang AE, Karnell LH, Menck HR. The National Cancer Data Base report on cutaneous and noncutaneous melanoma: a summary of 84,836 cases from the past decade. Cancer. 1998;83:1664–78.
9. Egan KM, Seddon JM, Glynn RJ, Gragoudas ES, Albert DM. Epidemiologic aspects of uveal melanoma. Surv Ophthalmol. 1988;32:239–51.
10. Margo CE, Mulla Z, Billiris K. Incidence of surgically treated uveal melanoma by race and ethnicity. Ophthalmology. 1998;105:1087–90.
11. Marshall JC, Gordon KD, Mccauley CS, De Souza Filho JP, Burnier MN. The effect of blue light exposure and use of intraocular lenses on human uveal melanoma cell lines. Melanoma Res. 2006;16:537–41.
12. Singh AD, Rennie IG, Seregard S, Giblin M, Mckenzie J. Sunlight exposure and pathogenesis of uveal melanoma. Surv Ophthalmol. 2004;49:419–28.
13. Vajdic CM, Kricker A, Giblin M, Mckenzie J, Aitken J, Giles GG, Armstrong BK. Sun exposure predicts risk of ocular melanoma in Australia. Int J Cancer. 2002;101:175–82.
14. Singh AD, Shields CL, De Potter P, Shields JA, Trock B, Cater J, Pastore D. Familial uveal melanoma. Clinical observations on 56 patients. Arch Ophthalmol. 1996a;114:392–9.
15. Jay M, Mccartney AC. Familial malignant melanoma of the uvea and p53: a Victorian detective story. Surv Ophthalmol. 1993;37:457–62.
16. Canning CR, Hungerford J. Familial uveal melanoma. Br J Ophthalmol. 1988;72:241–3.
17. Singh AD, Wang MX, Donoso LA, Shields CL, Potter PD, Shields JA, Elston RC, Fijal B. Familial uveal melanoma, III. Is the occurrence of familial uveal melanoma coincidental? Arch Ophthalmol. 1996b;114:1101–4.
18. Singh AD, Demirci H, Shields CL, Shields JA, Smith AF. Concurrent choroidal melanoma in son and father. Am J Ophthalmol. 2000;130:679–80.
19. Singh AD, De Potter P, Fijal BA, Shields CL, Shields JA, Elston RC. Lifetime prevalence of uveal melanoma in white patients with oculo(dermal) melanocytosis. Ophthalmology. 1998;105:195–8.
20. Van Raamsdonk CD, Bezrookove V, Green G, Bauer J, Gaugler L, O'brien JM, Simpson EM, Barsh GS, Bastian BC. Frequent somatic mutations of GNAQ in uveal melanoma and blue naevi. Nature. 2009;457:599–602.

21. NIH Consensus conference. Diagnosis and treatment of early melanoma. JAMA. 1992;268:1314–9.
22. Naeyaert JM, Brochez L. Clinical practice. Dysplastic nevi. N Engl J Med. 2003;349:2233–40.
23. Feng J, Kim ST, Liu W, Kim JW, Zhang Z, Zhu Y, Berens M, Sun J, Xu J. An integrated analysis of germline and somatic, genetic and epigenetic alterations at 9p21.3 in glioblastoma. Cancer. 2012;118:232–40.
24. Soura E, Eliades PJ, Shannon K, Stratigos AJ, Tsao H. Hereditary melanoma: update on syndromes and management: genetics of familial atypical multiple mole melanoma syndrome. J Am Acad Dermatol, 2016; 74: 395–407; quiz 408–10.
25. Abramson DH, Rodriguez-Sains RS, Rubman R. B-K mole syndrome. Cutaneous and ocular malignant melanoma. Arch Ophthalmol. 1980;98:1397–9.
26. Bellet RE, Shields JA, Soll DB, Bernardino EA. Primary choroidal and cutaneous melanomas occurring in a patient with the B-K mole syndrome phenotype. Am J Ophthalmol. 1980;89:567–70.
27. Richtig E, Langmann G, Mullner K, Smolle J. Ocular melanoma: epidemiology, clinical presentation and relationship with dysplastic nevi. Ophthalmologica. 2004;218:111–4.
28. Rodriguez-Sains RS. Ocular findings in patients with dysplastic nevus syndrome. Ophthalmology. 1986;93:661–5.
29. Rodriguez-Sains RS. Ocular findings in patients with dysplastic nevus syndrome. An update. Dermatol Clin. 1991;9:723–8.
30. Singh AD, Shields CL, Shields JA, Eagle RC, De Potter P. Uveal melanoma and familial atypical mole and melanoma (FAM-M) syndrome. Ophthalmic Genet. 1995;16:53–61.
31. Smith JH, Padnick-Silver L, Newlin A, Rhodes K, Rubinstein WS. Genetic study of familial uveal melanoma: association of uveal and cutaneous melanoma with cutaneous and ocular nevi. Ophthalmology. 2007;114:774–9.
32. Toth-Molnar E, Olah J, Dobozy A, Hammer H. Ocular pigmented findings in patients with dysplastic naevus syndrome. Melanoma Res. 2004;14:43–7.
33. Abdel-Rahman MH, Pilarski R, Massengill JB, Christopher BN, Noss R, Davidorf FH. Melanoma candidate genes CDKN2A/p16/INK4A, p14ARF, and CDK4 sequencing in patients with uveal melanoma with relative high-risk for hereditary cancer predisposition. Melanoma Res. 2011b;21:175–9.
34. Buecher B, Gauthier-Villars M, Desjardins L, Lumbroso-Le Rouic L, Levy C, De Pauw A, Bombled J, Tirapo C, Houdayer C, Bressac-De Paillerets B, Stoppa-Lyonnet D. Contribution of CDKN2A/P16 (INK4A), P14 (ARF), CDK4 and BRCA1/2 germline mutations in individuals with suspected genetic predisposition to uveal melanoma. Fam Cancer. 2010;9:663–7.
35. Mouriaux F, Maurage CA, Labalette P, Sablonniere B, Malecaze F, Darbon JM. Cyclin-dependent kinase inhibitory protein expression in human choroidal melanoma tumors. Invest Ophthalmol Vis Sci. 2000;41:2837–43.
36. Zeschnigk M, Tschentscher F, Lich C, Brandt B, Horsthemke B, Lohmann DR. Methylation analysis of several tumour suppressor genes shows a low frequency of methylation of CDKN2A and RARB in uveal melanomas. Comp Funct Genomics. 2003;4:329–36.
37. Friedman JM. Epidemiology of neurofibromatosis type 1. Am J Med Genet. 1999;89:1–6.
38. Foster WJ, Fuller CE, Perry A, Harbour JW. Status of the NF1 tumor suppressor locus in uveal melanoma. Arch Ophthalmol. 2003;121:1311–5.
39. Varley JM. Germline TP53 mutations and Li-Fraumeni syndrome. Hum Mutat. 2003;21:313–20.
40. Abdel-Rahman MH, Pilarski R, Cebulla CM, Massengill JB, Christopher BN, Boru G, Hovland P, Davidorf FH. Germline BAP1 mutation predisposes to uveal melanoma, lung adenocarcinoma, meningioma, and other cancers. J Med Genet. 2011a;48:856–9.
41. Wiesner T, Fried I, Ulz P, Stacher E, Popper H, Murali R, Kutzner H, Lax S, Smolle-Juttner F, Geigl JB, Speicher MR. Toward an improved definition of the tumor spectrum associated with BAP1 germline mutations. J Clin Oncol. 2012;30:e337–40.
42. Wiesner T, Obenauf AC, Murali R, Fried I, Griewank KG, Ulz P, Windpassinger C, Wackernagel W, Loy S, Wolf I, Viale A, Lash AE, Pirun M, Socci ND, Rutten A, Palmedo G, Abramson D,

Offit K, Ott A, Becker JC, Cerroni L, Kutzner H, Bastian BC, Speicher MR. Germline mutations in BAP1 predispose to melanocytic tumors. Nat Genet. 2011;43:1018–21.
43. Hoiom V, Edsgard D, Helgadottir H, Eriksson H, All-Ericsson C, Tuominen R, Ivanova I, Lundeberg J, Emanuelsson O, Hansson J. Hereditary uveal melanoma: a report of a germline mutation in BAP1. Genes Chromosomes Cancer. 2013;52:378–84.
44. Maerker DA, Zeschnigk M, Nelles J, Lohmann DR, Worm K, Bosserhoff AK, Krupar R, Jagle H. BAP1 germline mutation in two first grade family members with uveal melanoma. Br J Ophthalmol. 2014;98:224–7.
45. Harbour JW, Onken MD, Roberson ED, Duan S, Cao L, Worley LA, Council ML, Matatall KA, Helms C, Bowcock AM. Frequent mutation of BAP1 in metastasizing uveal melanomas. Science. 2010;330:1410–3.
46. Damato B, Eleuteri A, Taktak AF, Coupland SE. Estimating prognosis for survival after treatment of choroidal melanoma. Prog Retin Eye Res. 2011;30:285–95.
47. Damato BE, Heimann H, Kalirai H, Coupland SE. Age, survival predictors, and metastatic death in patients with choroidal melanoma: tentative evidence of a therapeutic effect on survival. JAMA Ophthalmol. 2014;132:605–13.
48. Kujala E, Makitie T, Kivela T. Very long-term prognosis of patients with malignant uveal melanoma. Invest Ophthalmol Vis Sci. 2003;44:4651–9.
49. Rietschel P, Panageas KS, Hanlon C, Patel A, Abramson DH, Chapman PB. Variates of survival in metastatic uveal melanoma. J Clin Oncol. 2005;23:8076–80.
50. Virgili G, Gatta G, Ciccolallo L, Capocaccia R, Biggeri A, Crocetti E, Lutz JM, Paci E, Group EW. Survival in patients with uveal melanoma in Europe. Arch Ophthalmol. 2008;126:1413–8.
51. Mclean IW, Foster WD, Zimmerman LE, Gamel JW. Modifications of Callender's classification of uveal melanoma at the Armed Forces Institute of Pathology. Am J Ophthalmol. 1983;96:502–9.
52. Mclean IW, Foster WD, Zimmerman LE. Uveal melanoma: location, size, cell type, and enucleation as risk factors in metastasis. Hum Pathol. 1982;13:123–32.
53. Mudhar HS, Parsons MA, Sisley K, Rundle P, Singh A, Rennie IG. A critical appraisal of the prognostic and predictive factors for uveal malignant melanoma. Histopathology. 2004;45:1–12.
54. Bronkhorst IH, Maat W, Jordanova ES, Kroes WG, Schalij-Delfos NE, Luyten GP, Jager MJ. Effect of heterogeneous distribution of monosomy 3 on prognosis in uveal melanoma. Arch Pathol Lab Med. 2011b;135:1042–7.
55. Scholes AG, Damato BE, Nunn J, Hiscott P, Grierson I, Field JK. Monosomy 3 in uveal melanoma: correlation with clinical and histologic predictors of survival. Invest Ophthalmol Vis Sci. 2003;44:1008–11.
56. Matatall KA, Agapova OA, Onken MD, Worley LA, Bowcock AM, Harbour JW. BAP1 deficiency causes loss of melanocytic cell identity in uveal melanoma. BMC Cancer. 2013;13:371.
57. Folberg R, Rummelt V, Parys-Van Ginderdeuren R, Hwang T, Woolson RF, Pe'er J, Gruman LM. The prognostic value of tumor blood vessel morphology in primary uveal melanoma. Ophthalmology. 1993;100:1389–98.
58. Makitie T, Summanen P, Tarkkanen A, Kivela T. Microvascular loops and networks as prognostic indicators in choroidal and ciliary body melanomas. J Natl Cancer Inst. 1999b;91:359–67.
59. Maniotis AJ, Folberg R, Hess A, Seftor EA, Gardner LM, Pe'er J, Trent JM, Meltzer PS, Hendrix MJ. Vascular channel formation by human melanoma cells in vivo and in vitro: vasculogenic mimicry. Am J Pathol. 1999;155:739–52.
60. Damato B, Duke C, Coupland SE, Hiscott P, Smith PA, Campbell I, Douglas A, Howard P. Cytogenetics of uveal melanoma: a 7-year clinical experience. Ophthalmology. 2007;114:1925–31.
61. Bronkhorst IH, Ly LV, Jordanova ES, Vrolijk J, Versluis M, Luyten GP, Jager MJ. Detection of M2-macrophages in uveal melanoma and relation with survival. Invest Ophthalmol Vis Sci. 2011a;52:643–50.

62. Makitie T, Summanen P, Tarkkanen A, Kivela T. Microvascular density in predicting survival of patients with choroidal and ciliary body melanoma. Invest Ophthalmol Vis Sci. 1999a;40:2471–80.
63. Whelchel JC, Farah SE, Mclean IW, Burnier MN. Immunohistochemistry of infiltrating lymphocytes in uveal malignant melanoma. Invest Ophthalmol Vis Sci. 1993;34:2603–6.
64. Damato B, Eleuteri A, Fisher AC, Coupland SE, Taktak AF. Artificial neural networks estimating survival probability after treatment of choroidal melanoma. Ophthalmology. 2008;115:1598–607.
65. Gamel JW, Mclean IW, Mccurdy JB. Biologic distinctions between cure and time to death in 2892 patients with intraocular melanoma. Cancer. 1993;71:2299–305.
66. Walter SD, Chao DL, Feuer W, Schiffman J, Char DH, Harbour JW. Prognostic implications of tumor diameter in association with gene expression profile for uveal melanoma. JAMA Ophthalmol. 2016;134:734–40.
67. Schmittel A, Bechrakis NE, Martus P, Mutlu D, Scheibenbogen C, Bornfeld N, Foerster MH, Thiel E, Keilholz U. Independent prognostic factors for distant metastases and survival in patients with primary uveal melanoma. Eur J Cancer. 2004;40:2389–95.
68. Shields CL, Kaliki S, Shah SU, Luo W, Furuta M, Shields JA. Iris melanoma: features and prognosis in 317 children and adults. J AAPOS. 2012;16:10–6.
69. Shields CL, Shields JA, Materin M, Gershenbaum E, Singh AD, Smith A. Iris melanoma: risk factors for metastasis in 169 consecutive patients. Ophthalmology. 2001;108:172–8.
70. Shields CL, Ramasubramanian A, Ganguly A, Mohan D, Shields JA. Cytogenetic testing of iris melanoma using fine needle aspiration biopsy in 17 patients. Retina. 2011;31:574–80.
71. Rey JA, Bello MJ, De Campos JM, Ramos MC, Benitez J. Cytogenetic findings in a human malignant melanoma metastatic to the brain. Cancer Genet Cytogenet. 1985;16:179–83.
72. Prescher G, Bornfeld N, Hirche H, Horsthemke B, Jockel KH, Becher R. Prognostic implications of monosomy 3 in uveal melanoma. Lancet. 1996;347:1222–5.
73. Coupland SE, Lake SL, Zeschnigk M, Damato BE. Molecular pathology of uveal melanoma. Eye (Lond). 2013;27:230–42.
74. Naus NC, Van Drunen E, De Klein A, Luyten GP, Paridaens DA, Alers JC, Ksander BR, Beverloo HB, Slater RM. Characterization of complex chromosomal abnormalities in uveal melanoma by fluorescence in situ hybridization, spectral karyotyping, and comparative genomic hybridization. Genes Chromosomes Cancer. 2001;30:267–73.
75. Singh N, Singh AD, Hide W. Inferring an evolutionary tree of uveal melanoma from genomic copy number aberrations. Invest Ophthalmol Vis Sci. 2015;56:6801–9.
76. Damato B, Dopierala JA, Coupland SE. Genotypic profiling of 452 choroidal melanomas with multiplex ligation-dependent probe amplification. Clin Cancer Res. 2010;16:6083–92.
77. Horsman DE, White VA. Cytogenetic analysis of uveal melanoma. Consistent occurrence of monosomy 3 and trisomy 8q. Cancer. 1993;71:811–9.
78. Naus NC, Verhoeven AC, Van Drunen E, Slater R, Mooy CM, Paridaens DA, Luyten GP, De Klein A. Detection of genetic prognostic markers in uveal melanoma biopsies using fluorescence in situ hybridization. Clin Cancer Res. 2002;8:534–9.
79. Singh AD, Boghosian-Sell L, Wary KK, Shields CL, De Potter P, Donoso LA, Shields JA, Cannizzaro LA. Cytogenetic findings in primary uveal melanoma. Cancer Genet Cytogenet. 1994;72:109–15.
80. Sisley K, Parsons MA, Garnham J, Potter AM, Curtis D, Rees RC, Rennie IG. Association of specific chromosome alterations with tumour phenotype in posterior uveal melanoma. Br J Cancer. 2000;82:330–8.
81. Wiltshire RN, Elner VM, Dennis T, Vine AK, Trent JM. Cytogenetic analysis of posterior uveal melanoma. Cancer Genet Cytogenet. 1993;66:47–53.
82. Sisley K, Rennie IG, Parsons MA, Jacques R, Hammond DW, Bell SM, Potter AM, Rees RC. Abnormalities of chromosomes 3 and 8 in posterior uveal melanoma correlate with prognosis. Genes Chromosomes Cancer. 1997;19:22–8.
83. Trolet J, Hupe P, Huon I, Lebigot I, Decraene C, Delattre O, Sastre-Garau X, Saule S, Thiery JP, Plancher C, Asselain B, Desjardins L, Mariani P, Piperno-Neumann S, Barillot E, Couturier

J. Genomic profiling and identification of high-risk uveal melanoma by array CGH analysis of primary tumors and liver metastases. Invest Ophthalmol Vis Sci. 2009;50:2572–80.
84. Damato B, Dopierala J, Klaasen A, Van Dijk M, Sibbring J, Coupland SE. Multiplex ligation-dependent probe amplification of uveal melanoma: correlation with metastatic death. Invest Ophthalmol Vis Sci. 2009;50:3048–55.
85. Blasi MA, Roccella F, Balestrazzi E, Del Porto G, De Felice N, Roccella M, Rota R, Grammatico P. 3p13 region: a possible location of a tumor suppressor gene involved in uveal melanoma. Cancer Genet Cytogenet. 1999;108:81–3.
86. Cross NA, Ganesh A, Parpia M, Murray AK, Rennie IG, Sisley K. Multiple locations on chromosome 3 are the targets of specific deletions in uveal melanoma. Eye (Lond). 2006;20:476–81.
87. Parrella P, Fazio VM, Gallo AP, Sidransky D, Merbs SL. Fine mapping of chromosome 3 in uveal melanoma: identification of a minimal region of deletion on chromosomal arm 3p25.1-p25.2. Cancer Res. 2003;63:8507–10.
88. Tschentscher F, Prescher G, Horsman DE, White VA, Rieder H, Anastassiou G, Schilling H, Bornfeld N, Bartz-Schmidt KU, Horsthemke B, Lohmann DR, Zeschnigk M. Partial deletions of the long and short arm of chromosome 3 point to two tumor suppressor genes in uveal melanoma. Cancer Res. 2001;61:3439–42.
89. Landreville S, Agapova OA, Harbour JW. Emerging insights into the molecular pathogenesis of uveal melanoma. Future Oncol. 2008;4:629–36.
90. Prescher G, Bornfeld N, Friedrichs W, Seeber S, Becher R. Cytogenetics of twelve cases of uveal melanoma and patterns of nonrandom anomalies and isochromosome formation. Cancer Genet Cytogenet. 1995;80:40–6.
91. Mooy CM, De Jong PT. Prognostic parameters in uveal melanoma: a review. Surv Ophthalmol. 1996;41:215–28.
92. White VA, Chambers JD, Courtright PD, Chang WY, Horsman DE. Correlation of cytogenetic abnormalities with the outcome of patients with uveal melanoma. Cancer. 1998;83:354–9.
93. Hammond DW, Al-Shammari NS, Danson S, Jacques R, Rennie IG, Sisley K. High-resolution array CGH analysis identifies regional deletions and amplifications of chromosome 8 in uveal melanoma. Invest Ophthalmol Vis Sci. 2015;56:3460–6.
94. Aalto Y, Eriksson L, Seregard S, Larsson O, Knuutila S. Concomitant loss of chromosome 3 and whole arm losses and gains of chromosome 1, 6, or 8 in metastasizing primary uveal melanoma. Invest Ophthalmol Vis Sci. 2001;42:313–7.
95. Horsman DE, Sroka H, Rootman J, White VA. Monosomy 3 and isochromosome 8q in a uveal melanoma. Cancer Genet Cytogenet. 1990;45:249–53.
96. Parada LA, Maranon A, Hallen M, Tranberg KG, Stenram U, Bardi G, Johansson D. Cytogenetic analyses of secondary liver tumors reveal significant differences in genomic imbalances between primary and metastatic colon carcinomas. Clin Exp Metastasis. 1999;17:471–9.
97. Chana JS, Wilson GD, Cree IA, Alexander RA, Myatt N, Neale M, Foss AJ, Hungerford JL. c-myc, p53, and Bcl-2 expression and clinical outcome in uveal melanoma. Br J Ophthalmol. 1999;83:110–4.
98. Ehlers JP, Harbour JW. NBS1 expression as a prognostic marker in uveal melanoma. Clin Cancer Res. 2005;11:1849–53.
99. Ehlers JP, Worley L, Onken MD, Harbour JW. DDEF1 is located in an amplified region of chromosome 8q and is overexpressed in uveal melanoma. Clin Cancer Res. 2005;11:3609–13.
100. Royds JA, Sharrard RM, Parsons MA, Lawry J, Rees R, Cottam D, Wagner B, Rennie IG. C-myc oncogene expression in ocular melanomas. Graefes Arch Clin Exp Ophthalmol. 1992;230:366–71.
101. Singh AD, Sisley K, Xu Y, Li J, Faber P, Plummer SJ, Mudhar HS, Rennie IG, Kessler PM, Casey G, Williams BG. Reduced expression of autotaxin predicts survival in uveal melanoma. Br J Ophthalmol. 2007;91:1385–92.

102. Boutin JA, Ferry G. Autotaxin. Cell Mol Life Sci. 2009;66:3009–21.
103. Onken MD, Worley LA, Harbour JW. A metastasis modifier locus on human chromosome 8p in uveal melanoma identified by integrative genomic analysis. Clin Cancer Res. 2008a;14:3737–45.
104. Harbour JW. A prognostic test to predict the risk of metastasis in uveal melanoma based on a 15-gene expression profile. Methods Mol Biol. 2014;1102:427–40.
105. Hoglund M, Gisselsson D, Hansen GB, White VA, Sall T, Mitelman F, Horsman D. Dissecting karyotypic patterns in malignant melanomas: temporal clustering of losses and gains in melanoma karyotypic evolution. Int J Cancer. 2004;108:57–65.
106. Singh AD, Tubbs R, Biscotti C, Schoenfield L, Trizzoi P. Chromosomal 3 and 8 status within hepatic metastasis of uveal melanoma. Arch Pathol Lab Med. 2009;133:1223–7.
107. Sisley K, Tattersall N, Dyson M, Smith K, Mudhar HS, Rennie IG. Multiplex fluorescence in situ hybridization identifies novel rearrangements of chromosomes 6, 15, and 18 in primary uveal melanoma. Exp Eye Res. 2006;83:554–9.
108. Parrella P, Sidransky D, Merbs SL. Allelotype of posterior uveal melanoma: implications for a bifurcated tumor progression pathway. Cancer Res. 1999;59:3032–7.
109. Speicher MR, Prescher G, Du Manoir S, Jauch A, Horsthemke B, Bornfeld N, Becher R, Cremer T. Chromosomal gains and losses in uveal melanomas detected by comparative genomic hybridization. Cancer Res. 1994;54:3817–23.
110. Van Gils W, Kilic E, Bruggenwirth HT, Vaarwater J, Verbiest MM, Beverloo B, Van Til-Berg ME, Paridaens D, Luyten GP, De Klein A. Regional deletion and amplification on chromosome 6 in a uveal melanoma case without abnormalities on chromosomes 1p, 3 and 8. Melanoma Res. 2008;18:10–5.
111. Hausler T, Stang A, Anastassiou G, Jockel KH, Mrzyk S, Horsthemke B, Lohmann DR, Zeschnigk M. Loss of heterozygosity of 1p in uveal melanomas with monosomy 3. Int J Cancer. 2005;116:909–13.
112. Kilic E, Naus NC, Van Gils W, Klaver CC, Van Til ME, Verbiest MM, Stijnen T, Mooy CM, Paridaens D, Beverloo HB, Luyten GP, De Klein A. Concurrent loss of chromosome arm 1p and chromosome 3 predicts a decreased disease-free survival in uveal melanoma patients. Invest Ophthalmol Vis Sci. 2005;46:2253–7.
113. Kilic E, Bruggenwirth HT, Meier M, Naus NC, Beverloo HB, Meijerink JP, Luyten GP, De Klein A. Increased expression of p73Deltaex2 transcript in uveal melanoma with loss of chromosome 1p. Melanoma Res. 2008;18:208–13.
114. Adhikary S, Eilers M. Transcriptional regulation and transformation by Myc proteins. Nat Rev Mol Cell Biol. 2005;6:635–45.
115. Cotterman R, Jin VX, Krig SR, Lemen JM, Wey A, Farnham PJ, Knoepfler PS. N-Myc regulates a widespread euchromatic program in the human genome partially independent of its role as a classical transcription factor. Cancer Res. 2008;68:9654–62.
116. Zeller KI, Jegga AG, Aronow BJ, O'donnell KA, Dang CV. An integrated database of genes responsive to the Myc oncogenic transcription factor: identification of direct genomic targets. Genome Biol. 2003;4:R69.
117. Ross DA, Wilson GD. Expression of c-myc oncoprotein represents a new prognostic marker in cutaneous melanoma. Br J Surg. 1998;85:46–51.
118. Esser P, Grisanti S, Bartz-Schmidt K. TGF-beta in uveal melanoma. Microsc Res Tech. 2001;52:396–400.
119. Siegel PM, Massague J. Cytostatic and apoptotic actions of TGF-beta in homeostasis and cancer. Nat Rev Cancer. 2003;3:807–21.
120. Derynck R, Akhurst RJ, Balmain A. TGF-beta signaling in tumor suppression and cancer progression. Nat Genet. 2001;29:117–29.
121. Koli K, Saharinen J, Hyytiainen M, Penttinen C, Keski-Oja J. Latency, activation, and binding proteins of TGF-beta. Microsc Res Tech. 2001;52:354–62.
122. El-Shabrawi Y, Ardjomand N, Radner H, Ardjomand N. MMP-9 is predominantly expressed in epithelioid and not spindle cell uveal melanoma. J Pathol. 2001;194:201–6.
123. Lai K, Conway RM, Crouch R, Jager MJ, Madigan MC. Expression and distribution of MMPs and TIMPs in human uveal melanoma. Exp Eye Res. 2008;86:936–41.

124. Woodward JK, Rennie IG, Burn JL, Sisley K. A potential role for TGFbeta in the regulation of uveal melanoma adhesive interactions with the hepatic endothelium. Invest Ophthalmol Vis Sci. 2005;46:3473–7.
125. Hanahan D, Weinberg RA. The hallmarks of cancer. Cell. 2000;100:57–70.
126. Pardali E, Ten Dijke P. Transforming growth factor-beta signaling and tumor angiogenesis. Front Biosci (Landmark Ed). 2009;14:4848–61.
127. Ziemssen F, Wegner R, Wegner J, Tatar O, Susskind D, Gelisken F, Rohrbach M, Bartz-Schmidt KU, Grisanti S. Analysis of neovasculature in uveal melanoma by targeting the TGFbeta-binding receptor endoglin: is there prognostic relevance of proliferating endothelium? Graefes Arch Clin Exp Ophthalmol. 2006;244:1124–31.
128. Maat W, Ly LV, Jordanova ES, De Wolff-Rouendaal D, Schalij-Delfos NE, Jager MJ. Monosomy of chromosome 3 and an inflammatory phenotype occur together in uveal melanoma. Invest Ophthalmol Vis Sci. 2008;49:505–10.
129. Ijland SA, Jager MJ, Heijdra BM, Westphal JR, Peek R. Expression of angiogenic and immunosuppressive factors by uveal melanoma cell lines. Melanoma Res. 1999;9:445–50.
130. Ma D, Niederkorn JY. Transforming growth factor-beta down-regulates major histocompatibility complex class I antigen expression and increases the susceptibility of uveal melanoma cells to natural killer cell-mediated cytolysis. Immunology. 1995;86:263–9.
131. Canovas D, Rennie IG, Nichols CE, Sisley K. Local environmental influences on uveal melanoma: vitreous humor promotes uveal melanoma invasion, whereas the aqueous can be inhibitory. Cancer. 2008;112:1787–94.
132. Cory S, Huang DC, Adams JM. The Bcl-2 family: roles in cell survival and oncogenesis. Oncogene. 2003;22:8590–607.
133. Adams JM, Cory S. Bcl-2-regulated apoptosis: mechanism and therapeutic potential. Curr Opin Immunol. 2007;19:488–96.
134. Thornberry NA. Caspases: key mediators of apoptosis. Chem Biol. 1998;5:R97–103.
135. Hussein MR. Analysis of Bcl-2 protein expression in choroidal melanomas. J Clin Pathol. 2005;58:486–9.
136. Jay V, Yi Q, Hunter WS, Zielenska M. Expression of bcl-2 in uveal malignant melanoma. Arch Pathol Lab Med. 1996;120:497–8.
137. Mooy CM, Luyten GP, De Jong PT, Luider TM, Stijnen T, Van De Ham F, Van Vroonhoven CC, Bosman FT. Immunohistochemical and prognostic analysis of apoptosis and proliferation in uveal melanoma. Am J Pathol. 1995;147:1097–104.
138. Druker BJ. Perspectives on the development of a molecularly targeted agent. Cancer Cell. 2002;1:31–6.
139. Triozzi PL, Eng C, Singh AD. Targeted therapy for uveal melanoma. Cancer Treat Rev. 2008;34:247–58.
140. Nemati F, De Montrion C, Lang G, Kraus Berthier L, Carita G, Sastre Garau X, Berniard A, Vallerand D, Geneste O, De Plater L, Pierre A, Lockhart B, Desjardins L, Piperno-Neumann S, Depil S, Decaudin D. Targeting Bcl-2/Bcl-XL induces antitumor activity in uveal melanoma patient-derived xenografts. PLoS One. 2014;9:e80836.
141. Harbour JW, Worley L, Ma D, Cohen M. Transducible peptide therapy for uveal melanoma and retinoblastoma. Arch Ophthalmol. 2002;120:1341–6.
142. Vogler M, Dinsdale D, Dyer MJ, Cohen GM. Bcl-2 inhibitors: small molecules with a big impact on cancer therapy. Cell Death Differ. 2009;16:360–7.
143. Brantley Jr MA, Harbour JW. Deregulation of the Rb and p53 pathways in uveal melanoma. Am J Pathol. 2000a;157:1795–801.
144. Janssen K, Kuntze J, Busse H, Schmid KW. p53 oncoprotein overexpression in choroidal melanoma. Mod Pathol. 1996;9:267–72.
145. Wang JY. DNA damage and apoptosis. Cell Death Differ. 2001;8:1047–8.
146. Brantley Jr MA, Worley L, Harbour JW. Altered expression of Rb and p53 in uveal melanomas following plaque radiotherapy. Am J Ophthalmol. 2002;133:242–8.
147. Prives C, Hall PA. The p53 pathway. J Pathol. 1999;187:112–26.
148. Chin L, Pomerantz J, Depinho RA. The INK4a/ARF tumor suppressor: one gene--two products--two pathways. Trends Biochem Sci. 1998;23:291–6.

149. Pomerantz J, Schreiber-Agus N, Liegeois NJ, Silverman A, Alland L, Chin L, Potes J, Chen K, Orlow I, Lee HW, Cordon-Cardo C, Depinho RA. The Ink4a tumor suppressor gene product, p19Arf, interacts with MDM2 and neutralizes MDM2's inhibition of p53. Cell. 1998;92:713–23.
150. Haupt Y, Maya R, Kazaz A, Oren M. Mdm2 promotes the rapid degradation of p53. Nature. 1997;387:296–9.
151. Honda R, Tanaka H, Yasuda H. Oncoprotein MDM2 is a ubiquitin ligase E3 for tumor suppressor p53. FEBS Lett. 1997;420:25–7.
152. Landers JE, Cassel SL, George DL. Translational enhancement of mdm2 oncogene expression in human tumor cells containing a stabilized wild-type p53 protein. Cancer Res. 1997;57:3562–8.
153. Coupland SE, Anastassiou G, Stang A, Schilling H, Anagnostopoulos I, Bornfeld N, Stein H. The prognostic value of cyclin D1, p53, and MDM2 protein expression in uveal melanoma. J Pathol. 2000;191:120–6.
154. Chene P, Fuchs J, Bohn J, Garcia-Echeverria C, Furet P, Fabbro D. A small synthetic peptide, which inhibits the p53-hdm2 interaction, stimulates the p53 pathway in tumour cell lines. J Mol Biol. 2000;299:245–53.
155. Vassilev LT, Vu BT, Graves B, Carvajal D, Podlaski F, Filipovic Z, Kong N, Kammlott U, Lukacs C, Klein C, Fotouhi N, Liu EA. In vivo activation of the p53 pathway by small-molecule antagonists of MDM2. Science. 2004;303:844–8.
156. Brantley Jr MA, Harbour JW. Inactivation of retinoblastoma protein in uveal melanoma by phosphorylation of sites in the COOH-terminal region. Cancer Res. 2000b;60:4320–3.
157. Pardo M, Pineiro A, De La Fuente M, Garcia A, Prabhakar S, Zitzmann N, Dwek RA, Sanchez Salorio M, Dominguez F, Capeans C. Abnormal cell cycle regulation in primary human uveal melanoma cultures. J Cell Biochem. 2004;93:708–20.
158. Scholes AG, Liloglou T, Maloney P, Hagan S, Nunn J, Hiscott P, Damato BE, Grierson I, Field JK. Loss of heterozygosity on chromosomes 3, 9, 13, and 17, including the retinoblastoma locus, in uveal melanoma. Invest Ophthalmol Vis Sci. 2001;42:2472–7.
159. Halaban R. Rb/E2F: a two-edged sword in the melanocytic system. Cancer Metastasis Rev. 2005;24:339–56.
160. Harbour JW. Eye cancer: unique insights into oncogenesis: the Cogan Lecture. Invest Ophthalmol Vis Sci. 2006;47:1736–45.
161. Laurie NA, Donovan SL, Shih CS, Zhang J, Mills N, Fuller C, Teunisse A, Lam S, Ramos Y, Mohan A, Johnson D, Wilson M, Rodriguez-Galindo C, Quarto M, Francoz S, Mendrysa SM, Guy RK, Marine JC, Jochemsen AG, Dyer MA. Inactivation of the p53 pathway in retinoblastoma. Nature. 2006;444:61–6.
162. El-Shabrawi Y, Radner H, Muellner K, Langmann G, Hoefler G. The role of UV-radiation in the development of conjunctival malignant melanoma. Acta Ophthalmol Scand. 1999;77:31–2.
163. Gear H, Williams H, Kemp EG, Roberts F. BRAF mutations in conjunctival melanoma. Invest Ophthalmol Vis Sci. 2004;45:2484–8.
164. Spendlove HE, Damato BE, Humphreys J, Barker KT, Hiscott PS, Houlston RS. BRAF mutations are detectable in conjunctival but not uveal melanomas. Melanoma Res. 2004;14:449–52.
165. Davies H, Bignell GR, Cox C, Stephens P, Edkins S, Clegg S, Teague J, Woffendin H, Garnett MJ, Bottomley W, Davis N, Dicks E, Ewing R, Floyd Y, Gray K, Hall S, Hawes R, Hughes J, Kosmidou V, Menzies A, Mould C, Parker A, Stevens C, Watt S, Hooper S, Wilson R, Jayatilake H, Gusterson BA, Cooper C, Shipley J, Hargrave D, Pritchard-Jones K, Maitland N, Chenevix-Trench G, Riggins GJ, Bigner DD, Palmieri G, Cossu A, Flanagan A, Nicholson A, Ho JW, Leung SY, Yuen ST, Weber BL, Seigler HF, Darrow TL, Paterson H, Marais R, Marshall CJ, Wooster R, Stratton MR, Futreal PA. Mutations of the BRAF gene in human cancer. Nature. 2002;417:949–54.
166. Cohen Y, Goldenberg-Cohen N, Parrella P, Chowers I, Merbs SL, Pe'er J, Sidransky D. Lack of BRAF mutation in primary uveal melanoma. Invest Ophthalmol Vis Sci. 2003; 44:2876–8.

167. Cruz 3rd F, Rubin BP, Wilson D, Town A, Schroeder A, Haley A, Bainbridge T, Heinrich MC, Corless CL. Absence of BRAF and NRAS mutations in uveal melanoma. Cancer Res. 2003;63:5761–6.
168. Edmunds SC, Cree IA, Di Nicolantonio F, Hungerford JL, Hurren JS, Kelsell DP. Absence of BRAF gene mutations in uveal melanomas in contrast to cutaneous melanomas. Br J Cancer. 2003;88:1403–5.
169. Rimoldi D, Salvi S, Lienard D, Lejeune FJ, Speiser D, Zografos L, Cerottini JC. Lack of BRAF mutations in uveal melanoma. Cancer Res. 2003;63:5712–5.
170. Zuidervaart W, Van Nieuwpoort F, Stark M, Dijkman R, Packer L, Borgstein AM, Pavey S, Van Der Velden P, Out C, Jager MJ, Hayward NK, Gruis NA. Activation of the MAPK pathway is a common event in uveal melanomas although it rarely occurs through mutation of BRAF or RAS. Br J Cancer. 2005;92:2032–8.
171. Van Raamsdonk CD, Fitch KR, Fuchs H, De Angelis MH, Barsh GS. Effects of G-protein mutations on skin color. Nat Genet. 2004;36:961–8.
172. Onken MD, Worley LA, Long MD, Duan S, Council ML, Bowcock AM, Harbour JW. Oncogenic mutations in GNAQ occur early in uveal melanoma. Invest Ophthalmol Vis Sci. 2008b;49:5230–4.
173. Van Raamsdonk CD, Griewank KG, Crosby MB, Garrido MC, Vemula S, Wiesner T, Obenauf AC, Wackernagel W, Green G, Bouvier N, Sozen MM, Baimukanova G, Roy R, Heguy A, Dolgalev I, Khanin R, Busam K, Speicher MR, O'brien J, Bastian BC. Mutations in GNA11 in uveal melanoma. N Engl J Med. 2010;363:2191–9.
174. O'hayre M, Vazquez-Prado J, Kufareva I, Stawiski EW, Handel TM, Seshagiri S, Gutkind JS. The emerging mutational landscape of G proteins and G-protein-coupled receptors in cancer. Nat Rev Cancer. 2013;13:412–24.
175. Oldham WM, Hamm HE. Heterotrimeric G protein activation by G-protein-coupled receptors. Nat Rev Mol Cell Biol. 2008;9:60–71.
176. Carvajal RD, Sosman JA, Quevedo JF, Milhem MM, Joshua AM, Kudchadkar RR, Linette GP, Gajewski TF, Lutzky J, Lawson DH, Lao CD, Flynn PJ, Albertini MR, Sato T, Lewis K, Doyle A, Ancell K, Panageas KS, Bluth M, Hedvat C, Erinjeri J, Ambrosini G, Marr B, Abramson DH, Dickson MA, Wolchok JD, Chapman PB, Schwartz GK. Effect of selumetinib vs chemotherapy on progression-free survival in uveal melanoma: a randomized clinical trial. JAMA. 2014;311:2397–405.
177. Carvajal RD, Schwartz GK, Tezel T, Marr B, Francis JH, Nathan PD. Metastatic disease from uveal melanoma: treatment options and future prospects. Br J Ophthalmol. 2017;101(1):38–44.
178. Johansson P, Aoude LG, Wadt K, Glasson WJ, Warrier SK, Hewitt AW, Kiilgaard JF, Heegaard S, Isaacs T, Franchina M, Ingvar C, Vermeulen T, Whitehead KJ, Schmidt CW, Palmer JM, Symmons J, Gerdes AM, Jonsson G, Hayward NK. Deep sequencing of uveal melanoma identifies a recurrent mutation in PLCB4. Oncotarget. 2016;7:4624–31.
179. Moore AR, Ceraudo E, Sher JJ, Guan Y, Shoushtari AN, Chang MT, Zhang JQ, Walczak EG, Kazmi MA, Taylor BS, Huber T, Chi P, Sakmar TP, Chen Y. Recurrent activating mutations of G-protein-coupled receptor CYSLTR2 in uveal melanoma. Nat Genet. 2016;48:675–80.
180. Testa JR, Cheung M, Pei J, Below JE, Tan Y, Sementino E, Cox NJ, Dogan AU, Pass HI, Trusa S, Hesdorffer M, Nasu M, Powers A, Rivera Z, Comertpay S, Tanji M, Gaudino G, Yang H, Carbone M. Germline BAP1 mutations predispose to malignant mesothelioma. Nat Genet. 2011;43:1022–5.
181. Jensen DE, Proctor M, Marquis ST, Gardner HP, Ha SI, Chodosh LA, Ishov AM, Tommerup N, Vissing H, Sekido Y, Minna J, Borodovsky A, Schultz DC, Wilkinson KD, Maul GG, Barlev N, Berger SL, Prendergast GC, Rauscher 3rd FJ. BAP1: a novel ubiquitin hydrolase which binds to the BRCA1 RING finger and enhances BRCA1-mediated cell growth suppression. Oncogene. 1998;16:1097–112.
182. Jensen DE, Rauscher 3rd FJ. BAP1, a candidate tumor suppressor protein that interacts with BRCA1. Ann N Y Acad Sci. 1999;886:191–4.
183. Bommi PV, Dimri M, Sahasrabuddhe AA, Khandekar J, Dimri GP. The polycomb group protein BMI1 is a transcriptional target of HDAC inhibitors. Cell Cycle. 2010;9:2663–73.

184. Landreville S, Agapova OA, Matatall KA, Kneass ZT, Onken MD, Lee RS, Bowcock AM, Harbour JW. Histone deacetylase inhibitors induce growth arrest and differentiation in uveal melanoma. Clin Cancer Res. 2012;18:408–16.
185. Harbour JW, Roberson ED, Anbunathan H, Onken MD, Worley LA, Bowcock AM. Recurrent mutations at codon 625 of the splicing factor SF3B1 in uveal melanoma. Nat Genet. 2013;45:133–5.
186. Martin M, Masshofer L, Temming P, Rahmann S, Metz C, Bornfeld N, Van De Nes J, Klein-Hitpass L, Hinnebusch AG, Horsthemke B, Lohmann DR, Zeschnigk M. Exome sequencing identifies recurrent somatic mutations in EIF1AX and SF3B1 in uveal melanoma with disomy 3. Nat Genet. 2013;45:933–6.
187. Furney SJ, Pedersen M, Gentien D, Dumont AG, Rapinat A, Desjardins L, Turajlic S, Piperno-Neumann S, De La Grange P, Roman-Roman S, Stern MH, Marais R. SF3B1 mutations are associated with alternative splicing in uveal melanoma. Cancer Discov. 2013;3:1122–9.
188. Harbour JW. Genomic, prognostic, and cell-signaling advances in uveal melanoma. Am Soc Clin Oncol Educ Book. 2013;2013:388–91.
189. Yavuzyigitoglu S, Koopmans AE, Verdijk RM, Vaarwater J, Eussen B, Van Bodegom A, Paridaens D, Kilic E, De Klein A, Rotterdam Ocular Melanoma Study, G. Uveal melanomas with SF3B1 mutations: a distinct subclass associated with late-onset metastases. Ophthalmology. 2016;123:1118–28.
190. Harbour JW. The genetics of uveal melanoma: an emerging framework for targeted therapy. Pigment Cell Melanoma Res. 2012;25:171–81.
191. Kilic E, Koopmans AE, Yavuzyigitoglu S, Vaarwater J, Van Ijcken WFJ, Paridaens D, De Klein J. SF3B1 and EIF1AX mutations in uveal melanoma: a protective factor, or not? Acta Ophthalmol, 2014;92.
192. Onken MD, Worley LA, Tuscan MD, Harbour JW. An accurate, clinically feasible multi-gene expression assay for predicting metastasis in uveal melanoma. J Mol Diagn. 2010;12:461–8.
193. Zuidervaart W, Van Der Velden PA, Hurks MH, Van Nieuwpoort FA, Out-Luiting CJ, Singh AD, Frants RR, Jager MJ, Gruis NA. Gene expression profiling identifies tumour markers potentially playing a role in uveal melanoma development. Br J Cancer. 2003;89:1914–9.
194. Tschentscher F, Husing J, Holter T, Kruse E, Dresen IG, Jockel KH, Anastassiou G, Schilling H, Bornfeld N, Horsthemke B, Lohmann DR, Zeschnigk M. Tumor classification based on gene expression profiling shows that uveal melanomas with and without monosomy 3 represent two distinct entities. Cancer Res. 2003;63:2578–84.
195. Onken MD, Worley LA, Davila RM, Char DH, Harbour JW. Prognostic testing in uveal melanoma by transcriptomic profiling of fine needle biopsy specimens. J Mol Diagn. 2006;8:567–73.
196. Onken MD, Worley LA, Ehlers JP, Harbour JW. Gene expression profiling in uveal melanoma reveals two molecular classes and predicts metastatic death. Cancer Res. 2004;64:7205–9.
197. Petrausch U, Martus P, Tonnies H, Bechrakis NE, Lenze D, Wansel S, Hummel M, Bornfeld N, Thiel E, Foerster MH, Keilholz U. Significance of gene expression analysis in uveal melanoma in comparison to standard risk factors for risk assessment of subsequent metastases. Eye (Lond). 2008;22:997–1007.
198. Coupland SE, Kalirai H, Ho V, Thornton S, Damato BE, Heimann H. Concordant chromosome 3 results in paired choroidal melanoma biopsies and subsequent tumour resection specimens. Br J Ophthalmol. 2015;99:1444–50.
199. Dogrusöz M, Kroes WGM, Van Duinen SG, Creutzberg CL, Versluis M, Bleeker JC, Marinkovic M, Luyten GPM, Jager MJ. Radiation treatment affects chromosome testing in uveal melanoma. Invest Ophthalmol Vis Sci. 2015;56:5956–64.
200. Hussain RN, Kalirai H, Groenewald C, Kacperek A, Errington RD, Coupland SE, Heimann H, Damato B. Prognostic biopsy of choroidal melanoma after proton beam radiation therapy. Ophthalmology. 2016;123:2264–5.
201. Wackernagel W, Tarmann L, Mayer C, Langmann G, Wedrich A. Genetic analysis of uveal melanoma by array comparative genomic hybridization before and after radiotherapy. Spektrum Augenheilkd. 2013;27:286–91.

Molecular Aspects of Thyroid Carcinogenesis

10

Rita Seeböck, Johannes Haybaeck, and Oleksiy Tsybrovskyy

Contents

Abstract

The thyroid is one of the largest endocrine glands in the body and of highest importance to healthy life by regulating energy metabolism, protein synthesis, and hormone sensitivity. Tumors of the thyroid are rare, accounting for only ~2% of all tumors being diagnosed worldwide. Remarkably, this cancer entity is more frequent in women than in men, with incidence ratios of approximately 3:1. This chapter will introduce the main features of thyroid cancer development, especially focusing on altered molecular signaling and epigenetic variations.

R. Seeböck (✉)
Department Life Sciences, IMC University of Applied Sciences, Krems, Austria
e-mail: rita.seeboeck@fh-krems.ac.at

J. Haybaeck
Department of Pathology, University Hospital, Medical Faculty, Magdeburg, Germany

Institute of Pathology, Medical University, Graz, Austria
e-mail: johannes.haybaeck@med.ovgu.de

Oleksiy Tsybrovskyy
Institute of Clinical Pathology, BHS Linz, Linz, Austria
e-mail: oleksiy.tsybrovskyy@bhs.at

J. Haybaeck (ed.), *Mechanisms of Molecular Carcinogenesis – Volume 1*,
DOI 10.1007/978-3-319-53659-0_10

10.1 Anatomy and Physiology of the Thyroid Gland

The thyroid gland is a key organ of the endocrine system and therefore an essential regulator of numerous physiological processes, including energy metabolism, protein synthesis, and hormone sensitivity. The principal hormones produced by the thyroid itself are triiodothyronine (T_3) and thyroxin (T_4), generated from the amino acid tyrosine and elemental iodine attachments [1]. The third major hormone secreted by the thyroid is calcitonin, involved in calcium homeostasis. Calcitonin is produced in the so-called C-cells of the thyroid gland [1]. The organ is built up of two lobes that are positioned on both sides of the trachea, closely underneath the larynx. The lobes are connected by a small band of thyroid tissue called isthmus. Each lobe has a size of 4–6 cm in length; the whole organ has a weight of 15–30 g in adults, but can be vastly increased in conditions of disease [2].

Microscopically, the thyroid tissue is made up of numerous follicles of varying sizes. The follicular lumen containing colloid is framed by one layer of follicular cells. The C-cells are located on top of or in between follicular cells; they are associated with a particular follicle. The name parafollicular cells for C-cells, which can be found frequently in the literature, is thus incorrect. Both normal and hyperplastic C-cells have an intrafollicular localization [3] (Fig. 10.1j).

Within the colloid the thyroid hormones, T_3 and T_4 are stored. These two hormones are the only iodine-containing compounds involved in physiologic processes [4]. Iodine taken up through nutrition is absorbed through the small intestine and shuttled to the thyroid via the bloodstream. The thyroid is very well supplied with blood, so that it only takes approximately one and a half hour for the whole blood to pass through the gland. Within the thyroid, iodine gets stored, oxidized, and finally incorporated in precursors of T_3 and T_4. When stored in colloid, the hormones are bound to thyroglobulin protein Tg; secreted hormones are first separated from Tg through proteolysis, but will interact with other proteins in circulation. In fact, 80% of circulating T_3 and T_4 are associated with thyroxine-binding globulin (TBG); 10% each are coupled to albumin or prealbumin [5].

Hormone release from the thyroid is tightly regulated in a negative feedback loop, called the hypothalamic-pituitary-thyroid axis. As soon as the hypothalamus recognizes low levels of circulating T_3 and especially T_4 hormone, it secretes thyrotropin-releasing hormone (TRH). TRH binding triggers the release of thyroid-stimulating hormone (TSH) in the pituitary. Consequently, TSH acts on the thyroid, which is stimulated to release T_3 and T_4, thereby increasing their concentration in the bloodstream and slowing down this regulatory circuit. The negative feedback acts directly on the hypothalamus but also signals to the pituitary gland [6].

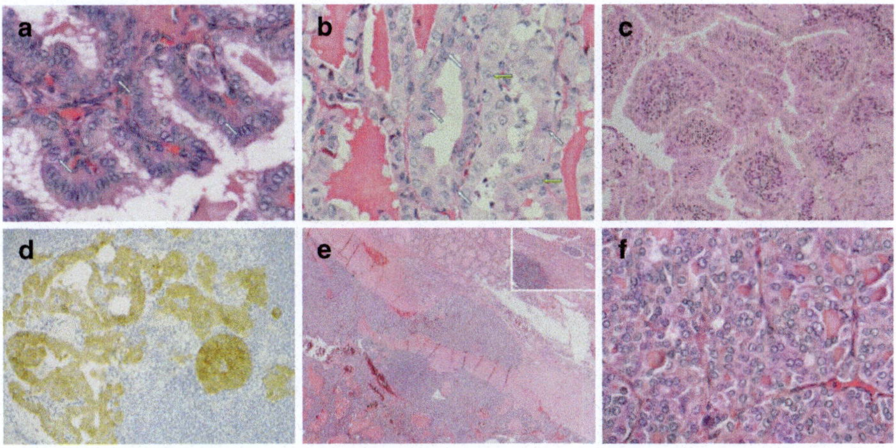

Fig. 10.1 (**a**) Papillary thyroid carcinoma (classic variant). The tumor is composed of papillary structures with gentle fibrovascular stalks. The cell nuclei display a number of characteristic alterations including elongation, overlapping, irregular contours, chromatin clearing, and numerous infoldings of the nuclear envelope, i.e., so-called grooves (*white arrows*). HE stain. (**b**) Papillary thyroid carcinoma, follicular variant. The tumor shows follicular architecture; no papillary structures are seen. However, the tumor cell nuclei display typical alterations of a papillary carcinoma including grooves (*white arrows*) and eosiniphilic intranuclear cytoplasmic inclusions (*yellow arrows*). HE stain. (**c**) Papillary thyroid carcinoma, tall cell variant. The cytoplasm of the tumor cells is larger and more eosinophilic than in conventional type. The height of the cytoplasm is approximately three times as much as its width. (**d**) Lymph node metastasis is typical for papillary carcinoma. The presence of BRAF mutation can be detected by means of immunohistochemistry (positive cytoplasmic +/− nuclear stain with anti-BRAF V600E VE1 monoclonal antibody, Ventana, cat.no. 790-4855). (**e**) Follicular thyroid carcinoma. The tumor usually consists of follicles of variable size. Penetration of the tumor capsule and blood vessel invasion (*inset*) resulting in distant hematogeneous metastasis is typical for this type of tumors. (**f**) Follicular thyroid carcinoma at high-power magnification. The cell nuclei do not display the typical features of a papillary carcinoma. (**g**) Anaplastic (undifferentiated) thyroid carcinoma. High cellularity, diffusely infiltrative solid growth pattern, as well as areas of necrosis and hemorrhage are typical. (**h**) Anaplastic (undifferentiated) thyroid carcinoma at high-power magnification. Primitively looking cell population with brisk mitotic activity (*white arrows*) and occasional multinucleated giant cells (*yellow arrow*) with overall sarcomatoid appearance underscore the loss of differentiation in this tumor type. (**i**) C-cell hyperplasia is barely visible on conventional HE-stain. (**j**) Immunohistochemistry for calcitonin reveals numerous C-cells occupying entire follicles or of parts of them (same location as in **i**). Note the intrafollicular location of the C-cells (the term "parafollicular cells" is actually a misnomer). (**k**) Medullary thyroid carcinoma. This tumor can show quite different growth patterns and cytologic features. Coarse "salt-and-pepper" chromatin is one of the most consistent features aiding the correct diagnosis. (**l**) Immunohistochemical stain for calcitonin is strongly positive in virtually all medullary carcinomas and is obligatory for definitive diagnosis. (**m**) Amyloid deposits in the tumor stroma are often seen and are positively stained with *congo red*. (**n**) Amyloid stained with *congo red* is birefringent and therefore shines *apple-green* in polarized light (same location as in **m**). This feature is essential in distinguishing true amyloid deposits from its mimics (e.g., hyalinized collagenous stroma) (color figure online)

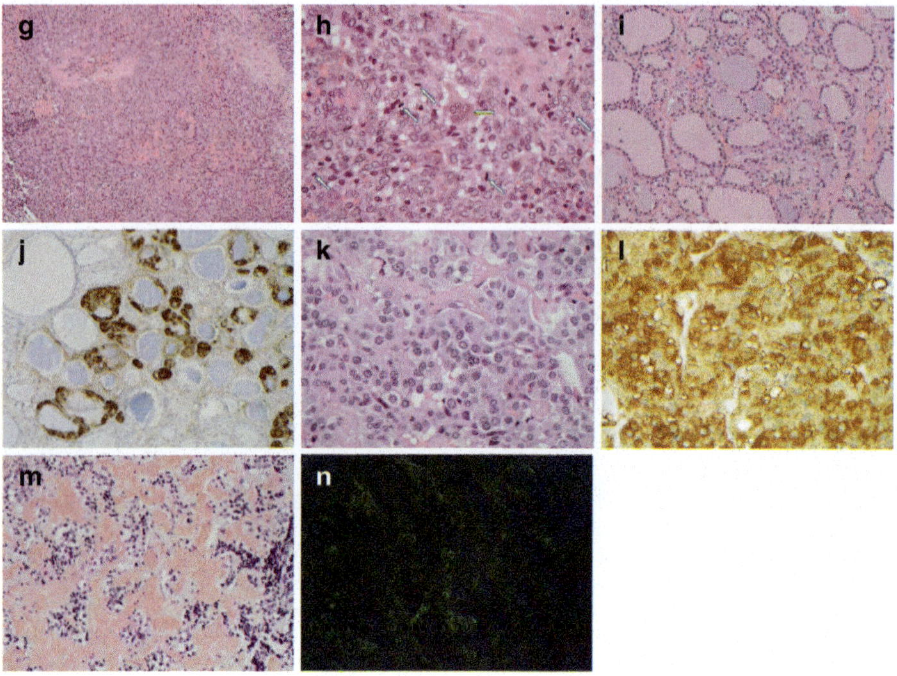

Fig. 10.1 (continued)

10.2 Epidemiology and Pathophysiology of Thyroid Cancer

Worldwide thyroid cancer accounts for less than 2% of all cancers diagnosed [7]. The distribution between female and male patients varies from country to country but is always higher in females, with an average ratio of 3:1 [8]. Even if thyroid cancer has a low prevalence and is associated with more than 95% survival with a very good survival rate, it must not be treated lightly. The incidence for thyroid cancer is significantly rising, including all tumor sizes and stages [9].

The majority of thyroid tumors originate from follicular cells as epithelial tumor. These lesions can further be divided in papillary thyroid carcinoma (PTC), follicular thyroid carcinoma (FTC), or anaplastic thyroid carcinoma (ATC). Tumors that arise from C-cells are of the category medullary thyroid carcinoma (MTC) [10]. Eighty-five to 90% of thyroid cancer cases present as PTC, 5–10% as FTC, and about 2% as MTC [10]. Further subtypes include carcinomas of mixed origin, squamous cell carcinoma, mucinous carcinoma, carcinosarcoma, and other less frequent variants [11]. Representative pictures of the various types of thyroid cancer are shown in Fig. 10.1.

A number of genetic alterations accompany the molecular carcinogenesis of the different tumor variants. A summary of these alterations and the associated signaling pathways, as well as a discussion of how they orchestrate tumorigenesis in the thyroid, is presented below.

10.3 MAP Kinase and PI3-Kinase Signaling Cascades

From all occurring mutations, most affect the signaling pathways of (mitogen-activated protein) MAP kinases and (phosphoinositide-3) PI3-kinase. Among these, we reckon mutations in the genes of *BRAF*, *HRAS*, *KRAS*, or *NRAS*, translocations of *RET*, as well as *PTEN* gene mutations or deletions.

The RAS/RAF/MEK/ERK pathway is among the most essential pathways of inter- and intracellular signaling and of vital importance in signal transductions regulating survival, growth, differentiation, migration, and cell-cell interactions [12]. The best described MAP kinases include ERK proteins, JNK, and p38 [13]. Each of them contains a three-tiered kinase cascade of a MAP kinase (MAPK), a MAP kinase kinase (MAPKK), and a MAP kinase kinase kinase (MAPKKK). Signals are received at the surface of the cell by different receptor molecules, activating signaling modules in the cells interior by phosphorylation events. The most intensively studied pathway is RAS/RAF/MEK/ERK (extracellular signal-regulated kinase) cascade. RAS is recruited to the intracellular domain of a common receptor tyrosine kinase (RTK), which has undergone dimerization upon binding of an extracellular stimulus. When activated by GTP at the cell membrane, RAS can mediate RAF activation. These kinases phosphorylate MEK kinases, and these are capable of activating ERK. ERK can target a wide portfolio of substrates in essentially every cellular compartment to regulate the appropriate cellular response [14].

There are three isoforms of the RAS protein, which occur in the thyroid HRAS, KRAS, and NRAS. Mutations may occur in any of the three, but most publications name *NRAS*'s codon 61 as the site being altered most often in case of *RAS* mutation [14]. Codon 61 encodes for the autocatalytic GTPase of RAS, and an alteration therefore could transform the RAS molecule to a constantly signaling trigger toward over- and mis-reaction. Another site that is frequently mutated is the GTP binding site, encoded in codons 12/13, which increases binding affinity or even locks GTP in the activating position. Mutant activation of RAS has been shown in vivo as well as in vitro to be able to induce thyroid neoplasia. Mutations are understood to occur at early stages, when cells are still well differentiated. At this stage, they were also identified as mutually exclusive with other genetic alterations in FTC and PTC. Mutations of one *RAS* isoform occur in up to 20% of PTC and 50% of FTC cases [14]. *RAS* mutations can also occur in follicular adenomas of the thyroid gland. These lesions are rather frequent and benign, but can also present a precursor lesion of FTC [14].

A point mutation at position 1799 from T to A in the *BRAF* gene is the most common gene mutation in PTC and may also occur in ATC. This mutation, also called V600E, results in constitutive activation of the serine/threonine kinase. Generally, the average rate of *BRAF* mutation in PTC and ATC is approximately 44% and 24%, respectively [15]. Especially the tall cell variant of PTC is characterized by *BRAF*^{V600E}, where up to 100% of cases bear the mutation. On the contrary, *BRAF*^{V600E} is rarely found in the follicular variant of PTC [16]. Increasing patients' age was identified as predisposing factor to sporadic *BRAF* mutation [17]. *BRAF*^{V600E} is associated with poor clinical outcome, aggressive pathological features, and higher recurrence rate [18]. Furthermore, the mutation is suggested to influence a patient's sensitivity to radiotherapy, as *BRAF*^{V600E} is described to cause a loss of avidity to radioiodine [19].

Charles et al. [20] have shown that *BRAF*^{V600E} is the driver mutation in adult PTC thyroid carcinogenesis, rather than mutations of *KRAS* (especially *KRAS*^{G12D}). A more aggressive mouse model was introduced by McFadden et al., who confirmed that *BRAF*^{V600E} is sufficient to initiate PTC in adult mice thyroid, but also showed that in advanced ATC, *BRAF*^{V600E} is not sufficient as therapeutic target. A combinatorial approach of MAPK pathway targeting by administration of MEK, as well as BRAF inhibitors, showed improved response rates in their respective mouse model. Furthermore, they could show that progression of *BRAF*^{V600E}-positive thyroid cancer to ATC is facilitated by loss of p53 [21].

As mentioned above, besides MAPK pathway regulators, effector proteins of the PI3K signaling cascade are frequently affected by the processes of molecular carcinogenesis in thyroid cancer. The PI3K cascade can be triggered by RAS signaling or other initiators like tyrosine kinase receptors or G protein-coupled receptors. PI3K catalyzes the transition of phosphatidylinositol bisphosphate (PIP2) to phosphatidylinositol trisphosphate (PIP3). PTEN antagonizes this reaction by dephosphorylating and inactivating PIP3. PIP3 in its phosphorylated form is capable of activating Akt. This serine-threonine kinase has multiple targets, of which TSC2, leading to downstream activation of mTOR, is the most popular and influential on protein synthesis and cell cycle progression. Mutations in *PI3K*, *PTEN*, and *AKT* itself are rare in early stages of thyroid cancer and frequently associated with disease progression or even metastasis [22]. In this context, it has to be noted that *PTEN* mutation and especially its deletion occur in approximately 30% of FTC cases. This is frequently in connection with Cowden's syndrome, a cancer predisposition syndrome with characteristic germ line mutation of *PTEN*. Besides mutation and deletion alterations, *PTEN* expression may also be modified by promoter hypermethylation, which occurs in FTC and ATC cases [23].

10.4 Gene Translocations and Fusions

Among the cell surface tyrosine kinase receptors that transduce extracellular signals to downstream signaling cascades as introduced above, the transmembrane protein rearranged during transfection (RET) is an important player in thyroid carcinogenesis. Mehlen and Bredesen [24] reported RET as belonging to the group of so-called dependence receptors. Unbound, RET possesses proapoptotic activity that is interrupted as soon as a ligand binds to the extracellular domain. This effect is the basis for the concept that *RET*-expressing cells might be controlled in a way that their growth and survival is limited to ligand co-localization. Effects on the development of cancer or other diseases are not elucidated in full detail [25]. Still, the cancer-associated mutant *RET*^{C634R} has no cleavage-dependent proapoptotic effect. In thyroid cancer, the most frequent alteration of *RET* is a genetic disruption of the gene leading to translocation and gene fusion with various heterologous genes. The resulting chimeric oncogenes are termed *RET/PTC* [25]. Prevalence of *RET/PTC* is highly dependent on the cohort under investigation and may range from 25% up to 70% in patient groups, including pediatric patients and individuals with high radioiodine isotope load [26]. Generally, *RET/PTC* formation enables constitutive RET kinase dimerization, activation independent of ligand binding, and autophosphorylation, which leads to steady downstream signaling.

Mouse models harboring *RET/PTC* rearrangements have shown a sufficiency to initiate thyroid carcinogenesis. Corresponding mouse lines were generated by two independent groups, where the transgene was expressed under different promoters and at varying copy numbers. All *RET/PTC* mice developed PTC, thus, at copy-number-dependent rates [27, 28].

RET/PTC rearrangement, other than *BRAF* mutation, which is a distinct tumor indicator, was also reported in benign nodules or healthy tissue surrounding tumor tissues [29]. Various studies report a rate of *RET/PTC* rearrangement in 13–15% of benign nodules. The rate is even higher in individuals with a history of irradiation. The rate of 52.4% *RET/PTC* rearrangement in post-Chernobyl benign nodules is almost identical to the rate of *RET/PTC* rearrangement in PTC [30, 31].

A further gene translocation occurring in up to 60% of FTC cases is the paired box 8 (PAX8)/peroxisome proliferator-activated receptor gamma (*PPARG*) gene fusion. It may also occur in approximately 30% of follicular variant PTC cases [32]. PAX8 plays an essential role in the terminal differentiation steps of thyrocyte development and, unlike other members of the *PAX* gene family, is a key regulator of terminally differentiated gene expression, including the sodium iodide symporter, thyroglobulin, and the TSH receptor [33]. It is discussed controversially how oncogenic the gene fusion *PAX8/PPARG* actually is. On the one hand, it was shown that the resulting fusion protein can act as a suppressor on PPARG-driven gene expression and thereby executes antiapoptotic features. On the other hand, it was shown that the fusion protein can disrupt PAX8 as a transcription factor and deregulate the expression of thyroid-specific genes [33].

10.5 Further Influential Molecular Alterations Contributing to Thyroid Carcinogenesis

Besides the genetic variations introduced above, this section summarizes further frequent and important alterations in thyroid cancer.

With a low involvement in the development but high impact on tumor progression, mutations on *p53* have to be mentioned here. P53 is the most commonly mutated tumor suppressor gene in all human cancers and associates with bad prognosis. Also in thyroid cancer, mutation of *p53* marks a malignant progression of an individual cancer. In accordance, *p53* mutation is detectable in 70–80% of ATC cases [34].

A similar frequency of 70–80% in ATC and 25% in poorly differentiated thyroid cancer is witnessed for mutations of the *CTNNB1* gene. *CTNNB1* encoding for β-catenin is a key regulator in the WNT pathway. Aberrant Wnt signaling is a hallmark of epithelial tumors' developmental phase. The family of Wnt proteins has a physiological role in embryonic development, controlling cell proliferation, cell fate specification, tissue patterning, and cell polarity. Later, in adult tissue, the proteins are involved in tissue homeostasis, as they control cell proliferation, stem cell activation, and self-renewal. Wnt proteins can execute their targeted signaling via three different pathways, including a canonical β-catenin-dependent and two β-catenin-independent pathways. The latter noncanonical pathways are a calcium pathway and planar cell polarity (PCP) signaling [35]. The canonical Wnt signaling, which is dependent on β-catenin, is extensively studied and known to contribute to cancer development and progression in various tumor entities [36]. This signaling cascade is triggered by extracellular binding of a Wnt protein to a frizzled receptor (Fzd) which consequently leads to recruitment of Axin to the cell membrane. This releases Axin's former interaction partners, including β-catenin, which will accumulate in the cytoplasm and enter the nucleus. Nuclear β-catenin acts as a transcription factor for Wnt target genes, like cyclin D1, c-Myc, and further potent regulators of cell proliferation [37]. In the absence of Wnt, β-catenin is tightly bound to E-cadherin in adherens junctions. For many years, Wnt pathway dysregulation was associated solely with ATC and as a feature of aggressive thyroid carcinoma. Throughout the last decade, evidence has been found that altered Wnt signaling is influential also on early stages of thyroid carcinogenesis. Indicators thereof are elevated levels of Wnt family member Wnt5a in FTC and PTC or the stabilization of cytoplasmic β-catenin by RET/PTC [38, 39].

10.6 Epigenetic Modifications in Thyroid Cancer Development and Progression

There are two main fields of epigenetic regulation reported to influence thyroid carcinogenesis. These are, on the one hand, microRNAs (miRNAs) and, on the other hand, aberrant methylation events.

MiRNAs are small noncoding RNAs of approximately 22 nucleotides length that bind to multiple mRNAs, initiating translational repression by cleavage of target mRNA. As for thyroid cancer, it was very early discovered that miRNAs might play an important role in PTC development and progression, as a set of miRNAs could be identified as significantly overexpressed, whereas, in general, cancers are associated with a global under-expression of miRNAs [40]. The most consistently overexpressed miRNAs reported in PTC are miR-221, miR-222, and miR-146b [41].

DNA methylation at cytosines, especially in CpG islands, is a powerful regulator of gene expression. Hypomethylation can cause genetic instability and activation of proto-oncogenes; on the contrary, hypermethylation may silence a gene and is frequently identified as causative for tumor suppressor gene downregulation. Affected tumor suppressor genes in thyroid cancer include *PTEN, RASSF1A, TIMP3, SLC5A8, DAPK, RAPβ2*, and *RAP1GAP* [42]. Aberrant DNA methylation of these genes is detectable in up to 100% of tumors analyzed.

References

1. Sarne D. Effects of the environment, chemicals and drugs on thyroid function. In: De Groot LJ, Beck-Peccoz P, Chrousos G, Dungan K, Grossman A, Hershman JM, Koch C, McLachlan R, New M, Rebar R, Singer F, Vinik A, Weickert MO, editors. Endotext. South Dartmouth, MA: MDText.com, Inc.; 2010.
2. Nussey S, Whitehead S, 2001, Chapter 3: The thyroid gland. Endocrinology: an integrated approach. Oxford: BIOS Scientific Publishers. Available from: http://www.ncbi.nlm.nih.gov/books/NBK28/.
3. DeLellis RA, Nunnemacher G, Wolfe HJ. C-cell hyperplasia. An ultrastructural analysis. Lab Invest. 1977;36:237–48.
4. Refetoff S. Thyroid hormone serum transport proteins. In: De Groot LJ, Chrousos G, Dungan K, et al., editors. Endotext [Internet]. South Dartmouth: MDText.com, Inc.; 2000. https://www.ncbi.nlm.nih.gov/books/NBK285566/. Accessed 7 Jun 2015.
5. Schweizer U, Johannes J, Bayer D, Braun D. Structure and function of thyroid hormone plasma membrane transporters. Eur Thyroid J. 2014;3(3):143–53.
6. Fekete C, Lechan RM. Central regulation of hypothalamic-pituitary-thyroid axis under physiological and pathophysiological conditions. Endocr Rev. 2014;35(2):159–94. doi:10.1210/er.2013-1087.
7. Ferlay J, Soerjomataram I, Ervik M, Dikshit R, Eser S, Mathers C, Rebelo M, Parkin DM, Forman D, Bray, F. GLOBOCAN 2012 v1.1, Cancer incidence and mortality worldwide: IARC CancerBase No. 11. Lyon: International Agency for Research on Cancer; 2014. Available from: http://globocan.iarc.fr. Accessed on 6 Jan 2016.
8. Jemal A, Bray F, Center MM, Ferlay J, Ward E, Forman D. Global cancer statistics. CA Cancer J Clin. 2011;61(2):69–90.
9. Enewold L, Zhu K, Ron E, Marrogi AJ, Stojadinovic A, Peoples GE, Devesa SS. Rising thyroid cancer incidence in the United States by demographic and tumor characteristics, 1980–2005. Cancer Epidemiol Biomark Prev. 2009;18(3):784–91.
10. Katoh H, Yamashita K, Enomoto T, Watanabe M. Classification and general considerations of thyroid cancer. Ann Clin Pathol. 2015;3:1045.
11. Pacini F, De Groot LJ. Thyroid nodules. In: De Groot LJ, Chrousos G, Dungan K, et al., editors. Endotext. South Dartmouth: MDText.com, Inc; 2000.
12. Cox AD, Der CJ. Ras history: the saga continues. Small GTPases. 2010;1:2–27.
13. Dhillon AS, Hagan S, Rath O, Kolch W. MAP kinase singaling pathways in cancer. Oncogene. 2007;26:3279–90.
14. Howell GM, Hodak SP, Yip L. RAS mutations in thyroid cancer. Oncologist. 2013;18:926–32.
15. Caronia LM, et al. Role of BRAF in thyroid oncogenesis. Clin Cancer Res. 2011;17(24):7511–7.
16. Nikiforova MN, Kimura ET, Gandhi M, Biddinger PW, Knauf JA, Basolo F, et al. BRAF mutations in thyroid tumors are restricted to papillary carcinomas and anaplastic or poorly differentiated carcinomas arising from papillary carcinomas. J Clin Endocrinol Metab. 2003;88:5399–404.
17. Ciampi R, Nikiforov YE. RET/PTC rearrangements and BRAF mutations in thyroid tumorigenesis. Endocrinology. 2007;148:936–41.
18. Xing M, Alzahrani AS, Carson KA, Violoa D, Elisei R, Bendlova B, et al. Association between BRAF V600E mutation and mortality in patients with papillary thyroid cancer. JAMA. 2013;309:1493–501.
19. Xing M. BRAF mutation in thyroid cancer. Endocr Relat Cancer. 2005;12:245–62.
20. Charles RP, Iezza G, Amendola E, Dankort D, McMahon M. Mutationally activated $BRAF^{V600E}$ elicits papillary thyroid cancer in the adult mouse. Cancer Res. 2011;71:3863–71.
21. McFadden DG, Vernon A, Santiago PM, Martinez-McFaline R, Bhutkar A, Crowley DM, McMahon M, Sdow PM, Jacks T. p53 constrains progression to anaplastic thyroid carcinoma in a Braf-mutant mouse model of papillary thyroid cancer. Proc Natl Acad Sci U S A. 2014;111(16):E1600–9.

22. Robbins HL, Hague A. The PI3K/Akt pathway in tumors of endocrine tissues. Front Endocrinol. 2016;6:188.
23. Hoiu P, Ji M, Xing M. Association of PTEN gene methylation with genetic alterations in the PI3K/AKT signaling pathway in thyroid tumors. Cancer. 2008;113:2440–7.
24. Mehlen P, Bredesen DE. Dependence receptors: from basic research to drug development. Sci Signal. 2011;4:mr2.
25. Santoro M, Carlomagno F. Central role of RET in thyroid cancer. Cold Spring Harb Perspect Biol. 2013;5:a009233.
26. Zhu Z, Ciampi R, Nikiforova MN, Gandhi M, Nikiforov YE. Prevalence of RET/PTC rearrangements in thyroid papillary carcinoma: effects of the detection methods and genetic heterogeneity. J Clin Endocrinol Metab. 2006;91:3603–10.
27. Jhiang SM, Sagartz JE, Tong Y, Parker-Thornburg J, Capen CC, Cho JY, Xing S, Ledent C. Targeted expression of the RET/PTC1 oncogene induces papillary thyroid carcinomas. Endocrinology. 1996;137:375–8.
28. Santoro M, Chiappetta G, Cerrato A, Salvatore D, Zhang L, Manzo G, Picone A, Portella G, Santelli G, Veccio G, Fusco A. Development of thyroid papillary carcinomas secondary to tissue-specific expression of the RET/PTC1 oncogene in transgenic mice. Oncogene. 1996;12:1821–6.
29. Marotta V, Guerra A, Sapio MR, Vitale M. RET/PTC rearrangement in benign and malignant thyroid disease: a clinical standpoint. Eur J Endocrinol. 2011;165:499–507.
30. Elisei R, Romei C, Vorontsova T, Cosci B, Veremeychik V, Kuchinskaya E, Basolo F, Demidchik EP, Miccoli P, Pinchera A, Pacini F. RET/PTC rearrangements in thyroid nodules: studies in irradiated and not irradiated, malignant and benign thyroid lesions in children and adults. J Clin Endocrinol Metab. 2001;86(7):3211 6.
31. Guerra A, Sapio MR, Marotta V, Campanile E, Moretti MA, Deandrea M, Motta M, Limone PP, Fenzi G, Rossi G, Vitale M. Prevalence of RET/PTC rearrangement in benign and malignant thyroid nodules and its clinical application. Endocr J. 2011;58:31–8.
32. Placzkowski KA, Reddi HV, Grebe SK, Eberardt NL, McIver B. The role of the PAX8/PPARgamma fusion oncogene in thyroid cancer. PPAR Res. 2008;2008:672829.
33. Eberhardt NL, Grebe SKG, McIver B, Reddi HV. The role of the PAX8/PPARγ fusion oncogene in the pathogenesis of follicular thyroid cancer. Mol Cell Endocrinol. 2010;321:50–6.
34. Walerych D, Lisek K, Del Sal G. Mutant p53: one, no one, and one hundred thousand. Front Oncol. 2015;5:289.
35. Sastre-Perona A, Santisteban P. Role of the Wnt pathway in thyroid cancer. Front Endocrinol. 2012;3:31.
36. Ducharte Y, Kim YM, Kahn M. The Wnt signaling pathway in cancer. Crit Rev Oncol Hematol. 2015;15:300093–7.
37. Tetsu O, McCormick F. Beta-catenin regulates expression of cyclin D1 in colon carcinoma cells. Nature. 1999;398:422–6.
38. Kremenevskaja N, von Wasielewski R, Rao AS, Schofl C, Andersson T, Barbant G. Wnt5a has tumor suppressor activity in thyroid carcinoma. Oncogene. 2005;24:2144–54.
39. Tartari CJ, Donadoni C, Manieri E, Mologni L, Mina PD, Villa A, Gambacorti-Passerini C. Dissection of the RET/beta-catenin interaction in the TPC1 thyroid cancer cell line. Am J Cancer Res. 2011;1:716–25.
40. He H, Jazdzewski K, Li W, et al. The role of microRNA genes in papillary thyroid carcinoma. Proc Natl Acad Sci U S A. 2005;102:19075–80.
41. Lee JC, Gundara JS, Glover A, Serpell J, Sidhu SB. MicroRNA expression profiles in the management of papillary thyroid cancer. Oncologist. 2014;19:1141–7.
42. Faam B, Ghaffari MA, Ghadiri A, Azizi F. Epigenetic modifications in human thyroid cancer. Biomed Rep. 2015;3:3–8.

Carcinogenesis in the Epithelium of the Upper Aerodigestive Tract

11

Dietmar Thurnher

Contents

Abstract

Cancer of the head and neck is one of the six most common malignancies, accounting for more than 600,000 case per year worldwide.

Risk factors associated with carcinogenesis of the squamous epithelium of the upper aerodigestive tract are smoking and alcohol consumption, which, when taken together, act highly synergistic. More recently, infection with human papillomavirus (HPV) was noted.

Interestingly, patients with HPV positive oropharyngeal cancer have a significantly better prognosis than patients without HPV infection, regardless of the type of treatment. Currently, it still needs to be clarified whether the HPV vaccination program for cervical cancer will have a future impact on the incidence of oropharyngeal cancer.

D. Thurnher
Department of Otorhinolaryngology, Medical University of Graz, Auenbruggerplatz 26, Graz 8036, Austria
e-mail: dietmar.thurnher@medunigraz.at

J. Haybaeck (ed.), *Mechanisms of Molecular Carcinogenesis – Volume 1*,
DOI 10.1007/978-3-319-53659-0_11

11.1 Introduction

Head and neck squamous cell carcinoma (HNSCC) is one of the six most common malignancies, accounting for more than 600,000 cases per year worldwide [1]. The incidence of cancers of the head and neck area is about 6% of all tumors, the annual mortality being about 6/100,000 worldwide [2].

Head and neck cancer refers to malignancies located in the oral cavity, oropharynx, nasopharynx, hypopharynx, and larynx, of which more than 90% are squamous cell carcinomas of the mucosa.

Risk factors associated with carcinogenesis of head and neck cancers include tobacco and alcohol abuse [3] and, more recently, human papillomavirus infection.

Alcohol and tobacco have a highly synergistic carcinogenic effect, which is possibly due to the fact that alcohol acts as a solvent for the tobacco carcinogens [4].

In addition, women seem to have a higher genetic tobacco-associated susceptibility [5]. In a recent Swedish study, it could be shown that when women inhale lower amounts of tobacco smoke than men, they nevertheless have a higher risk for oral cancer [6]. This gender-specific susceptibility was also demonstrated for lung cancer [7].

Overall, tobacco-induced gene expression in the upper and lower respiratory tract is very similar; the involved molecular processes are, however, already explored much better in lung carcinomas [8].

11.2 Mechanistic Model of Cancer Development as a Result of Tobacco Use

The mechanistic model currently used for carcinoma development by tobacco smoke, also published in the American Surgeon General's Report 2010, was introduced by S. Hecht (Fig. 11.1, modified according to S. Hecht) [9].

With each inhalation of cigarette smoke, more than 5000 substances—including over 70 known carcinogens—are transported into the aerodigestive tract. These very diverse substances belong to different chemical classes: polycyclic aromatic hydrocarbons (polycyclic aromatic hydrocarbons (PAHs)), the "tobacco-specific"

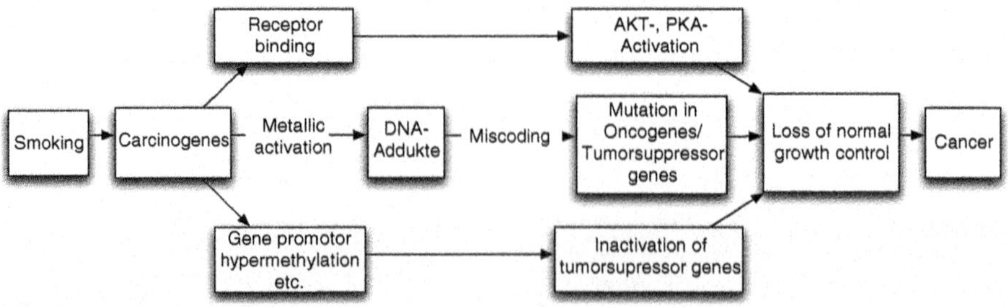

Fig. 11.1 (Modified according to S. Hecht [9])

N-nitrosamine 4-(methylnitrosamino)-1-(3-pyridyl)-1-butanone (NNK) and *N*-nitrosonornicotine (NNN), aromatic amines, aldehydes (formaldehyde), soluble hydrocarbons, and metals (cadmium and the radioactive polonium). Of all known smoking-related carcinogens, NNK and NNN are by far the most potent. Although a subject of controversy in the past, nicotine itself is not seen as a carcinogen in the recent literature [10].

The first step is the conversion and metabolic activation of carcinogens, a process that is significantly catalyzed by isoenzymes of cytochrome P-450, in particular P-450 1A1 and 1B1. This produces metabolites that can bind covalently to DNA, thus forming DNA adducts. This second step, the formation of DNA adducts, is seen as the central process in the molecular carcinogenesis caused by tobacco smoke. If these DNA adducts are not removed by repair mechanisms, it subsequently leads to coding defects during DNA replication and thus to the permanent mutations. In the lung tissue of smokers, thousands of mutations could be found, often in critical state. The proliferation of cells is controlled by genes, such as the oncogene K-Ras or tumor suppressor gene TP53 [11]. Frequently mutated genes, such as CDKN2A or STK11, and mutant genes of the EGFR-RAS-RAF-MEK-ERK signaling pathway have been described [12]. These mutations, therefore, ultimately lead to the formation of cancer.

Some nitrosamines bind without prior metabolism directly to cellular nicotinic receptors and, thereby, are able to activate different molecular pathways, including the AKT or the PKA signaling pathway [13].

Furthermore, the inhalation of tobacco smoke leads to the activation of signal pathways that have a crucial role in inflammatory processes, e.g., the NF-KB signaling pathway, and are involved in the carcinogenesis of lung cancers [14].

In addition to the accumulation of mutations by DNA adduct formation, hypermethylation of gene promoters, which, e.g., might prevent transcription of tumor suppressor genes, is another major process in the development of cancer. An often described example is the tumor suppressor gene CDKN2A (p16). Mutations of p16 in lung cancers are very rare; however, this gene is inactivated by hypermethylation with a prevalence of up to 70% [15]. In recent years, in addition to p16, more than 50 genes have been described in lung cancers, which are inactivated by hypermethylation, including genes of cell cycle regulation (PAX5), DNA repair (AGT), apoptosis (FAS), or invasiveness (E-cadherin) [16]. The same could be shown for head and neck cancers [17].

11.3 Human Papillomavirus and Oropharyngeal Cancer

Human papillomaviruses (HPV) are evolutionarily conserved good and stable viruses with a low mutation rate. Currently, about 120 of these viruses are fully characterized, and at least 30 of these exclusively infect the skin and mucous membranes of the anogenital region. These HPV viruses are classified according to their carcinogenic potential viruses in “low risk” and “high risk.”

The “low-risk” viruses include, for example, HPV serotypes 6 and 11, which are responsible for causing genital warts or recurrent respiratory papillomatosis, but also the serotype HPV 40, 42, 43, 44, etc.

The “high-risk” viruses include HPV 16, 18, 31, 33, etc., with HPV 16 occurring most frequently [18].

In recent years, HPV has been identified as a carcinogen for various carcinomas. These include cervical and anal cancers and also affect the vulva, vagina, penis, and oropharynx.

The association of HPV infection and the emergence of cervical cancers have long been known [19]. In recent years, it was shown that infection with the HPV serotypes 16 and 18 plays a role in the development of some head and neck tumors [20]. In the literature, the prevalence of HPV in head and neck tumors ranges from 11% to 44%, most studies reporting a prevalence of about 20–25% [21]. For oropharyngeal cancer, a distinct change in the epidemiology is recorded. The “classic” tonsillar carcinoma, which is associated with nicotine and alcohol abuse, low socioeconomic status, and typically occurring at an older age, is declining. On the other hand, in oropharyngeal carcinomas of young patients [22], a very high prevalence of HPV was demonstrated (45–90%), particularly in carcinomas of the palate and the base of tongue [23]. These patients with HPV-positive oropharyngeal carcinomas represent a clinical subgroup. They are usually younger, nonsmoking patients who appear to have a better prognosis than patients with HPV-negative oropharyngeal carcinomas. Additional risk factors, such as smoking behavior and/or the expression of associated biomarkers, like p16 [24] and non-mutated p53 or EGFR [25], will subsequently worsen the favorable prognosis.

11.4 HPV-Positive Oropharyngeal Cancer and Response to Therapy

Patients with HPV-positive oropharyngeal cancer have a significantly better response in all currently established standard therapies, i.e., surgical resection with adjuvant therapy or primary radiation therapy with or without chemotherapy [26].

Clinically, HPV-positive oropharyngeal tumors have a typical appearance. In contrast to HPV-negative tumors, which grow ulcerative and highly invasive, HPV-positive oropharyngeal cancers have a typical exophytic and “mulberry-like” growth. HPV-positive oropharyngeal cancer is internationally on the rise. Some authors estimate for the US population that in 2020, the number of HPV-associated oropharyngeal cancers in men and women together will exceed the number of cervical cancers, and thus the otolaryngologist could be the first contact person for patients with an HPV-related disease [27].

Ongoing studies will show whether patients with HPV-positive oropharyngeal cancers can be treated with modified standard therapies, for example, whether a dose reduction of the radio (chemo)therapy would be recommendable [28].

11.5 HPV-Positive Oropharyngeal Cancer and Staging

Furthermore, the significant difference in overall survival of HPV+ positive and – negative oropharyngeal cancers affects not only the clinical trial design and outcomes research but interferes with clinical decision-making because prognosis is not reflected by the 7th edition of the TNM staging system [29]. Therefore, the International Collaboration on Oropharyngeal cancer Network for Staging (ICON-S) aims at developing a TNM classification specific to HPV+ oropharyngeal cancer, which was also proposed for the upcoming 8th edition of the UICC/AJCC TNM classification [30].

11.6 The Potential Impact of Prophylactic Human Papillomavirus Vaccination on Oropharyngeal Cancer

Many countries, in government-sponsored programs, started prophylactic HPV vaccination in girls to prevent cervical cancer [31]. However, current evidence is insufficient to determine the efficacy of these vaccines within the context of oral or oropharyngeal HPV infection or subsequent development of oropharyngeal cancer, especially when only girls are vaccinated [32].

Austria was the first European Country to offer a government-sponsored HPV vaccination program to both girls and boys. Since February 2014, for all children living in Austria, the vaccine is administered in two separate injections at a minimum distance of 6 months in the fourth grade, in the hope that subsequently not only the incidence of cervical cancer but also oropharyngeal cancer will decline [33].

References

1. Jemal A, Bray F, Center MM, et al. Global cancer statistics. CA Cancer J Clin. 2011;61:69–90.
2. Parkin DM, Bray F, Ferlay J, et al. Global cancer statistics, 2002. CA Cancer J Clin. 2005;55:74–108.
3. Marur S, Forastiere AA. Head and neck cancer: changing epidemiology, diagnosis, and treatment. Mayo Clin Proc. 2008;83:489–501.
4. Menvielle G, Luce D, Goldberg A, et al. Smoking, alcohol drinking and cancer risk for various sites of the larynx and hypopharynx. A case control study in France. Eur J Cancer Prev. 2004;13:165–72.
5. Muscat J, Richie JJ, Thompson S. Gender differences in smoking and risk for oral cancer. Cancer Res. 1996;56:5192–7.
6. Rosenquist K, Wennerberg J, Schildt EB, et al. Use of Swedish moist snuff, smoking and alcohol consumption in the aetiology of oral and oropharyngeal squamous cell carcinoma. A population based case-control study in southern Sweden. Acta Otolaryngol. 2005;125:991–8.
7. Joshua E. Muscat, John P. et al. Gender differences in smoking and risk for oral cancer. Cancer Res. 1996;56:5192–7.
8. Zhang X, Sebastini P, Liu G, et al. Similarities and Differences between smoking related gene expression in nasal and bronchial epithelium. Physiol Genomics. 2010;41:1–8.

9. Hecht SS. Tobacco smoke carcinogens and lung cancer. J Natl Cancer Inst. 1999;91: 1194–210.
10. Maier CR, Hollander MC, Hobbs EA, et al. Nicotine does not enhance tumorigenesis in mutant K-ras-driven mouse models of lung cancer. Cancer Prev Res (Phila). 2011;4(11):1743–51.
11. Hecht SS. Progress and challenges in selected areas of tobacco carcinogenesis. Chem Res Toxicol. 2008;21:160–71.
12. Lee W, Jiang Z, Liu J, et al. The mutation spectrum revealed by paired genome sequences from a lung cancer patient. Nature. 2010;465(7297):473–7.
13. Chen RJ, Chang LW, Lin P, et al. Epigenetic effects and molecular mechanisms of tumorigenesis induced by cigarette smoke: an overview. J Oncol. 2011;2011:654931.
14. Lee JM, Yanagawa J, Peebles KA, et al. Inflammation in lung carcinogenesis: new targets for lung cancer chemoprevention and treatment. Crit Rev Oncol Hematol. 2008;66(3):208–17.
15. Zöchbauer-Müller S, Fong KM, Virmani AK, et al. Aberrant promoter methylation of multiple genes in non-small cell lung cancers. Cancer Res. 2001;61(1):249–55.
16. Belinsky SA. Gene-promoter hypermethylation as a bio- marker in lung cancer. Nat Rev Cancer. 2004;4(9):707–17.
17. Demokan S, Dalay N. Role of DNA methylation in head and neck cancer. Clin Epigenetics. 2011;2:123–50.
18. Muñoz N, Bosch FX, de Sanjosé S, et al. Epidemiologic classification of human papillomavirus types associated with cervical cancer. N Engl J Med. 2013;348:518–27.
19. Walboomers JMM, Jacobs MV, Manos MM, et al. Human papillomavirus is a necessary cause of invasive cervical cancer world- wide. J Pathol. 1999;189:12–9.
20. Gillison ML, Lowy DR. A causal role for human papillomavirus in head and neck cancer. Lancet. 2004;363:1488–9.
21. Gillison M. HPV-associated head and neck cancer is a distinct epidemiologic, clinical and molecular entity. Semin Oncol. 2004;31:744–54.
22. Young D, Xiao CC, Murphy B, et al. Increase in head and neck cancer in younger patients due to human papillomavirus (HPV). Oral Oncol. 2015;8:727–30.
23. Chaturvedi AK, Engels EA, Pfeiffer RM, et al. Human papillomavirus and rising oropharyngeal cancer incidence in the United States. J Clin Oncol. 2011;29:4294–301.
24. Ferris RL, Martinez I, Sirianni N, et al. Human papillomavirus-16 associated squamous cell carcinoma of the head and neck (SCCHN): a natural disease model provides insights into viral carcinogenesis. Eur J Cancer. 2005;41:807–15.
25. Reimers N, Kasper HU, Weissenborn SJ, et al. Combined analysis of HPV-DNA, p16 and EGFR expression to predict prognosis in oropharyngeal cancer. Int J Cancer. 2007;120:1731–8.
26. Ang KK, Harris J, Wheeler R, Weber R, et al. Human papillomavirus and survival of patients with oropharyngeal cancer. N Engl J Med. 2010;363:24–35.
27. Zandberg DP, et al. The role of human papilloma virus in nongenital cancers. CA Cancer J Clin. 2013;1:57–81.
28. Owadally W, Hurt C, Timmins H, et al. PATHOS: a phase II/III trial of risk-stratified, reduced intensity adjuvant treatment in patients undergoing transoral surgery for Human papillomavirus (HPV) positive oropharyngeal cancer. BMC Cancer. 2015;15:602.
29. Brizel DM. Different strokes for different folks: new paradigms for staging oropharynx cancer. J Clin Oncol. 2015;33:817–8.
30. O'Sullivan B, Huang HS, Su J, et al. Development and validation of a staging system for HPV-related oropharyngeal cancer by the International Collaboration on Oropharyngeal cancer Network for Staging (ICON-S): a multicentre cohort study. Lancet Oncol. 2016;17(4):440–51.
31. Joura EA, Giuliano AR, Iversen OE, et al. Broad Spectrum HPV Vaccine Study. A 9-valent HPV vaccine against infection and intraepithelial neoplasia in women. N Engl J Med. 2015;372:711–23.
32. Guo T, Eisele DW, Fakry C. The potential impact of prophylactic human papillomavirus vaccination on oropharyngeal cancer. Cancer. 2016;122(15):2313–23.
33. www.bmg.gv.at.

Carcinogenesis of Sinonasal Carcinomas 12

P.V. Tomazic

Contents

Abstract

Sinonasal carcinomas are rare lesions, comprising 1% of all malignancies. They develop in a variety of tissues ranging from epithelial to neuroendocrine origin. The most prevalent ones are squamous cell carcinomas followed by adenoid cystic and adenocarcinomas. Generally speaking, ongoing inflammatory processes in the mucosa and exogenic noxa, such as smoking, might trigger their development. A high risk exists for adenocarcinoma, when patients are exposed to wood dust. Research on signaling pathways and genetic studies are still ongoing, and the first promising results may lead to future development of targeted tumor therapy, where today surgery combined with radio-(chemo)therapy is still the primary choice of treatment (Nat Rev Clin Oncol 11(8):460–472, 2014).

P.V. Tomazic, M.D.
Department of General ORL, H&NS, ENT-University Hospital Graz, Medical University of Graz, Auenbruggerplatz 26, Graz 8036, Austria
e-mail: peter.tomazic@medunigraz.at

J. Haybaeck (ed.), *Mechanisms of Molecular Carcinogenesis – Volume 1*,
DOI 10.1007/978-3-319-53659-0_12

Sinonasal carcinomas are rare tumors, comprising 1% of all malignancies, 3% of all upper respiratory tract malignancies and only 3–5% of all head and neck malignancies. The incidence in European countries (e.g. Italy) is around 0.5–1 new cases/100,000 inhabitants and similar to the United States [1–3]. There is a variety of tumors with respect to their tissue origin: epithelial (squamous cell carcinoma, sinonasal undifferentiated carcinoma, adenocarcinoma, salivary gland-type tumors), soft tissue, neuroendocrine tumors, bone/cartilage soft tissue tumors, neuroectodermal tumors (sinonasal primary melanoma, olfactory neuroblastoma), hematolymphoid tumors, germ cell tumors, borderline tumors, and secondary tumors [2, 3]. The vast majority of these are squamous cell carcinomas, with a 50–80% prevalence, followed by intestinal-type adenocarcinomas with 10–20%. The remaining tumors are very rare, and almost no data exist regarding their carcinogenesis [1].

12.1 Sinonasal Squamous Cell Carcinoma

The development of sinonasal squamous cell carcinoma (SNSCC) may be associated with chronic inflammation due to inhalant irritants and smoking. The ongoing stimulus of TNF and IL-1β might activate downstream transcription of factor NF-κB, playing a major role in tumorigenesis. Cyclooxygenase 2 was also found to be elevated in SNSCC, further underlining the connection between inflammation and carcinogenesis, which was also true for sinonasal inverted papilloma as precursor lesion to SNSCC [4]. Another impact is seen by reactive oxygen species and reactive nitrogen species generating DNA mutations. The production of nitric oxide through inducible nitric oxide synthase may lead to G > A nucleotide transitions in *TP53* and *KRAS*. Similar findings were made with HRAS and NRAS. Another factor influencing this pathway in SNSCC is EGFR being overexpressed in around 40% of SNSCCs. A weaker but still noticeable expression of HER2 was detected in approximately 10% of cases. FGFR1 and SOX2 were detected in 20% and 37%, respectively. Genomic profiling revealed further potential target oncogenes, such as CD44, CCND1/CTTN, and ERBB2 [1, 5]. A rare finding in around 10 cases was a fusion gene of nuclear protein in testis (*NUT*) and *BRD4,* resulting in a BRD4-NUT fusion gene. Histologically, these tumors appear as undifferentiated basaloid cells with focally abrupt squamous differentiation. These tumors are coined NUT midline carcinomas [5, 6].

Precursor lesions, such as inverted papilloma, may be associated with HPV infections and may lead, via metaplasia and cytokeratin switch, to the development of SNSCC, given a high expression of p16. In an HPV-negative tumor, *SYNE1* and *NOTCH3* were found to be mutated. Those genes are responsible for nuclear polarity and terminal differentiation of squamous epithelia [1].

12.2 Carcinoma Ex-Schneiderian/Inverted Papilloma

Inverted (IP) or Schneiderian papillomas are benign tumors; however, they show expansive and destructive growth and can become malignant in up to 27% of cases. As mentioned above, they are associated with HPV infection and p16 protein. Ki-67 proliferation was found to be associated with dysplasia and invasion [7, 8]. Another positive correlation with malignancy was found for *Msx2*, *topoll-α*, and *VEGF* [9]. Jung et al. [10] found that the Wnt signaling pathway was prone to malignant transformation of IP. The overexpression of the signaling proteins beta-catenin, cyclin D1, and Dvl-1 may play a significant role in malignant transformation of IP. Yu et al. proposed that the downregulation of tissue factor pathway inhibitor-2 (TFPI-2) could lead to malignant transformation in IP [11] as well as SMAC, survivin and a decreased expression of E-cadherin and catenin [12].

12.3 Sinonasal Undifferentiated Carcinoma

Sinonasal undifferentiated carcinomas (SNUC) were first described by Frierson as high-grade malignancy with or without neuroendocrine differentiation but without glandular or squamous histological appearance. Compared to SNSCC, they lack the expression of cytokeratins 5/6 and 13. In addition, they present a limited expression of p63. The small cell carcinomas express NSE, synaptophysin, and chromogranin, similar to their pulmonary counterparts [5]. Very little is known about this distinct tumor entity as regards carcinogenesis. Takahashi et al. were able to establish a cell line of SNUC, showing 12 chromosomal translocations [13]. They found out that HER2/neu could be a potential target for SNUC treatment; moreover, lapatinib could induce apoptosis in their cell line [14]. Another potential predictive marker was VEGF [15]. Ansari et al. [16] proposed a successful chemotherapeutic induction scheme (metformin, doxorubicin, and etoposide based on morphoproteomic structure of SNUCs). In two cases, CD133, FASN, topoisomerase II alpha, and mTOR (p-mTOR [Ser 2448]) had an impact on treatment response.

12.4 Adenoid Cystic Carcinoma

Adenoid cystic carcinomas (ACC) are the second largest group of sinonasal tumors after SNSCC [2]. They are characterized by slow growth, perineural growth, and intracranial extension. Cribriform, tubular, and solid subtypes are classified. Little is known about sinonasal as compared to salivary gland carcinogenesis of ACC. A potential marker for tumor development could be the fusion of *MYB-NFIB* proto-oncogenes in ACC [17, 18]. A high expression of Ki-67 correlated with poor outcomes in salivary and non-salivary gland ACC [19].

12.5 Adenocarcinoma

Adenocarcinoma of the sinonasal tract is subdivided into an intestinal-type (ITAC) and a non-intestinal-type adenocarcinoma, where ITACs express cytokeratin 20, villin, and CDX-2. ITAC is associated to wood dust exposure, which was first postulated in the 1960s due to a rising incidence in the furniture industry. The risk for woodworkers is around 900 times higher than in the normal population. A less prominent but significant association was seen in workers of the shoe and leather industry, with a ten-fold elevated risk of developing adenocarcinoma in the sinonasal tract. Similar to SNSCC, chronic inflammation and elevated COX2 levels were found to be associated with tumor development. Smoking, in contrast to SNSCC, was not correlated to ITAC development. Due to this association, ITACs were studied more intensively than non-ITACs. Moreover, they represent the third most common sinonasal malignancy after squamous cell carcinoma and adenoid cystic carcinoma with 8–25% of all sinonasal tumors [1, 2, 5, 20]. ITACs are classified into various pathological subtypes: colonic (40%), solid (20%), papillary (18%), mucinous, and mixed types (22% together). LGALS4 encoding for galectin-4 was found to be upregulated in highly differentiated tumors, whereas clusterin was found to be downregulated. Moreover, a gain in function of genes encoding for growth factors like hepatocyte growth factor (HGF), MOS, and MYCC was found, whereas chromosomal losses were detected for tumor suppressor genes DCC, SMAD4, APC, and TP53 [5].

Promotor methylation of p14(ARF), p16(INK4a), and LOH was found in ITACs. Gene amplification of *CCND1*, *PIK3CA*, and *ERBB1* and *ERBB2* was detected at lower frequency. *EGRF*, *KRAS*, and *BRAF* mutation could be detected in ITACs, where KRAS was associated with a better prognosis [21]. In poorly differentiated sinonasal ITAC, moderate to strong cytoplasmic positivity for ADAM-9 in association with moderate membrane staining for c-erbB-2 oncoprotein was shown by Caltabiano et al. [22].

For *mucoepidermoid* and *acinic cell* carcinoma, no relevant data exists as of today.

12.6 Sinonasal Malignant Melanoma

They account for only 1% of all melanomas [2]. In sinonasal melanoma, *cKIT*, *BRAF*, *NRAS*, and *TERT* mutations are described. However, there exists heterogeneity in the expression of these biomarkers [23, 24].

Chraybi et al. [25] suggested that *NRAS* or *KIT* mutations and *cyclin D1* amplification are significant in these tumors and that targeted therapy should not focus on *BRAF* as its mutation was less frequently seen in his sample. Turri-Zanoni et al. [26] also support that *BRAF* would have low clinical efficacy, but they proposed targeting *RAS* and *KIT* mutations or inhibiting *PI3K-Akt-mTOR* pathways.

More studies are needed to confirm sensitivity to targeted therapies as reported in single cases [27]. In metastatic melanomas, keratins are more strongly expressed than in primary melanomas, particularly K8, K18, and MNF-116 [28].

12.7 Olfactory Neuroblastoma

Olfactory neuroblastoma (ON) arises from cells of the olfactory epithelium, accounting for around 2% of sinonasal neoplasms with peak incidences in the second and sixth decade of life [2, 29]. Trk-A, Trk-B, GRP78, and p75NRT were found to be expressed in ON but did not influence prognosis or outcome [30]. Kim et al. [31] suggested Bcl-2 expression to be associated with better therapeutic response to neoadjuvant chemotherapy but also showed worse survival. Diensthuber et al. [32] suggested that the expression of HIF-1alpha, Epo, Epo-R, and bcl-2 may play a functional role in ON pathogenesis. They suggested bcl-2 acting as a stimulator of angiogenesis in ON, and, thus, it might be targeted in antiangiogenic treatment approaches in ON.

12.8 Neuroendocrine Carcinoma

Neuroendocrine carcinoma (NC) of the sinonasal tract is extremely rare and has poor prognosis [2, 33]. Expression of GLUT-1, HIF-1α, PI3K, and p-Akt was seen in sinonasal NC, as well as in laryngeal NC, and was higher than in precancerous lesions. However, the cohort consisted of only three sinonasal NCs [34]. Achaete-scute homolog 1 (ASH1) seems to be associated with tumor grade where high-grade tumors show an increased protein expression [35].

12.9 Clival Chordoma

Clival chordomas (CC) originate from the notochord and are aggressively growing tumors in the clivus. Due to their lacking response to chemotherapy and difficult localization as regards surgical removal, radiation after maximum tumor debulking is therapy of choice. Rinner et al. [36] described gene hyper- and hypomethylation in chordoma, where the following genes were found to have a potential impact on tumor development: *C3, XIST, TACSTD2, FMR1, HIC1, RARB, DLEC1, KL*, and *RASSF1*. Miozzo et al. [37] found that a tumor suppressor gene on locus 1p36 could be associated with spontaneous but also familial inherited clivus chordoma.

Diaz et al. [38] report on the deletion at 9p involving *CDKN2A*, *CDKN2B*, and *MTAP*, which has been previously reported at higher rates. Moreover, they suggest that chromosome 3 aneuploidy and epigenetic regulation of FHIT contribute to loss of the FHIT tumor suppressor in chordoma. Recently established cell lines would open the possibility of testing these potential therapeutic targets [39].

References

1. Llorente JL, et al. Sinonasal carcinoma: clinical, pathological, genetic and therapeutic advances. Nat Rev Clin Oncol. 2014;11(8):460–72.
2. Haerle SK, et al. Sinonasal carcinomas: epidemiology, pathology, and management. Neurosurg Clin N Am. 2013;24(1):39–49.
3. Lund VJ, et al. European position paper on endoscopic management of tumours of the nose, paranasal sinuses and skull base. Rhinol Suppl. 2010;(22):1–143.
4. Lee GH, et al. Pattern of expression of cyclooxygenase-2 in malignant transformation of sinonasal inverted papilloma. Am J Otolaryngol. 2012;33(5):585–9.

5. Franchi A, et al. Sinonasal carcinomas: recent advances in molecular and phenotypic characterization and their clinical implications. Crit Rev Oncol Hematol. 2011;79(3):265–77.
6. Bell D, Hanna EY. Sinonasal undifferentiated carcinoma: morphological heterogeneity, diagnosis, management and biological markers. Expert Rev Anticancer Ther. 2013;13(3):285–96.
7. Nudell J, Chiosea S, Thompson LD. Carcinoma ex-Schneiderian papilloma (malignant transformation): a clinicopathologic and immunophenotypic study of 20 cases combined with a comprehensive review of the literature. Head Neck Pathol. 2014;8(3):269–86.
8. Tsou YA, et al. Evaluation of correlation of cell cycle proteins and Ki-67 interaction in paranasal sinus inverted papilloma prognosis and squamous cell carcinoma transformation. Biomed Res Int. 2014;2014:634945.
9. Zhang G, et al. Outcomes of the extended endoscopic approach for management of inverted papilloma. J Otolaryngol. 2007;36(2):83–7.
10. Jung YG, et al. Role of Wnt signaling pathway in progression of sinonasal inverted papilloma to squamous cell carcinoma. Am J Rhinol Allergy. 2015;29(3):e81–6.
11. Yu H, et al. The role of tissue factor pathway inhibitor-2 in malignant transformation of sinonasal inverted papilloma. Eur Arch Otorhinolaryngol. 2014;271(8):2191–6.
12. Koo BS, et al. Altered expression of E-cadherin and β-catenin in malignant transformation of sinonasal inverted papillomas. Rhinology. 2011;49(4):479–85.
13. Takahashi Y, et al. Establishment and characterization of novel cell lines from sinonasal undifferentiated carcinoma. Clin Cancer Res. 2012;18(22):6178–87.
14. Takahashi Y, et al. Human epidermal growth factor receptor 2/neu as a novel therapeutic target in sinonasal undifferentiated carcinoma. Head Neck. 2016;38(Suppl 1):E1926–34.
15. Gelbard A, et al. Molecular profiling of sinonasal undifferentiated carcinoma. Head Neck. 2014;36(1):15–21.
16. Ansari M, et al. Sinonasal undifferentiated carcinoma (SNUC): morphoproteomic-guided treatment paradigm with clinical efficacy. Ann Clin Lab Sci. 2013;43(1):45–53.
17. Di Palma S, et al. Primary sinonasal adenoid cystic carcinoma presenting with skin metastases—genomic profile and expression of the MYB-NFIB fusion biomarker. Histopathology. 2014;64(3):453–5.
18. Dillon PM, et al. Adenoid cystic carcinoma: a review of recent advances, molecular targets, and clinical trials. Head Neck. 2016;38(4):620–7.
19. Lin YC, et al. Clinicopathological features of salivary and non-salivary adenoid cystic carcinomas. Int J Oral Maxillofac Surg. 2012;41(3):354–60.
20. Llorente JL, et al. Genetic and clinical aspects of wood dust related intestinal-type sinonasal adenocarcinoma: a review. Eur Arch Otorhinolaryngol. 2009;266(1):1–7.
21. Szablewski V, et al. EGFR expression and KRAS and BRAF mutational status in intestinal-type sinonasal adenocarcinoma. Int J Mol Sci. 2013;14(3):5170–81.
22. Caltabiano R, et al. ADAM-9 expression in intestinal-type adenocarcinoma of the sinonasal tract. Appl Immunohistochem Mol Morphol. 2011;19(3):283–7.
23. Mikkelsen LH, et al. Mucosal malignant melanoma—a clinical, oncological, pathological and genetic survey. Acta Pathol Microbiol Immunol Scand. 2016;124(6):475–86.
24. Zebary A, et al. KIT, NRAS and BRAF mutations in sinonasal mucosal melanoma: a study of 56 cases. Br J Cancer. 2013;109(3):559–64.
25. Chraybi M, et al. Oncogene abnormalities in a series of primary melanomas of the sinonasal tract: NRAS mutations and cyclin D1 amplification are more frequent than KIT or BRAF mutations. Hum Pathol. 2013;44(9):1902–11.
26. Turri-Zanoni M, et al. Sinonasal mucosal melanoma: molecular profile and therapeutic implications from a series of 32 cases. Head Neck. 2013;35(8):1066–77.
27. Rapisuwon S, et al. Novel somatic KIT exon 8 mutation with dramatic response to imatinib in a patient with mucosal melanoma: a case report. Melanoma Res. 2014;24(5):509–11.
28. Safadi RA, et al. Immunohistochemical expression of keratins 6, 7, 8, 14, 16, 18, 19, and MNF-116 pancytokeratin in primary and metastatic melanoma of the head and neck. Oral Surg Oral Med Oral Pathol Oral Radiol. 2016;121(5):510–9.

29. Thompson LD. Olfactory neuroblastoma. Head Neck Pathol. 2009;3(3):252–9.
30. Weinreb I, et al. Expression patterns of Trk-A, Trk-B, GRP78, and p75NRT in olfactory neuroblastoma. Hum Pathol. 2009;40(9):1330–5.
31. Kim JW, et al. Expression of Bcl-2 in olfactory neuroblastoma and its association with chemotherapy and survival. Otolaryngol Head Neck Surg. 2008;139(5):708–12.
32. Diensthuber M, et al. Expression of bcl-2 is associated with microvessel density in olfactory neuroblastoma. J Neurooncol. 2008;89(2):131–9.
33. Montone KT. The differential diagnosis of sinonasal/nasopharyngeal neuroendocrine/neuroectodermally derived tumors. Arch Pathol Lab Med. 2015;139(12):1498–507.
34. Chai L, et al. Clinical features and hypoxic marker expression of primary sinonasal and laryngeal small-cell neuroendocrine carcinoma: a small case series. World J Surg Oncol. 2014;12:199.
35. Taggart MW, et al. Achaete-scute homolog 1 expression closely correlates with endocrine phenotype and degree of differentiation in sinonasal neuroendocrine tumors. Ann Diagn Pathol. 2015;19(3):154–6.
36. Rinner B, et al. Chordoma characterization of significant changes of the DNA methylation pattern. PloS One. 2013;8(3):e56609.
37. Miozzo M, et al. A tumor suppressor locus in familial and sporadic chordoma maps to 1p36. Int J Cancer. 2000;87(1):68–72.
38. Diaz RJ, et al. High-resolution whole-genome analysis of skull base chordomas implicates FHIT loss in chordoma pathogenesis. Neoplasia (New York, N.Y.). 2012;14(9):788–98.
39. Gellner V, et al. Establishment of clival chordoma cell line MUG-CC1 and lymphoblastoid cells as a model for potential new treatment strategies. Sci Rep. 2016;6:24195.

Parotid Cancer

13

Axel Wolf

Contents

Abstract

Salivary gland carcinomas are a heterogeneous group of tumors that comprise about 5–10% of all oropharyngeal cancers. Tumor classification and stage have a significant impact on patient survival. Primary treatment of salivary gland carcinomas is surgical resection in combination with postoperative radiotherapy and/or chemotherapies when required. Molecular abnormalities as potential therapeutic targets differ between certain tumor types.

13.1 Introduction

Salivary glands comprise three pairs of the major glands (parotid, submandibular, or sublingual) and hundreds of the minor salivary glands throughout the mucosa of the respiratory tract. Salivary gland carcinomas (SGCs) are relatively rare malignancies that occur in around 1 per 100,000 people per year counting, accounting for 5–10% of all oral and pharyngeal cancers. The majority of carcinomas occur in the parotid

A. Wolf, M.D.
Department of Otorhinolaryngology, Medical University of Graz,
Auenbruggerplatz 26, Graz 8036, Austria
e-mail: a.wolf@medunigraz.at

J. Haybaeck (ed.), *Mechanisms of Molecular Carcinogenesis – Volume 1*,
DOI 10.1007/978-3-319-53659-0_13

gland (70–75%) followed by the submandibular gland (10–15%), the sublingual gland (<1%), and minor salivary glands [1, 2].

Salivary gland neoplasms can be classified into 24 subtypes of malignant epithelial tumors. Mucoepidermoid carcinomas (MECs, 34%), adenoid cystic carcinomas (ACCs, 22%), and adenocarcinomas (ADNs, 18%) are the most common SGC although demographic and/or geographic aspects may have a significant impact on the incidence of tumors. Primary squamous cell carcinoma (SCC) of the salivary glands is described in the literature although some authors suggest that these tumors might be metastatic intraparotid lymph nodes originating from cutaneous SCC [3–5].

There is no clear evidence that SGC is associated with smoking or alcohol intake, while radiation exposure and UV light exposure are relevant risk factors. Nitroso compounds in rubbers may explain a higher incidence of SGC in rubber industrial workers [6–8].

SGC usually presents as painless swelling. The presence of facial nerve palsy, skin fixation, and cervical lymphadenopathy is highly suspect for malignancies. Clinical examination can be amended with fine-needle aspiration biopsy for cytological examination. Magnetic resonance imaging, computer tomography, and ultrasound are important for diagnosis although definitive diagnosis can only be performed with histologic examination.

The vast majority of malignant salivary gland tumors originate from acinar/ductal epithelial cells and/or myoepithelial/basal cells. Monophasic tumors have one cellular component (e.g., ACN, salivary duct carcinoma, and myoepithelioma), while biphasic tumors originate from both cell types (e.g., ADN, ACC, and epithelial-myoepithelial carcinoma) or demonstrate specific cellular differentiation (e.g., MEC) [5].

Primary treatment of salivary gland carcinomas is surgical resection in combination with postoperative radiotherapy and/or chemotherapies when required. Some authors have described positive effects of neutron-beam radiation due to its reduced toxic effects of the surrounding tissue [1, 9, 10].

SGC are a very heterogeneous group of neoplasms. Numerous general carcinogenic molecular mechanisms, including c-kit, EGF/EGFR, VEGF/VEGFR ErbB-1, ErbB-2, and Her2, are being examined in SGC. Furthermore, loss of the vascular protein sorting-associated protein 4b homolog (VPS4B) might promote carcinogenesis, leading to a prolonging effect of EGFR [11].

Molecular abnormalities, including Her2, c-kit, and EGFR as potential therapeutic targets in certain SGC types, are being investigated and will be discussed below in detail according to the most frequent tumor entities [1].

13.2 Mucoepidermoid Carcinoma

MECs originate from the main ducts and are composed of basal, intermediate, and differentiated cells and may also develop in the lung, skin, breast, cervix, and thyroid [3]. Carcinomas can be graded into low-, intermediate-, and high-grade tumors according to architectural formation and cellular and cytological features, which is of essential prognostic importance [3]. MECs are usually positive for CK5, CK6, CK7, CK8, CK14, CK18, CK 19, EMA, carcinoma antigen, and p63, while negative for CK20, SMA, muscle-specific actin (MSA), and S100. Especially p63 is an important marker to differentiate MEC from ACC and low-grade MEC from mucous retention cysts and papillary adenomas [1].

A t(11;19)(q21-22;p13) tumor-specific gene alteration occurs in 40–70% of MEC. The occurrence of this translocation has been described in other organs, such as the lung and thyroid, but not in other SGCs. It involves the MECT1 gene and the MAML2 gene leading to a fusion gene. The effect on signaling pathways is largely unknown, although it has been shown that it leads to an upregulation of AREG as a ligand of EGFR. This autocrine process promotes MEC growth and survival [12, 13].

Similar to mammary carcinomas, Her2 is a biomarker overexpressed in about 20–40% of MECs. In contrast, rare cases of ACC express Her2. EGFR is overexpressed in about 50% of MECs. Agulnik et al. investigated the effect of lapatinib, an oral tyrosine kinase inhibitor of EGFR and Her2, but no responses were observed [2, 3, 14].

13.3 Adenoid Cystic Carcinoma

ACC develops in both, the major and minor salivary gland, as well as in other sites, e.g., the bronchial tree, breast, cervix, and skin. As in sinonasal carcinomas, three tumor patterns (tubular, cribriform {Swiss cheese}, and solid forms) have been described. Characteristically, tumors grow slowly with early perineural invasion (e.g., leading to paresis of facial nerve in parotid gland tumors). They show a high level of hematogenic metastasis; thus, these biphasic tumors are typically associated with poor outcome [3].

ACC typically expresses CK7, CAM 5.2, calponin, SMA, SMMHC, p53, SOX10, and S100. Overexpression of Ki-67, p53, and H3K9me3 and low expression of H3K9AC are associated with poor survival [1, 15, 16].

Persson et al. described a specific t(6;9) (q22-23; p23-24) gene translocation fusing the *Myb* oncogene to the transcription factor gene *NFIB* and consequent potential activation of *Myb* targets occurring in approximately 60% of ACC. Furthermore, *Myb* overexpression seems to be triggered by other unknown pathways because it occurs in approximately 90%. Due to the frequent expression in ACC, *Myb* appears to be a useful marker for the differential diagnosis of other salivary gland tumors [13, 17, 18].

C-kit is a transmembrane cell surface receptor encoded by the c-kit gene associated with cell migration, differentiation, and proliferation. Hotte et al. observed c-kit overexpression in 90% of ACC, but did not find any mutation or amplifications in the corresponding gene loci. In contrast to ACC, c-kit overexpression was not found in MEC. Especially the predominant c-kit expression in its inner ductal cells is useful to differentiate the ACC from its mimics, although it may also be expressed in low-grade adenocarcinoma. The use of Imatinib as a c-kit inhibitor is a new therapeutic approach but first clinical trials did not show evidence of tumour response. Alternative target therapies may show more promising effects in the future [1, 19, 20].

EGFR overexpression appears in about 30–40%, but, similar to c-kit, no gene mutations or amplifications have been described; thus, the therapeutic impact of EGFR antagonists remains uncertain.

NF-kB being expressed in some ACC may be antagonized by bortezomib used for myeloma treatment. Argiris et al. reported disease stabilization in advanced tumors but no objective responses in a clinical trial using treatment protocols including bortezumib [21].

13.4 Acinic Cell Carcinoma

ACNs make up about 7–17% of all malignant tumors of SGC affecting the parotid gland in the vast majority of cases. Clinically, they often lead to local pain perception and occur bilaterally in 30% of the cases. "High-grade" and "low-grade" tumors can be differentiated, leading to dramatic differences in therapeutic outcome [1, 22].

ACN is a biphasic tumor mostly expressing CK7 and CAM 2.2. ACNs are usually negative for p63, SMA/SMMHC/calponin, and CK20. Originally described in gastrointestinal stromal tumors, the DOG1 protein is expressed in ACN and may be used to be distinguished from mimics, e.g., mammary analogue secretory carcinoma [23].

Very little is known about the genetic profile of ACC. Mitelman et al. reported 11 cases with abnormal karyotypic profile, while common changes were described in three trisomy eight cases only [24, 25].

The complex PI3K axis plays an important role in tumorigenesis, leading to upregulation of several regular tumor growth factors, including EGFR, HER2, and VEGF. Diegel et al. reported on the activation of the pathway in adenomatous polyposis coli (APC)/PTEN transgenic mice that lead to the formation of AZN [26].

13.5 Adenocarcinoma

ADNs show a very aggressive behavior. ADNs express hormonal and growth factors similar to mammary and epipharyngeal adenocarcinomas.

In contrast to ACC, EGFR overexpression (40%) and *EGFR* mutations have been described in ADN; consequently, EGFR inhibitors may play a therapeutic role in this subset of patient. Results of clinical trials using lapatinib, gefitinib, cetuximab and trastuzumab, all EGFR and Her2 antagonist showed limited success. Alternative treatment protocols may show more beneficial effects in the future [3, 27, 28].

Androgen receptors are a pathologic marker for salivary duct carcinomas found in about 20–40% of cases. Similar to antiestrogen therapies, the identification of these receptors might be therapeutically useful.

References

1. Ho K, et al. An overview of the rare parotid gland cancer. Head Neck Oncol. 2011;3:40.
2. Prenen H, Kimpe M, Nuyts S. Salivary gland carcinomas: molecular abnormalities as potential therapeutic targets. Curr Opin Oncol. 2008;20(3):270–4.
3. Adelstein DJ, et al. Biology and management of salivary gland cancers. Semin Radiat Oncol. 2012;22(3):245–53.
4. Spiro RH. Salivary neoplasms: overview of a 35-year experience with 2,807 patients. Head Neck Surg. 1986;8(3):177–84.
5. Zhu S, Schuerch C, Hunt J. Review and updates of immunohistochemistry in selected salivary gland and head and neck tumors. Arch Pathol Lab Med. 2015;139(1):55–66.
6. Mancuso TF, Brennan MJ. Epidemiological considerations of cancer of the gallbladder, bile ducts and salivary glands in the rubber industry. J Occup Med. 1970;12(9):333–41.
7. Muscat JE, Wynder EL. A case/control study of risk factors for major salivary gland cancer. Otolaryngol Head Neck Surg. 1998;118(2):195–8.
8. Whatley WS, Thompson JW, Rao B. Salivary gland tumors in survivors of childhood cancer. Otolaryngol Head Neck Surg. 2006;134(3):385–8.
9. Douglas JG, et al. Treatment of locally advanced adenoid cystic carcinoma of the head and neck with neutron radiotherapy. Int J Radiat Oncol Biol Phys. 2000;46(3):551–7.
10. Huber PE, et al. Radiotherapy for advanced adenoid cystic carcinoma: neutrons, photons or mixed beam? Radiother Oncol. 2001;59(2):161–7.
11. Chen N, Debnath J. Autophagy and tumorigenesis. FEBS Lett. 2010;584(7):1427–35.
12. Jee KJ, et al. Genomic profiles and CRTC1-MAML2 fusion distinguish different subtypes of mucoepidermoid carcinoma. Mod Pathol. 2013;26(2):213–22.
13. Pusztaszeri MP, Faquin WC. Update in salivary gland cytopathology: recent molecular advances and diagnostic applications. Semin Diagn Pathol. 2015;32(4):264–74.
14. Agulnik M, et al. Phase II study of lapatinib in recurrent or metastatic epidermal growth factor receptor and/or erbB2 expressing adenoid cystic carcinoma and non adenoid cystic carcinoma malignant tumors of the salivary glands. J Clin Oncol Off J Am Soc Clin Oncol. 2007;25(25):3978–84.
15. Xia R, et al. High expression of H3K9me3 is a strong predictor of poor survival in patients with salivary adenoid cystic carcinoma. Arch Pathol Lab Med. 2013;137(12):1761–9.
16. Xia R-H, et al. Low expression of endothelin receptor B (EDNRB) is related to H3K9me3 binding with the EDNRB promoter region and is associated with the clinical T tumor stage in salivary adenoid cystic carcinoma. Oral Surg Oral Med Oral Pathol Oral Radiol. 2015;120(2):258–68.
17. Persson M, et al. Recurrent fusion of MYB and NFIB transcription factor genes in carcinomas of the breast and head and neck. Proc Natl Acad Sci U S A. 2009;106(44):18740–4.
18. West RB, et al. MYB expression and translocation in adenoid cystic carcinomas and other salivary gland tumors with clinicopathologic correlation. Am J Surg Pathol. 2011;35(1):92–9.
19. Hotte SJ, et al. Imatinib mesylate in patients with adenoid cystic cancers of the salivary glands expressing c-kit: a Princess Margaret Hospital phase II consortium study. J Clin Oncol Off J Am Soc Clin Oncol. 2005;23(3):585–90.
20. Pfeffer MR, et al. A phase II study of Imatinib for advanced adenoid cystic carcinoma of head and neck salivary glands. Oral Oncol. 2007;43(1):33–6.
21. Argiris A, et al. A phase 2 trial of bortezomib followed by the addition of doxorubicin at progression in patients with recurrent or metastatic adenoid cystic carcinoma of the head and neck: a trial of the Eastern Cooperative Oncology Group (E1303). Cancer. 2011;117(15):3374–82.
22. Barnes EL, Eveson JW, Reichart P, Si-dransky D. Tumors of the salivary glands: introduction. WHO classification of tumours: pathology & genetics. Head and neck tumours. Lyon: IARCPress; 2005. p. 221–2.

23. Miettinen M, Wang Z-F, Lasota J. DOG1 antibody in the differential diagnosis of gastrointestinal stromal tumors: a study of 1840 cases. Am J Surg Pathol. 2009;33(9):1401–8.
24. Mitelman F, Johansson B, Mertens F. Database of chromosome aberrations and gene fusions in cancer. 2014. http://cgap.nci.nih.gov/Chromosomes/Mitelman.
25. Stenman G, Persson F, Andersson MK. Diagnostic and therapeutic implications of new molecular biomarkers in salivary gland cancers. Oral Oncol. 2014;50(8):683–90.
26. Diegel CR, et al. Mammalian target of rapamycin-dependent acinar cell neoplasia after inactivation of Apc and Pten in the mouse salivary gland: implications for human acinic cell carcinoma. Cancer Res. 2010;70(22):9143–52.
27. Murrah VA, Batsakis JG. Salivary duct carcinoma. Ann Otol Rhinol Laryngol. 1994;103(3):244–7.
28. Williams MD, et al. Differential expression of hormonal and growth factor receptors in salivary duct carcinomas: biologic significance and potential role in therapeutic stratification of patients. Am J Surg Pathol. 2007;31(11):1645–52.

Carcinogenesis of Laryngeal Tumors 14

G.P. Hammer

Contents

Abstract

Laryngeal cancer is a rare disease comprising 1–2% of all malignancies with an incidence of 1.1% (men 2.1%, women 0.3%) and mortality of 1.0% worldwide with a geographical variability. Nevertheless, it is the most common of all malignant tumors in the upper aerodigestive tract. Similar to other head and neck regions, it develops in a variety of tissues from epithelial to neuroendocrine origin. In general, squamous carcinogenesis results from successive accumulation of molecular genetic alterations in the squamous epithelium lining the upper aerodigestive tract. Proteins related to cell proliferation and cell cycle regulation have been used (p53, p16, and cyclin D1) as markers of potential genetic results. Squamous cell carcinoma develops from the squamous mucosal lining of the upper aerodigestive tract mainly in patients with a history of abusing risk factors like cigarette smoking, alcohol abuse, or human papillomavirus. Only 20% of individuals with evident risk factors in the personal history, however, develop squamous cell carcinoma. As of today, extensive global research concerning signaling pathways, genetic studies, or carcinogenic mediators is ongoing with the aim of targeting tumor therapy in the future.

G.P. Hammer, M.D.
Division of Phoniatrics, Department of Otorhinolaryngology, Medical University of Graz, Speech and Swallowing, Auenbruggerplatz 26, Graz 8036, Austria
e-mail: GeorgPhilipp.Hammer@klinikum-graz.at

J. Haybaeck (ed.), *Mechanisms of Molecular Carcinogenesis – Volume 1*,
DOI 10.1007/978-3-319-53659-0_14

14.1 Introduction

It is estimated that there were around 156,000 new cases (incidence 1.1%) with laryngeal cancer and 83,000 deaths (mortality 1.0%) from laryngeal cancer worldwide in 2012 [1]. Men have a four times higher risk having a head and neck (H&N) cancer in comparison with women; focusing on the larynx alone, this ratio grows up to a seven times higher risk in men (incidence in men 2.1%, in women 0.3%) [1]. There is a large geographical variability in disease frequency. In Central Europe, it is with 1–2% of all cancers, a relatively rare tumor entity [1, 2]. Nevertheless, it is the most common of all malignancies of the upper respiratory and digestive tracts [3]. In the United States, there were an estimated 12,000 new cases in 2013, resulting in more than 3600 deaths [4]. In an analysis of 250,000 patients with H&N cancer in Europe, a 5-year relative survival was the poorest for the hypopharynx (25%) and the highest for the larynx (59%) [5]. The outcome was significantly better in female than in male patients. Age-standardized 5-year survival remained stable from 1999–2001 to 2005–2007 for laryngeal cancer, while it is increased for all the other H&N cancers [6].

The main risk factors for cancer of the larynx are tobacco and alcohol, which together have a synergistic effect on the risk of laryngeal cancer [7–9]. For smoking, the reported effect estimates have been generally strong and consistent and show a pronounced exposure-response relationship [10]. The etiological role of alcohol is less clear, but evidence suggests both a weaker association and a joint effect with tobacco. It is most common among males and is especially common among those with a history of smoking [11, 12], with some suggesting that the larynx is the organ most susceptible to the deleterious effects of chronic carcinogen inhalation [13]. The incidence in males declined between 1980 and 2005 and increased in females during the same time, most likely due to the changing patterns of tobacco use [11]. Squamous cell carcinoma (SCC) is the most common type of cancer involving the larynx, comprising more than 95% of these cancers [11, 14].

14.2 Squamous Cell Carcinoma

Squamous carcinogenesis results from successive accumulation of molecular genetic alterations in the squamous epithelium lining the upper aerodigestive tract. Although the temporal occurrence and the order of these events are mainly unknown, some certainly precede the phenotypic changes associated with preinvasive dysplastic lesions. The progression of a late-stage dysplasia to invasive carcinoma is a complex one and comprised of both cellular and structural changes as a result of dysregulation of key pathways triggered by interaction of the epithelium and the host stromal elements [15, 16].

The molecular biological analysis and understanding of squamous tumorigenesis of the larynx are mainly based on the concept of field cancerization conceived by Slaughter et al. in 1953 [17]. This concept assumes that risk factors render the entire aerodigestive mucosal surface susceptible to squamous cell carcinoma development. In the small subset of patients with no history of risk factors and/or short temporal exposure to these factors, an inherent genetic predisposition may play a role [15]. According to Szyfter et al. [18], a model of multistep carcinogenesis preceding clinically recognized head and neck squamous cell carcinoma (HNSCC) can be presented by the following scheme:

```
Carcinogens (of different origin) →
   →desoxyribonucleic acid (DNA) damage →
      → chromosome aberrations/gene mutations →
         → altered gene function →
             → loss of control over cell cycle →
                  → increased cell proliferation →
                       → histopathological changes →
                          → squamous cell carcinoma of the larynx
```

A few years later, Almadori et al. [19] published an extensive review of the most recent knowledge on head and neck cancer, focusing on this multistep laryngeal carcinogenesis. The presence of both inciting oncogenes and defective tumor suppression is often necessary for a laryngeal lesion to progress to cancer, with generally 4–10 mutations necessary for abnormal cells to become invasive [5, 13, 19]. Histologic changes have been noted to correlate with these progressive genetic events [13]. Immunohistochemistry utilizes genetic markers to extend traditional histopathology. In general, proteins related to cell proliferation and cell cycle regulation have been used (p53, p16, and cyclin D1) as markers of potential genetic results, similar to other lesions of the upper aerodigestive tract [5]. The tumor protein p53 is encoded in humans by the *TP53* gene, which is located on the short arm of chromosome 17. It is the most frequently mutated tumor suppressor gene in HNSCC in approximately 50% of the cases. Tumors from patients with long histories of risk factor exposure are more frequently mutated. Most of *TP53* mutations are transversion in type (G:T), but missense mutations can also be found and are clustered between exons 5 and 9 [15]. p53 is a well-studied transcription factor that induces cell cycle arrest or apoptosis in response to cellular stress or DNA damage and has been attributed the roles of "guardian of the genome" and "policeman of the oncogenes." The first role consists in sensing and reacting to DNA damage through the ATM/ATR and Chk1/Chk2 kinases and the second in responding to oncogenic signaling through the p53-stabilizing protein ARF [20]. Another tumor suppressor protein is p16, which is encoded by the *CDKN2A* gene on chromosome 9p21. It is a potent inhibitor of the cell cycle; its loss leads to uncontrolled proliferation. In contrast to p53, mutations of p16 are infrequent events in HNSCC. Instead, hypermethylation of the p16 promoter and the first exon is the major mechanism for the loss of function [15, 21–23]. The cyclin-D1 protein is encoded by the *CCND1* gene, which is a critical cell cycle gene within the chromosome 11p amplicon. It has also been found to be highly amplified in advanced premalignant and malignant lesions of the larynx. Polymorphism at this gene has been associated with the high risk of developing squamous cell carcinoma [15, 22, 24]. Additionally, various studies and utilization reports have been published dealing with markers of proliferation (Ki67, PNCA), angiogenesis (VEGF or vascular endothelial growth factor; FGF or fibroblast growth factor), apoptosis dysregulation (Bcl-2), transmembrane receptor dysfunction (EGFR or epidermal growth factor), and cell adhesion (osteopontin, cortactin, and CD44); the results were varying [5, 25–33]. Many molecular factors are reported to be involved in the mechanisms of carcinogenesis in the larynx, such as *RECQL5*, *nucleotide excision repair (NER)* genes, and *NOD2* and *GSTM1* genes [25–29, 33]. DNA repair systems play a pivotal role in maintaining the stability and integrity of the genome, which include nucleotide excision repair, base excision repair (BER), mismatch repair (MMR), and double-strand break repair (DSBR) [30]. Polymorphisms of several *NER* genes have previously been studied in relation to the development of laryngeal cancer [25–27, 31, 33], but the results are conflicting. A recent study by Sun et al. [32] showed that *ERCC1 rs11615* and *ERCC2 rs50871* polymorphisms could influence the risk of laryngeal cancer in Chinese population, particularly among smokers.

Recently, the International Head and Neck Scientific Group (IHNSG) has published a review on literature focusing on biomarkers predicting malignant progression of laryngeal epithelial precursor lesions; reported studies of related markers in laryngeal epithelial precursor lesions (e.g., proliferation markers or cell cycle markers) yield inconsistent conclusions or stand in direct contradiction. A variety of problems could explain these discrepancies, such as general methodological differences, poor study design, assays that are not standardized or lacking reproducibility, and inappropriate or misleading statistical analyses that are often based on sample sizes which are too small to draw meaningful conclusions. In conclusion, various biomarkers have suggested in preliminary investigations that they might ultimately prove to have prognostic value and could be clinically relevant. Focal adhesion kinase (FAK) and cortactin, in particular, have shown the strongest association with laryngeal cancer risk [34].

14.2.1 Carcinogenic Factors in Laryngeal Tumorigenesis

SCC develops from the squamous mucosal lining of the upper aerodigestive tract mainly in patients with a history of risk factors like cigarette smoking, alcohol abuse, or human papillomavirus (Table 14.1). Only 20% of individuals with evident risk factors in the personal history, however, develop squamous cell carcinoma [15].

14.2.1.1 Tobacco

The effect of individual exogenous agents in tobacco carcinogenesis is difficult to assess at the molecular level because there is chronic exposure to a complex mixture of carcinogens, tumor promoters, and cocarcinogens. Apart from lung cancers, several common neoplasms are strongly associated with tobacco use. This is the case for squamous cell carcinomas of the oral cavity, larynx, and esophagus and for cancers of the bladder (both squamous cell carcinomas and transitional cell carcinomas). It is important to note that these cancers occur at variable incidences in

Table 14.1 Preventable exposures associated with laryngeal cancer, as identified by the IACR

Carcinogenic agents with *sufficient* evidence in humans	Agents with *limited* evidence in humans
• Tobacco smoking	• Tobacco smoke, secondhand
• Alcoholic beverages	• *Human papillomavirus* type 16
• Asbestos (all forms)	• Mate drinking, hot
• Acid mists, strong inorganic	• Rubber production industry
	• Sulfur mustard

Adapted from Table 4 in Cogliano et al. [35] (this table does not include factors not covered in the IARC monographs, notably genetic traits, reproductive status, and some nutritional factors)

Table 14.2 Classes of chemicals with sufficient evidence for carcinogenicity in either laboratory animals or humans, according to Hoffmann et al. [37]

Polycyclic aromatic hydrocarbons (PAH) (ten compounds)
• Azaarenes (3)
• *N*-nitrosamines (8)
• Aromatic amines (4)
• Heterocyclic amines (8)
• Aldehydes (2)
• Volatile hydrocarbons (4)
• Nitro compounds (3)
• Miscellaneous organic compounds (12)
• Metals and other inorganic compounds (9)

different regions of the world and that not all of these cancers are a direct consequence of tobacco use [1, 36]. Cigarette smoke is an aerosol which contains about 10^{10} particles/mL and 4800 compounds. According to the International Agency for Research on Cancer (IARC), it contains more than 60 carcinogenic combustion products [7, 37] (Table 14.2). Experimentally, vapor-phase components of the smoke can be separated from the particulate phase by a glass fiber filter. The vapor phase comprises over 90% of the mainstream smoke weight [37]. The main constituents of the vapor phase are nitrogen, oxygen, and carbon dioxide. Potentially, carcinogenic vapor-phase compounds include nitrogen oxides, isoprene, butadiene, benzene, styrene, formaldehyde, acetaldehyde, acrolein, and furan. The particulate phase contains at least 3500 compounds and many carcinogens including polycyclic aromatic hydrocarbons (PAH), *N*-nitrosamines, aromatic amines, and metals. Cigarette smoke condensates reproducibly and robustly causes tumors when applied to a mouse skin and implanted in a rodent lung [38]. Fractions of the condensate which contain PAH also induce tumors in these models, but the concentrations of the PAH are too low to explain the carcinogenicity [39]. Other fractions of the condensate have tumor-promoting and cocarcinogenic activities which enhance the carcinogenicity of the PAH-containing fractions. Inhalation experiments using Syrian golden hamsters demonstrate that whole cigarette smoke and its particulate phase consistently induce preneoplastic lesions and benign and malignant tumors of the larynx [38]. This model system has been widely applied and is the most reliable one for induction of tumors by inhalation of cigarette smoke. Tumors are also observed in hamsters exposed only to the particulate phase of smoke [38]. Studies in A/J mice exposed to environmental tobacco smoke (comprised of 89% mainstream smoke and 11% sidestream smoke in this experimental model) by inhalation demonstrate a small but reproducible increase in lung tumor multiplicity [40]. Tumor induction in this model is due to vapor-phase constituents of cigarette smoke. Thus, there is reliable evidence that both particulate-phase and vapor-phase constituents of cigarette smoke cause tumors in laboratory animals and that tumor promoters and cocarcinogens are also involved in the observed response.

The potential disease burden of the long-term use of electronic cigarettes (ECs) is unclear because ECs are a novel commodity [41]. A study by McAuley et al. [42] comparing the particles and components found in EC vapor and tobacco cigarette (TC) smoke in indoor air samples showed that EC vapor contains significantly less carcinogens and carcinogenic agents than TC, thus posing a significantly lower carcinogenic risk than TC. EC use has the potential to effectively allow TC smokers to quit or decrease TC use, thereby eliminating combustion of carcinogenic TC components and subsequent active and passive exposure to carcinogens exposed directly to smokers, secondhand smokers, and the environment.

During the past decades, the mechanisms by which tobacco smoke causes different cancers and other health effects have been studied intensively. One mechanism involves the mutagenic activity of tobacco smoke, which has been demonstrated clearly and reviewed in the past years [7, 43, 44]. Some reviews have summarized the studies on smoking-related DNA and protein adducts in human tissues [36, 45] as well as the chemical biomarkers associated with tobacco smoke exposure [46]. Tobacco smoking (as well as quid chewing) causes oxidative stress to tissues, that is, the sustained presence of reactive oxygen species (ROS), which initiate free radical reactions. ROS can damage proteins, lipids, carbohydrates, and DNA. Minor DNA damage can result in mutations that can be part of the causal chain for malignant transformation, while sustained DNA damage can result in further perturbations of cell cycle control. In addition to an extensive literature on the carcinogenicity of tobacco smoke in cell and animal models, numerous case-control and cohort studies affirm this key role in human and the super-multiplicative synergism with alcohol drinking [47]. The mechanisms through which cigarette smoking induces tumorigenesis and promotes the development of cancer in the larynx include the formation of bulky DNA adducts, DNA single-/double-strand breaks, chromosome fragmentation/pulverization and the induction of oxidative DNA damages, the genome-wide changes in DNA methylation, as well as the stimulation of tumor angiogenesis [48]. Cigarette smoke and its active compounds impair the fundamental structure of the upper aerodigestive and even the whole gastrointestinal tract through the induction of cellular apoptosis and the inhibition of mucosal cell renewal. It also interferes with the protective mucosal mechanisms by decreasing the blood flow and modulating the mucosal immune system. Furthermore, cigarette smoke also inhibits the synthesis and release of epidermal growth factor (EGF) and polyamines and thereby mucus secretion, which plays an important role in protecting mucosal integrity. Chronic inflammation induced by cigarette smoke exposure releases various inflammatory components, including the cytokines TNF-α, IL-1, and IL-6 and the chemokines CXCL1 and CXCL8. These inflammatory components are capable of promoting tumor growth and tumor adhesion and invasion. Moreover, these mediators also induce angiogenesis and immune suppression in the tumor microenvironment [25, 33].

14.2.1.2 Alcohol

Aside from being a major factor in hepatocarcinogenesis, epidemiological studies show that chronic alcohol consumption is a strong risk factor for several forms of cancer of the oral cavity, pharynx, larynx, and esophagus [49, 50]. Alcohol drinking is second in importance only to smoking as a proven cause of cancer [51]. Therefore, alcoholic beverages have been classified as group 1 carcinogens (carcinogenic to humans) by the IARC. Alcoholic beverages contain many different substances derived from fermentation, such as ethyl carbamate, which may be formed naturally during fermentation and have been proved probably carcinogenic to humans (group 2A) [50]. The ingestion of all types of alcoholic beverages is associated with an increased cancer risk (Fig. 14.1), suggesting an etiological role for ethanol and its primary oxidative metabolite acetaldehyde [8, 9]. A meta-analysis of a large number of studies showed an association of alcohol drinking with an approximately twofold increase in the risk of laryngeal cancer; while light alcohol drinking is not associated with the risk of laryngeal cancer, moderate and heavy drinking is associated with a 1.5-fold and 2.5-fold increased risk, respectively [9].

The mechanisms by which alcoholic drinks exert their carcinogenic effect are not fully understood and probably differ by its target organ, as do other

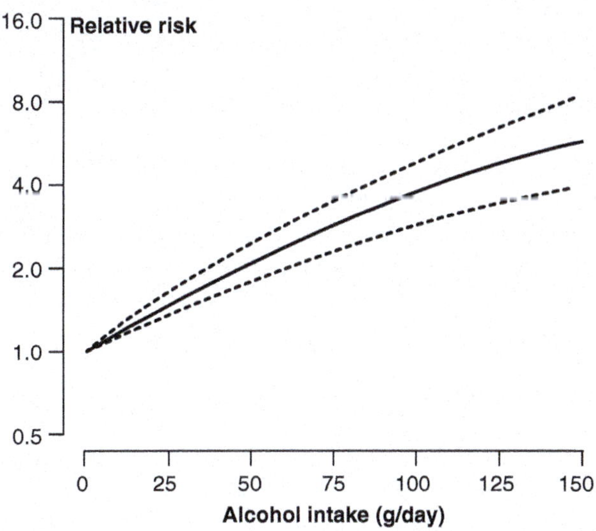

Fig. 14.1 Relative risk function and the corresponding 95% confidence interval, based on the best-fitting dose-risk relationship between alcohol consumption and risk of laryngeal cancer (from [9])

carcinogens that act at many sites [52]. Pure ethanol does not act as a carcinogen in animal studies [53, 54]. The primary metabolite of ethanol-acetaldehyde-is a plausible candidate for the carcinogenic effect of alcoholic drinks, although evidence for acetaldehyde as a direct cause of cancer in human beings is weak [50]. Various theories on the etiology of alcohol-related cancer have been developed. One theory is that exposure to high levels of acetaldehyde, as mentioned before, is responsible for the carcinogenic effect of ethanol, owing to its multiple mutagenic effects on DNA (e.g., forming DNA adducts and therefore triggering the occurrence of replication errors or mutations in oncogenes or tumor suppressor genes) [47]. Another theory suggests that alcohol may act as a solvent that enhances the penetration of carcinogenic compounds, especially tobacco carcinogens, into the mucosa [55, 56]. A third theory is that production of ROS by ethanol induces expression of the cytochrome P450 2E1 (CYP2E1) [57]. Finally, alcohol is highly calorific and lessens the protective effect of beneficial foods such as fruits and vegetables by depressing hunger [47].

14.2.1.3 Pollution

The mechanism of coal- and woodsmoke-mediated carcinogenesis of upper respiratory tract cancers has recently been summarized by Ding et al. [58]. Polycyclic aromatic hydrocarbons with inhalable particles, volatile organic compounds, and some metals are the main carcinogenic components released from solid fuel. While the insoluble particles entering the extra-thoracic or trachea-thoracic regions are cleared by mucociliary mechanisms or via exhalation, those in alveolar regions likely undergo chemical transformations and lead to tumor formation following the uptake of particles by phagocytes and other cells. The deposited particles potentially initiate sustained inflammation, cell injury, cell proliferation, depletion of antioxidants, elevated production of reactive oxygen species, and gene mutation. Polycyclic aromatic hydrocarbons absorbed through the respiratory tract get distributed to most tissues and are metabolized to epoxides, phenols, dihydrodiols, phenol dihydrodiol epoxides, quinines, and tetrols, which are known to bind in the nitrogen bases in DNA and cause deleterious mutations and, eventually, transformation of the cell to a cancerous phenotype. It has been suggested that inflammation is part of the etiology underlying cancer and that measuring inflammation using a marker such as suPAR (plasma-soluble urokinase plasminogen activator receptor) along with the established risk factors, such as age, sex, smoking, and alcohol consumption, could improve cancer risk stratification [58].

14.2.1.4 Human Papillomavirus (HPV)

The most recent review concerning the molecular mechanisms through which HPVs induce carcinogenesis has been published by Zaravinos [59]: The HPV genome is composed of six early (E1, E2, E4, E5, E6, and E7) and two late (L1 and L2) open reading frames and a noncoding long control region (LCR) [60]. E5, E6, and E7 genes encode three viral oncoproteins. E6/E7 proteins function as the dominant oncoproteins of high-risk HPVs inactivating the tumor suppressor proteins, p53 and pRb, respectively. E6 and E7 genes can modify the cell cycle so as to retain the differentiating host keratinocyte in a state that is favorable to the amplification of viral genome replication and consequent late gene expression [61]. HPV E6 in association with host ubiquitin ligase E6-associated protein (E6AP) acts to ubiquitinate p53, leading to its proteasomal degradation [62]. While in most cancers, p53 malfunction is determined by p53 mutations, in HPV-associated carcinomas, wild-type functional p53 is degraded by E6 oncoprotein. Moreover, cells expressing HPV-16 E6 show chromosomal instability [63, 64]. HPV E7 on the other hand inactivates pRb, which controls the G1-S phase transition of the cell cycle by binding the transcription factor E2F. As a consequence, E2F is released with consequent promotion of cell G1-S phase transition and transcription of genes of required proteins for cell cycle progression, such as cyclin E and cyclin A [65, 66]. This functional inactivation of pRb results in a reciprocal overexpression of p16INK4A. The HPV(+) tonsillar SCC shares a disruption of the pRb pathway as a common biological marker. By immunohistochemistry (IHC), most HPV(+) HNSCCs show p16INK4A overexpression. In non-HPV-related HNSCC, continuous tobacco and alcohol exposure can lead to mutational loss of the p16INK4A and p53 genes. These early neoplastic events are detected in 80% of HNSCCs and cause uncontrolled cellular growth [67]. The expression of p53 and Bcl-2 is not associated with HPV(+) oral cavity SCC [68], and mutations in p53 are rarely seen in HPV(+) tumors compared with HPV(−) tumors [69]. Furthermore, there seems to be an inverse relationship between epidermal growth factor receptor (EGFR) expression and HPV status. For patients with oral SCC, high p16INK4A and low EGFR were associated with improved outcome, suggesting a predictive role in surgically treated patients [70]. All HPVs can induce transient proliferation, but only HPV-16 and HPV-18 can immortalize cell lines in vitro. Carcinogenic mechanisms in HPV-associated OSCCs may be similar to those in cervical cancers. However, since the oral cavity and the oropharynx are exposed to higher levels of chemical carcinogens compared to the genital tract, it is likely that different mechanisms are implicated in cervical and oropharyngeal carcinogenesis. Although HPV infection (particularly types 16 and 18) may play a role in the development of laryngeal cancer, there does not appear to be a strong causal association as in oropharyngeal cancer [71]. Three to seven percent of the cases of respiratory papillomatosis undergo malignant change to SCC [72], and-what is interesting-HPV types 6 and 11 prevail in these cases [73].

14.2.1.5 Gastroesophageal Reflux Disease (GERD)

According to the recent extensive review of literature by Herbella et al. [74], a correlation between GERD and carcinogenesis of laryngopharyngeal tumors is evident, especially in nonsmokers. GERD has long been considered as a risk factor for laryngeal/pharyngeal cancer [75, 76]. Gastric contents reach the larynx/pharynx in healthy volunteers and in patients with GERD. This has been proven by different methods, such as dual-probe pH monitoring [77], multichannel intraluminal impedance [78], and aerosolized reflux detection [79]. Few studies did not show GERD as an independent risk factor for cancer in multivariate analysis when tobacco and alcohol consumption are considered [80]; however, other studies, including a meta-analysis, do show GERD as an independent risk factor especially in nonsmokers

[81–84]. Also, the incidence of these tumors is increasingly parallel to GERD. Another piece of evidence that links GERD and laryngeal/pharyngeal cancer is the putative higher risk in patients with heterotopic acid-producing gastric mucosa in the proximal esophagus (inlet patch) [85]. The literature on oral cancer and GERD is scarce even though they also may be associated [86]. GERD induces chronic inflammation and consequent oxidative stress leading to DNA damage. Both acid and bile are active on oncogenic pathways. Acid induces DNA damage, decreases proliferation, and increases apoptosis. Bile salts induce DNA damage, affect proliferation in a pH-dependent manner, and cause resistance to apoptosis. More detailed molecular mechanisms are available in several reviews [87–89].

There is still a lack of a definitive evidence for a causal relation of reflux with laryngeal cancer. The association of reflux with tobacco and alcohol use makes it difficult to separate the relative contribution of reflux to cancer development. Difficulty in diagnosing reflux in a consistent fashion also makes it difficult to draw firm conclusions. In addition, it is necessary to distinguish between reflux that involves the upper esophagus, pharynx, and larynx (so-called laryngopharyngeal reflux) and reflux confined to the lower esophagus (GERD). Most reports fail to make this distinction. Carefully designed studies using differential pH measurement instruments and control groups, matched for lifestyle-related risk factors in particular, are needed to conclusively determine the relationship between reflux and laryngopharyngeal cancer [90].

14.2.1.6 Asbestos

Asbestos is an important nonmetallic mineral raw material. The most common types of asbestos are chrysolite (white) asbestos, amosite (brown), and crocidolite (blue) asbestos. Due to the properties of high intensity, flexibility, heat resistance, electrical nonconductivity, and spinnability, asbestos is widely used in various industries, such as the building, vehicle, and textile industries [91]. Since asbestos was listed in the First Annual Report on Carcinogens by the IARC in the 1970s and 1980s of the twentieth century [92–94], the evidence of carcinogenicity of asbestos was reevaluated in 2009 [95]. IARC concluded that exposure to all forms of asbestos is associated with an increased risk of lung cancer and mesothelioma. In addition, it concluded that there was sufficient evidence from epidemiological studies that asbestos also caused cancer of the larynx and ovary, as well as limited evidence that it caused cancer of the colorectum, pharynx, and stomach [96].

The role of asbestos in the etiology of laryngeal cancer has been investigated over the past years, but the results are inconsistent. The larynx lies directly in the path of an inhaled air stream; thus, asbestos fibers can easily become lodged in the laryngeal mucosa. Asbestos could affect the human immune system and make people sensitive to the development of malignancies. A review by Kumagai-Takei et al. [97] summarizes possible mechanisms for asbestos resulting in cancer; the authors proposed that asbestos fibers having iron produce reactive oxygen/nitrogen species that cause DNA damage to nearby cells and moreover that fibers are directly inserted into the cells and injure chromosomes. A recent meta-analysis by Peng et al. [91] provided an association between asbestos and laryngeal cancer. Workers exposed to asbestos have 1.69 times the likelihood of suffering from laryngeal cancer compared with the general population. Although all forms of asbestos are thought to be harmful to humans, different types of asbestos contributed to different health risks. Crocidolite was more harmful than chrysotile and amosite. It has to be mentioned that in this study, significant association of asbestos with laryngeal cancer was only observed in male cohort, which is related to a larger exposure to risk factors related to laryngeal cancer, such as smoking and alcohol.

Cigarette smoke and asbestos both are considered by most authorities to have a synergistic effect on lung cancer induction, and both are complex carcinogens that can affect multiple steps in the multistage process of carcinogenesis. The combined effect of cigarette smoke and asbestos involves an interactive effect, whereby the joint effect is greater than the sum of the two separate effects [98]. At least four mechanisms have been proposed as potential explanations for the synergy between cigarette smoke and asbestos [99–101]: (1) Cigarette smoke may facilitate penetration of asbestos fibers into the bronchial walls. (2) Carcinogens in cigarette smoke such as benzopyrene may be absorbed onto asbestos fibers with subsequent delivery of the carcinogens into cells at high concentration. (3) Cigarette smoke may interfere with the clearance of asbestos from the lungs. Churg and Stevens recorded elevated concentrations of asbestos fibers in the airway tissues of smokers in comparison to nonsmokers, for both amosite (~sixfold) and chrysotile (~50-fold), especially for short fibers [99]. (4) Free fatty acids in cigarette smoke may translocate iron into cell membranes, with enhancement of cell sensitivity to oxidants such as active oxygen species.

14.2.1.7 Strong Inorganic Acid Mists Containing Sulfuric Acid/Rubber

An increased risk of lung and laryngeal cancer has been suggested in workers exposed to strong inorganic acid, for example, sulfuric acid, in a number of industries, including production of isopropanol and ethanol, steel pickling, battery manufacture and sulfuric acid production, as well as manufacture of soaps and detergents [102]. A strong association was found overall for laryngeal cancer with exposure to sulfuric acid [103].

Rubber processes, dusts, and fumes may cause exposure to many chemicals, including PAHs, chromium (VI) compounds, lead and lead compounds, crystalline silica, cadmium and cadmium compounds, cobalt and cobalt compounds, acrylonitrile, styrene, 1,3-butadiene, and N-nitrosodimethylamine [92, 94]. Mortality and laryngeal cancer morbidity in workers employed in the rubber industry were published in several cohort studies and meta-analyses [104, 105].

14.2.1.8 Genetic Predisposition

An extensive meta-analysis by Guha et al. [106] pooled individual-level data across 12 case-control studies and showed, after adjusting for potential cofactors, an increased risk (OR = 1.7, 95% CI 1.2–2.3) with a family history of H&N cancer in first-degree relatives; the risk was higher when the affected relative was a sibling rather than a parent and for more distal H&N sites such as the hypopharynx and larynx. The highest odds ratio of 7.2 (95% CI 5.5–9.5) was shown among subjects with a positive family history and a chronic tobacco and alcohol abuse. No association was observed for family history of nontobacco-related neoplasms and the risk of HNSCC [47].

14.3 Neuroendocrine Tumors

Neuroendocrine neoplasms of the larynx include both epithelial (carcinomas) and neural-type lesions (paragangliomas). The nomenclature of these tumors has changed quite a bit over time, but recently, clearer categories have emerged that are biologically meaningful. Neuroendocrine neoplasms of the larynx can be clearly categorized into the five tumor types: typical carcinoid, atypical carcinoid, small cell neuroendocrine carcinoma, large cell neuroendocrine carcinoma, and paraganglioma [107]. The diffuse neuroendocrine system is composed of a wide variety of cells from the central nervous system, peripheral nervous system, and virtually all organs, all of which have the common phenotype of producing biologically active amines or peptides such as calcitonin, chromogranin A, bombesin, serotonin, and cholecystokinin [108]. Neuroendocrine cells in the laryngeal mucosa appear to have diverse functions ranging from oxygen sensing to regulation of local epithelial cell growth and regeneration to affecting nearby vasculature and/or autonomic nerve terminals. The term "neuroendocrine neoplasm" has been accepted to encompass a variety of different tumors arising in the larynx and having these features. Although neuroendocrine neoplasms are uncommon tumors of the larynx, they represent the most common non-squamous types of neoplasms arising in this area [109].

According to the most recent reviews by the IHNSG [107, 109], more than 700 neuroendocrine neoplasms of the larynx have been reported in the literature, and approximately 500 publications deal with this relatively uncommon yet intriguing family of laryngeal tumors. These tumors are identified as such by histochemistry showing argyrophilic cytoplasmic granules; by immunohistochemistry showing expression of a variety of neuroendocrine markers, such as synaptophysin, chromogranin, Leu-7, neuron-specific enolase, CD56 (NCAM), CD57, and neurofilament protein; and by ultrastructural examination showing dense-core granules. By producing amines, other molecules, or inducing autoantibodies, they also can, rarely, produce different paraneoplastic syndromes, including Schwartz-Bartter, Cushing, Lambert-Eaton, and carcinoid syndromes [110]. Detailed knowledge of special carcinogenesis of laryngeal neuroendocrine tumors is still poor. In large cell NET, for example, some cases showed a *TP53* point mutation: CAT to AAT transition in codon 179 on exon 5, resulting in a His substitution for Asn. Some authors have suggested that Bcl-2 may regulate cellular NE differentiation; two cases showed diffuse Bcl-2 expression, which may support this notion [111, 112].

14.4 Chondrosarcomas

In 2014, the International Head and Neck Scientific Group (IHNSG) published a summary of the current state of information about biology, diagnosis, and management of chondrosarcomas of the head and neck [113]. According to the definition endorsed by the World Health Organization (WHO), head and neck chondrosarcoma (HNCS) is a malignant tumor characterized by the formation of cartilage, but not of the bone, by tumor cells [114]. Although it is the most common malignant mesenchymal tumor of the larynx, primary laryngeal chondrosarcoma is rare, comprising only 0.2% of head and neck malignancies and 1% of laryngeal malignancies. It shows a predilection for male gender with mean age at presentation between 60 and 64 years [113, 115, 116].

The tumorigenesis of chondrosarcomas is controversial, and several theories are suggested. One theory of origin argues that remnants of the cartilage from failures of the ossification of chondrocranium may persist at the skull base (e.g., at the temporo-occipital junction, middle cranial fossa, sphenoethmoid complex, anterior cranial fossa, and clivus) and give rise to chondrosarcomas [117]. Histologically, it is well known that islands of hyaline cartilage are often present in the area of the nasopalatine duct in adults, and these may account for chondrosarcomas in the anterior maxilla. It is, however, difficult to envisage the origin of mandibular chondrosarcomas from remnants of Meckel's cartilage, an embryonic structure that does not persist after birth [113].

Laryngeal chondrosarcomas usually arise in the cricoid and thyroid cartilages, but are very rare in the epiglottis. This distribution corresponds with a second theory endorsing the development of chondrosarcoma from ossified cartilage. On the other hand, the significant under-representation of chondrosarcomas in the mandibular condyle, a site where calcified cartilage is common in adults, argues against this being the only mechanism of tumorigenesis [113]. Another widely propagated theory is that mesenchymal pluripotential cells undergo malignant transformation and differentiate toward a chondrocytic phenotype [117]. This aspect is used to ascribe the origin of periosteal chondrosarcoma to the periosteum. Despite the popularity of this theory, there is a paucity of good "markers" to establish the existence and confident morphological recognition of such cells [113]. Another theory suggests that chondrosarcomas arise from solitary enchondromas. It is known that chondrosarcomas can develop in nonhereditary skeletal disorders characterized by multiple enchondromas (Ollier disease and Maffucci syndrome), and these conditions are associated with *IDH1* and *IDH2* mutations [118, 119]. The major issue in proving this theory has been the difficulty in histologically distinguishing between chondroma and grade I chondrosarcoma. Other syndromes and disorders in which chondrosarcoma may arise (including Paget's disease) were reviewed by Helliwell [120].

References

1. GLOBOCAN. Estimated cancer incidence, mortality and prevalence worldwide in 2012. 2012. www.globocan.iarc.fr/Pages/fact_sheets_population.aspx. Assessed 23 Feb 2016.
2. Jahnke V, Strange R, Matthias C, Fryer AA. Initial results of glutathione-S-transferase GSTM1 and GSTT1 genotypes and genetic predisposition for laryngeal carcinoma. Laryngorhinootologie. 1995;74(11):691–4.
3. Maier H, Tisch M. Epidemiology of laryngeal cancer: results of the Heidelberg case-control study. Acta Otolaryngol Suppl. 1997;527:160–4.
4. Howlader N, Noone AM, Krapcho M, Garshell J, Miller D, Altekruse SF, Kosary CL, Yu M, Ruhl J, Tatalovich Z, Mariotto A, Lewis DR, Chen HS, Feuer EJ, Cronin KA, editors. SEER cancer statistics review. Bethesda: National Cancer Institute; 1975–2012. http://seer.cancer.gov/csr/1975_2012/, based on November 2014 SEER data submission, posted to the SEER web site, April 2015. Accessed 23 Feb 2016.
5. Russell JO, Scharf J. Premalignant and early malignant lesions of the larynx. In: Sataloff RT, Series editor, Benninger MS, volume editor. Sataloff's comprehensive textbook of otolaryngology: head and neck surgery: laryngology, vol. 4). New Delhi: Jaypee Brothers Medical Publishers (P) Ltd; 2016.
6. Gatta G, Botta L, Sánchez MJ, Anderson LA, Pierannunzio D, Licitra L. EUROCARE Working Group. Prognoses and improvement for head and neck cancers diagnosed in Europe in early 2000s: the EUROCARE-5 population-based study. Eur J Cancer. 2015;51(15):2130–43.
7. IARC Working Group on the Evaluation of Carcinogenic Risks to Humans. Tobacco smoke and involuntary smoking. IARC Monogr Eval Carcinog Risk Chem Hum. 2004;83:1–1438.
8. Pelucchi C, Gallus S, Garavello W, Bosetti C, La Vecchia C. Cancer risk associated with alcohol and tobacco use: focus on upper aero-digestive tract and liver. Alcohol Res Health. 2006;29(3):193–8.
9. Islami F, Tramacere I, Rota M, Bagnardi V, Fedirko V, Scotti L, Garavello W, Jenab M, Corrao G, Straif K, Negri E, Boffetta P, La Vecchia C. Alcohol drinking and laryngeal cancer: overall and dose-risk relation—a systematic review and meta-analysis. Oral Oncol. 2010;46(11):802–10.
10. Olshan A. Cancer of the larynx. In: Schottenfeld D, Fraumeni J, editors. Cancer epidemiology and prevention. 3rd ed. New York: Oxford University Press; 2006.
11. Schultz P. Vocal fold cancer. Eur Ann Otorhinolaryngol Head Neck Dis. 2011;128(6):301–8.
12. Lubin JH, Purdue M, Kelsey K, Zhang ZF, Winn D, Wei Q, Talamini R, Szeszenia-Dabrowska N, Sturgis EM, Smith E, Shangina O, Schwartz SM, Rudnai P, Neto JE, Muscat J, Morgenstern H, Menezes A, Matos E, Mates IN, Lissowska J, Levi F, Lazarus P, La Vecchia C, Koifman S, Herrero R, Franceschi S, Wünsch-Filho V, Fernandez L, Fabianova E, Daudt AW, Maso LD, Curado MP, Chen C, Castellsague X, Brennan P, Boffetta P, Hashibe M, Hayes RB. Total exposure and exposure rate effects for alcohol and smoking and risk of head and neck cancer: a pooled analysis of case-control studies. Am J Epidemiol. 2009;170(8):937–47.
13. Loyo M, Pai SI. The molecular genetics of laryngeal cancer. Otolaryngol Clin North Am. 2008;41(4):657–72.
14. Merati AL, Bielamowicz SA, editors. Textbook of laryngology. San Diego: Plural Publishing; 2007.
15. El-Naggar AK. Cellular and molecular pathology of head and neck tumors. In: Bernier J, editor. Head and neck cancer. New York: Springer; 2011.
16. Mandal M, Myers JN, Lippman SM, Johnson FM, Williams MD, Rayala S, Ohshiro K, Rosenthal DI, Weber RS, Gallick GE, El-Naggar AK. Epithelial to mesenchymal transition in head and neck squamous carcinoma: association of Src activation with E-cadherin down-regulation, vimentin expression, and aggressive tumor features. Cancer. 2008;112(9):2088–100.

17. Braakhuis BJ, Tabor MP, Kummer JA, Leemans CR, Brakenhoff RH. A genetic explanation of Slaughter's concept of field cancerization: evidence and clinical implications. Cancer Res. 2003;63(8):1727–30.
18. Szyfter K, Szmeja Z, Szyfter W, Hemminki K, Banaszewski J, Jaskula-Sztul R, Louhelainen J. Molecular and cellular alterations in tobacco smoke-associated larynx cancer. Mutat Res. 1999;445(2):259–74.
19. Almadori G, Bussu F, Cadoni G, Galli J, Rigante M, Artuso A, Maurizi M. Multistep laryngeal carcinogenesis helps our understanding of the field cancerization phenomenon: a review. Eur J Cancer. 2004;40(16):2383–8.
20. Efeyan A, Serrano M. p53: guardian of the genome and policeman of the oncogenes. Cell Cycle. 2007;6(9):1006–10.
21. Coombes MM, Briggs KL, Bone JR, Clayman GL, AK EN, Dent SY. Resetting the histone code at CDKN2A in HNSCC by inhibition of DNA methylation. Oncogene. 2003;22(55):8902–11.
22. Papadimitrakopoulou VA, Izzo J, Mao L, Keck J, Hamilton D, Shin DM, El-Naggar A, den Hollander P, Liu D, Hittelman WN, Hong WK. Cyclin D1 and p 16 alterations in advanced premalignant lesions of the upper aerodigestive tract: role in response to chemoprevention and cancer development. Clin Cancer Res. 2001;7(10):3127–34.
23. Wang D, Grecula JC, Gahbauer RA, Schuller DE, Jatana KR, Biancamano JD, Lang JC. p16 gene alterations in locally advanced squamous cell carcinoma of the head and neck. Oncol Rep. 2006;15(3):661–5.
24. Nakahara Y, Shintani S, Mihara M, Kiyota A, Ueyama Y, Matsumura T. Alterations of Rb, p16(INK4A) and cyclin D1 in the tumorigenesis of oral squamous cell carcinomas. Cancer Lett. 2000;160(1):3–8.
25. Li X, Xu J, Yang X, Wu Y, Cheng B, Chen D, Bai B. Association of single nucleotide polymorphisms of nucleotide excision repair genes with laryngeal cancer risk and interaction with cigarette smoking and alcohol drinking. Tumour Biol. 2014;35(5):4659–65.
26. Lu B, Li J, Gao Q, Yu W, Yang Q, Li X. Laryngeal cancer risk and common single nucleotide polymorphisms in nucleotide excision repair pathway genes ERCC1, ERCC2, ERCC3, ERCC4, ERCC5 and XPA. Gene. 2014;542(1):64–8.
27. Qi Y, Zhou X. Haplotype analysis of RECQL5 gene and laryngeal cancer. Tumour Biol. 2014;35(3):2669–73.
28. Liu J, He C, Xu Q, Xing C, Yuan Y. NOD2 polymorphisms associated with cancer risk: a meta-analysis. PLoS One. 2014;9(2):e89340.
29. Zhang Y, Chen W, Ji JF, Wang ZY, Wu MH, Zhang K, Wang QP. GSTM1 null polymorphisms is associated with laryngeal cancer risk: a meta-analysis. Tumour Biol. 2014;35(7):6303–9.
30. Goode EL, Ulrich CM, Potter JD. Polymorphisms in DNA repair genes and associations with cancer risk. Cancer Epidemiol Biomarkers Prev. 2002;11(12):1513–30.
31. Abbasi R, Ramroth H, Becher H, Dietz A, Schmezer P, Popanda O. Laryngeal cancer risk associated with smoking and alcohol consumption is modified by genetic polymorphisms in ERCC5, ERCC6 and RAD23B but not by polymorphisms in five other nucleotide excision repair genes. Int J Cancer. 2009;125(6):1431–9.
32. Sun Y, Tan L, Li H, Qin X, Liu J. Association of NER pathway gene polymorphisms with susceptibility to laryngeal cancer in a Chinese population. Int J Clin Exp Pathol. 2015;8(9):11615–21.
33. Li LF, Chan RL, Lu L, Shen J, Zhang L, Wu WK, Wang L, Hu T, Li MX, Cho CH. Cigarette smoking and gastrointestinal diseases: the causal relationship and underlying molecular mechanisms (review). Int J Mol Med. 2014;34(2):372–80.
34. Rodrigo JP, García-Pedrero JM, Suárez C, Takes RP, Thompson LD, Slootweg PJ, Woolgar JA, Westra WH, Brakenhoff RH, Rinaldo A, Devaney KO, Williams MD, Gnepp DR, Ferlito A. Biomarkers predicting malignant progression of laryngeal epithelial precursor lesions: a systematic review. Eur Arch Otorhinolaryngol. 2012;269(4):1073–83.

35. Cogliano VJ, Baan R, Straif K, Grosse Y, Lauby-Secretan B, El Ghissassi F, Bouvard V, Benbrahim-Tallaa L, Guha N, Freeman C, Galichet L, Wild CP. Preventable exposures associated with human cancers. J Natl Cancer Inst. 2011;103(24):1827–39.
36. Pfeifer GP, Denissenko MF, Olivier M, Tretyakova N, Hecht SS, Hainaut P. Tobacco smoke carcinogens, DNA damage and p53 mutations in smoking-associated cancers. Oncogene. 2002;21(48):7435–51.
37. Hoffmann D, Hoffmann I, El-Bayoumy K. The less harmful cigarette: a controversial issue. A tribute to Ernst L. Wynder. Chem Res Toxicol. 2001;14(7):767–90.
38. IARC Working Group on the Evaluation of Carcinogenic Risks to Humans. Tobacco smoking. IARC Monogr Eval Carcinog Risk Chem Hum. 38:35–394;1986.
39. Rubin H. Synergistic mechanisms in carcinogenesis by polycyclic aromatic hydrocarbons and by tobacco smoke: a bio-historical perspective with updates. Carcinogenesis. 2001;22(12):1903–30.
40. Witschi H, Uyeminami D, Moran D, Espiritu I. Chemoprevention of tobacco-smoke lung carcinogenesis in mice after cessation of smoke exposure. Carcinogenesis. 2000;21(5):977–82.
41. Oh AY, Kacker A. Do electronic cigarettes impart a lower potential disease burden than conventional tobacco cigarettes? Review on E-cigarette vapor versus tobacco smoke. Laryngoscope. 2014;124(12):2702–6.
42. McAuley TR, Hopke PK, Zhao J, Babaian S. Comparison of the effects of e-cigarette vapor and cigarette smoke on indoor air quality. Inhal Toxicol. 2012;24(12):850–7.
43. DeMarini DM. Genotoxicity of tobacco smoke and tobacco smoke condensate. Mutat Res. 1983;114(1):59–89.
44. DeMarini DM. Genotoxicity of tobacco smoke and tobacco smoke condensate: a review. Mutat Res. 2004;567(2 3):447–74.
45. Phillips DH. Smoking-related DNA and protein adducts in human tissues. Carcinogenesis. 2002;23(12):1979–2004.
46. Hecht SS. Tobacco carcinogens, their biomarkers and tobacco-induced cancer. Nat Rev Cancer. 2003;3(10):733–44.
47. Johnson NW, Amarasinghe HK. Epidemiology and aetiology of head and neck cancers. In: Bernier J, editor. Head and neck cancer. New York: Springer; 2011.
48. Starska K, Forma E, Nowacka-Zawisa M, Lewy-Trenda I, Ciesielski P, Pietruszewska W, Skóra M, Bryś M. The c.*229C > T gene polymorphism in 3'UTR region of the topoisomerase IIβ binding protein 1 gene and LOH in BRCA1/2 regions and their effect on the risk and progression of human laryngeal carcinoma. Tumour Biol. 2016;37(4):4541–57.
49. Yu HS, Oyama T, Isse T, Kitagawa K, Pham TT, Tanaka M, Kawamoto T. Formation of acetaldehyde-derived DN adducts due to alcohol exposure. Chem Biol Interact. 2010;188(3):367–75.
50. IARC. Alcohol consumption and ethyl carbamate. IARC Monogr Eval Carcinog Risks Hum. 2010;96:3–1383.
51. Austoker J. Reducing alcohol intake. BMJ. 1994;308(6943):1549–52.
52. Boffetta P, Hashibe M. Alcohol and cancer. Lancet Oncol. 2006;7(2):149–56.
53. World Health Organization. Global status report on alcohol. Geneva: World Health Organization; 1999.
54. Boyle P, Autier P, Bartelink H, Baselga J, Boffetta P, Burn J, Burns HJ, Christensen L, Denis L, Dicato M, Diehl V, Doll R, Franceschi S, Gillis CR, Gray N, Griciute L, Hackshaw A, Kasler M, Kogevinas M, Kvinnsland S, La Vecchia C, Levi F, McVie JG, Maisonneuve P, Martin-Moreno JM, Bishop JN, Oleari F, Perrin P, Quinn M, Richards M, Ringborg U, Scully C, Siracka E, Storm H, Tubiana M, Tursz T, Veronesi U, Wald N, Weber W, Zaridze DG, Zatonski W, zur Hausen H. European code against cancer and scientific justification: third version. Ann Oncol. 2003;14(7):973–1005.
55. Wright AJ, Ogden GR. Possible mechanisms by which alcohol may influence the development of oral cancer—a review. Oral Oncol. 1998;34(6):441–7.
56. Brennan P, Boffetta P. Mechanistic considerations in the molecular epidemiology of head and neck cancer. IARC Sci Publ. 2004;157:393–414.

57. Seitz HK, Stickel F. Molecular mechanisms of alcohol-mediated carcinogenesis. Nat Rev Cancer. 2007;7(8):599–612.
58. Ding N, Zhou N, Zhou M, Ren GM. Respiratory cancers and pollution. Eur Rev Med Pharmacol Sci. 2015;19(1):31–7.
59. Zaravinos A. An updated overview of HPV-associated head and neck carcinomas. Oncotarget. 2014;5(12):3956–69.
60. Ganguly N, Parihar SP. Human papillomavirus E6 and E7 oncoproteins as risk factors for tumorigenesis. J Biosci. 2009;34(1):113–23.
61. Munger K, Howley PM. Human papillomavirus immortalization and transformation functions. Virus Res. 2002;89(2):213–28.
62. Ault KA. Epidemiology and natural history of human papillomavirus infections in the female genital tract. Infect Dis Obstet Gynecol. 2006;2006:5.
63. Lee D, Kwon JH, Kim EH, Kim ES, Choi KY. HMGB2 stabilizes p53 by interfering with E6/E6AP-mediated p53 degradation in human papillomavirus-positive HeLa cells. Cancer Lett. 2010;292(1):125–32.
64. Tomaic V, Pim D, Thomas M, Massimi P, Myers MP, Banks L. Regulation of the human papillomavirus type 18 E6/E6AP ubiquitin ligase complex by the HECT domain-containing protein EDD. J Virol. 2011;85(7):3120–7.
65. Classon M, Harlow E. The retinoblastoma tumour suppressor in development and cancer. Nat Rev Cancer. 2002;2(12):910–7.
66. Masciullo V, Khalili K, Giordano A. The Rb family of cell cycle regulatory factors: clinical implications. Int J Oncol. 2000;17(5):897–902.
67. Califano J, van der Riet P, Westra W, Nawroz H, Clayman G, Piantadosi S, Corio R, Lee D, Greenberg B, Koch W, Sidransky D. Genetic progression model for head and neck cancer: implications for field cancerization. Cancer Res. 1996;56(11):2488–92.
68. Oliveira MC, Soares RC, Pinto LP, Souza LB, Medeiros SR, Costa Ade L. High-risk human papillomavirus (HPV) is not associated with p53 and bcl-2 expression in oral squamous cell carcinomas. Auris Nasus Larynx. 2009;36(4):450–6.
69. Fallai C, Perrone F, Licitra L, Pilotti S, Locati L, Bossi P, Orlandi E, Palazzi M, Olmi P. Oropharyngeal squamous cell carcinoma treated with radiotherapy or radiochemotherapy: prognostic role of TP53 and HPV status. Int J Radiat Oncol Biol Phys. 2009;75(4):1053–9.
70. Liang C, Marsit CJ, McClean MD, Nelson HH, Christensen BC, Haddad RI, Clark JR, Wein RO, Grillone GA, Houseman EA, Halec G, Waterboer T, Pawlita M, Krane JF, Kelsey KT. Biomarkers of HPV in head and neck squamous cell carcinoma. Cancer Res. 2012;72(19):5004–13.
71. Upile NS, Shaw RJ, Jones TM, Goodyear P, Liloglou T, Risk JM, Boyd MT, Sheard J, Sloan P, Robinson M, Schache AG. Squamous cell carcinoma of the head and neck outside the oropharynx is rarely human papillomavirus related. Laryngoscope. 2014;124(12):2739–44.
72. Gallagher TQ, Derkay CS. Recurrent respiratory papillomatosis: update 2008. Curr Opin Otolaryngol Head Neck Surg. 2008;16(6):536–42.
73. Jeong WJ, Park SW, Shin M, Lee YJ, Jeon YK, Jung YH, Hah JH, Kwon TK, Song YS, Kim KH, Sung MW. Presence of HPV type 6 in dysplasia an carcinoma arising from recurrent respiratory papillomatosis. Head Neck. 2009;31(8):1095–101.
74. Herbella FA, Neto SP, Santoro IL, Figueiredo LC. Gastroesophageal reflux disease and non-esophageal cancer. World J Gastroenterol. 2015;21(3):815–9.
75. Ward PH, Hanson DG. Reflux as an etiological factor of carcinoma of the laryngopharynx. Laryngoscope. 1988;98(11):1195–9.
76. Koufman JA, Burke AJ. The etiology and pathogenesis of laryngeal carcinoma. Otolaryngol Clin North Am. 1997;30(1):1–19.
77. Neto SC, Herbella FA, Silva LC, Patti MG. Ratio between proximal/distal gastroesophageal reflux does not discriminate abnormal proximal reflux. World J Surg. 2014;38(4):890–6.
78. Oelschlager BK, Quiroga E, Isch JA, Cuenca-Abente F. Gastroesophageal and pharyngeal reflux detection using impedance and 24-hour pH monitoring in asymptomatic subjects: defining the normal environment. J Gastrointest Surg. 2006;10(1):54–62.

79. Sun G, Muddana S, Slaughter JC, Casey S, Hill E, Farrokhi F, Garrett CG, Vaezi MF. A new pH catheter for laryngopharyngeal reflux: normal values. Laryngoscope. 2009;119(8):1639–43.
80. Francis DO, Maynard C, Weymuller EA, Reiber G, Merati AL, Yueh B. Reevaluation of gastroesophageal reflux disease as a risk factor for laryngeal cancer. Laryngoscope. 2011;121(1):102–5.
81. Bacciu A, Mercante G, Ingegnoli A, Ferri T, Muzzetto P, Leandro G, Di Mario F, Bacciu S. Effects of gastroesophageal reflux disease in laryngeal carcinoma. Clin Otolaryngol Allied Sci. 2004;29(5):545–8.
82. Qadeer MA, Colabianchi N, Vaezi MF. Is GERD a risk factor for laryngeal cancer? Laryngoscope. 2005;115(3):486–91.
83. Vaezi MF, Qadeer MA, Lopez R, Colabianchi N. Laryngeal cancer and gastroesophageal reflux disease: a case-control study. Am J Med. 2006;119(9):768–76.
84. Langevin SM, Michaud DS, Marsit CJ, Nelson HH, Birnbaum AE, Eliot M, Christensen BC, McClean MD, Kelsey KT. Gastric reflux is an independent risk factor for laryngopharyngeal carcinoma. Cancer Epidemiol Biomarkers Prev. 2013;22(6):1061–8.
85. Chong VH. Clinical significance of heterotopic gastric mucosal patch of the proximal esophagus. World J Gastroenterol. 2013;19(3):331–8.
86. Mercante G, Bacciu A, Ferri T, Bacciu S. Gastroesophageal reflux as a possible co-promoting factor in the development of the squamous-cell carcinoma of the oral cavity, of the larynx and of the pharynx. Acta Otorhinolaryngol Belg. 2003;57(2):113–7.
87. Wang DH, Souza RF. Biology of Barrett's esophagus and esophageal adenocarcinoma. Gastrointest Endosc Clin N Am. 2011;21(1):25–38.
88. Denlinger CE, Thompson RK. Molecular basis of esophageal cancer development and progression. Surg Clin North Am. 2012;92(5):1089–103.
89. Fang Y, Chen X, Bajpai M, Verma A, Das KM, Souza RF, Garman KS, Donohoe CL, O'Farrell NJ, Reynolds JV, Dvorak K. Cellular origins and molecular mechanisms of Barrett's esophagus and esophageal adenocarcinoma. Ann N Y Acad Sci. 2013;1300:187–99.
90. Coca-Pelaz A, Rodrigo JP, Takes RP, Silver CE, Paccagnella D, Rinaldo A, Hinni ML, Ferlito A. Relationship between reflux and laryngeal cancer. Head Neck. 2013;35(12):1814–8.
91. Peng WJ, Mi J, Jiang YH. Asbestos exposure and laryngeal cancer mortality. Laryngoscope. 2016;126(5):1169–74.
92. IARC. Asbestos. In: Overall evaluations of carcinogenicity. IARC monographs on the evaluation of carcinogenic risks to humans, suppl 7. Lyon: International Agency for Research on Cancer; 1987. p. 106–16.
93. IARC. Asbestos. monographs on the evaluation of carcinogenic risks to humans, vol. 14. Lyon: International Agency for Research on Cancer; 1997. p. 106.
94. IARC. Overall evaluations of carcinogenicity: an updating of IARC monographs volumes 1 to 42. IARC Monogr Eval Carcinog Risks Hum Suppl. 1987;7:1–440.
95. Straif K, Benbrahim-Tallaa L, Baan R, Grosse Y, Secretan B, El Ghissassi F, Bouvard V, Guha N, Freeman C, Galichet L, Cogliano V, WHO International Agency for Research on Cancer Monograph Working Group. A review of human carcinogens—Part C: metals, arsenic, dusts, and fibres. Lancet Oncol. 10(5):453–4;2009.
96. Im S, Youn KW, Shin D, Lee MJ, Choi SJ. Review of carcinogenicity of asbestos and proposal of approval standards of an occupational cancer caused by asbestos in Korea. Ann Occup Environ Med. 2015;27:34.
97. Kumagai-Takei N, Maeda M, Chen Y, Matsuzaki H, Lee S, Nishimura Y, Hiratsuka J, Otsuki T. Asbestos induces reduction of tumor immunity. Clin Dev Immunol. 2011;2011:9.
98. Erren TC, Jacobsen M, Piekarski C. Synergy between asbestos and smoking on lung cancer risks. Epidemiology. 1999;10(4):405–11.
99. Churg A, Stevens B. Enhanced retention of asbestos fibers in the airways ofhuman smokers. Am J Respir Crit Care Med. 1995;151(5):1409–13.
100. Nelson HH, Kelsey KT. The molecular epidemiology of asbestos and tobacco in lung cancer. Oncogene. 2002;21(48):7284–8.

101. Bach PB, Kattan MW, Thornquist MD, Kris MG, Tate RC, Barnett MJ, Hsieh LJ, Begg CB. Variations in lung cancer risk among smokers. J Natl Cancer Inst. 2003;95(6):470–8.
102. IARC. Occupational exposures to mists and vapours from strong inorganic acids. IARC Monogr Eval Carcinog Risks Hum. 1992;54:41–130.
103. Soskolne CL, Jhangri GS, Siemiatycki J, Lakhani R, Dewar R, Burch JD, Howe GR, Miller AB. Occupational exposure to sulfuric acid in southern Ontario, Canada, in association with laryngeal cancer. Scand J Work Environ Health. 1992;18(4):225–32.
104. Alder N, Fenty J, Warren F, Sutton AJ, Rushton L, Jones DR, Abrams KR. Meta-analysis of mortality and cancer incidence among workers in the synthetic rubber-producing industry. Am J Epidemiol. 2006;164(5):405–20.
105. Brown T, Darnton A, Fortunato L, Rushton L, British Occupational Cancer Burden Study Group. Occupational cancer in Britain. Respiratory cancer sites: larynx, lung and mesothelioma. Br J Cancer. 2012;107(Suppl 1):56–70.
106. Guha N, Boffetta P, Wunsch Filho V, Eluf Neto J, Shangina O, Zaridze D, Curado MP, Koifman S, Matos E, Menezes A, Szeszenia-Dabrowska N, Fernandez L, Mates D, Daudt AW, Lissowska J, Dikshit R, Brennan P. Oral health and risk of squamous cell carcinoma of the head and neck and esophagus: results of two multicentric case-control studies. Am J Epidemiol. 2007;166(10):1159–73.
107. Lewis Jr JS, Ferlito A, Gnepp DR, Rinaldo A, Devaney KO, Silver CE, Travis WD, International Head and Neck Scientific Group. Terminology and classification of neuroendocrine neoplasms of the larynx. Laryngoscope. 2011;121(6):1187–93.
108. Pearse AG. The diffuse neuroendocrine system and the apud concept: related "endocrine" peptides in brain, intestine, pituitary, placenta, and anuran cutaneous glands. Med Biol. 1977;55(3):115–25.
109. Ferlito A, Silver CE, Bradford CR, Rinaldo A. Neuroendocrine neoplasms of the larynx: an overview. Head Neck. 2009;31(12):1634–46.
110. Ferlito A, Rinaldo A. Paraneoplastic syndromes in patients with cancer of the larynx and hypopharynx. Ann Otol Rhinol Laryngol. 2007;116(7):502–13.
111. Jiang SX, Kameya T, Shinada J, Yoshimura H. The significance of frequent and independent p53 and bcl-2 expression in large-cell neuroendocrine carcinomas of the lung. Mod Pathol. 1999;12(4):362–9.
112. Kusafuka K, Ferlito A, Lewis Jr JS, Woolgar JA, Rinaldo A, Slootweg PJ, Gnepp DR, Devaney KO, Travis WD, Barnes L. Large cell neuroendocrine carcinoma of the head and neck. Oral Oncol. 2012;48(3):211–5.
113. Coca-Pelaz A, Rodrigo JP, Triantafyllou A, Hunt JL, Fernández-Miranda JC, Strojan P, de Bree R, Rinaldo A, Takes RP, Ferlito A. Chondrosarcomas of the head and neck. Eur Arch Otorhinolaryngol. 2014;271(10):2601–9.
114. Fletcher CDM. The evolving classification of soft tissue tumours—an update based on the new 2013 WHO classification. Histopathology. 2013;64(1):2–11.
115. Koch BB, Karnell LH, Hoffman HT, Apostolakis LW, Robinson RA, Zhen W, Menck HR. National cancer database report on chondrosarcoma of the head and neck. Head Neck. 2000;22(4):408–25.
116. Fidai SS, Ginat DT, Langerman AJ, Cipriani NA. Dedifferentiated chondrosarcoma of the Larynx. Head Neck Pathol. 2004;97(3):369–75.
117. Neff B, Sataloff RT, Storey L, Hawkshaw M, Spiegel JR. Chondrosarcoma of the skull base. Laryngoscope. 2002;112(1):134–9.
118. Rosenberg A, Nielsen P, Keel SB, Renard LG, Fitzek MM, Munzenrider JE, Liebsch NJ. Chondrosarcomas of the base of skull: a clinicopathologic study of 200 cases with emphasis on its distinction from chordoma. Am J Surg Pathol. 1999;23(11):1370–8.
119. Lustig LR, Sciubba J, Holliday MJ. Chondrosarcomas of the skull base and temporal bone. J Laryngol Otol. 2007;121(8):725–35.
120. Helliwell TR. Central cartilaginous neoplasms of bone. In: Helliwell TR, editor. Pathology of bone and joint neoplasms, Major problems in pathology, vol. 37. Philadelphia: Saunders; 1999. p. 193–214.

Photodynamic Therapy for Esophageal Cancer: Functional, Clinical, and Immunological Aspects

15

Joerg Lindenmann and Nicole Fink-Neuboeck

Contents

J. Lindenmann, M.D. (✉) • N. Fink-Neuboeck
Division of Thoracic- and Hyperbaric Surgery, Department of Surgery, Medical University of Graz, Auenbruggerplatz 29, 8036 Graz, Austria
e-mail: jo.lindenmann@medunigraz.at

J. Haybaeck (ed.), *Mechanisms of Molecular Carcinogenesis – Volume 1*,
DOI 10.1007/978-3-319-53659-0_15

Abstract

Patients suffering from esophageal cancer ineligible for curative surgery still have a poor prognosis. Endoscopic photodynamic therapy (PDT) represents a feasible and effective approach beneath the different treatment options for this aggressive malignancy. The mechanism of function is based on the illumination of malignant tumor tissue with laser light after selective accumulation of the specific photosensitizer in these tumor cells. As a result of this photochemical reaction, distinct necrosis of the endoluminal tumor develops quickly resulting in reducing tumor load and increasing quality of life. In case of esophageal cancer, PDT is performed endoscopically under general anesthesia. Moreover, PDT does not only represent a local anticancer therapy limited to the carcinoma. By inducing a considerable systemic inflammatory response, PDT is able to interfere with the human immune system. On the one hand, PDT induces the synthesis of tumor-specific cytotoxic T cells which are able to eliminate distant untreated tumor cells. On the other hand, PDT leads to the development of antitumor memory immunity that may sustainably prevent tumor recurrence.

15.1 Introduction

Esophageal cancer, one of the least studied cancers, is the sixth-leading cause of death from cancer worldwide [1]. There are remarkable geographical variations in the incidence of esophageal carcinoma from approximately 10 cases per 100,000 persons in Europe and the USA to 139 cases per 100,000 persons in some areas of China [2–4]. Esophageal squamous cell carcinoma represents one of the two most common tumor subtypes, which is mainly associated with tobacco smoking [5, 6] and heavy alcohol use [5] but also with age >60 years, male gender, and heredity [1, 7, 8]. Esophageal adenocarcinoma represents the second histologic subtype, which increasingly gains in both frequency and importance. It is mainly associated with chronic gastroesophageal reflux disease and obesity [9].

Up to now the conventional treatment modality for the cure of esophageal cancer is still esophageal resection either with or without radio-chemotherapy. These radical procedures are, however, associated with a high morbidity and mortality in these very patients, who are mostly in a reduced nutritional as well as performance status often because of accompanying diseases such as limited cardiac and/or pulmonary function or impaired liver function.

In general, the prognosis of patients with operable esophageal cancer depends on the depth of tumor penetration (T-stage) and the extent of lymph node metastasis (N-stage) [10]. In patients with intraepithelial cancer, the risk of lymph node metastasis is low, and resection of the carcinoma carries an excellent prognosis [11]. Nevertheless, the overall 5-year survival rate of patients after curative esophagectomy for esophageal cancer is between 20 and 40% depending on the histologic subtype, respectively. However, these results still represent a distressing outcome although many efforts have been undertaken during the last decades to improve this poor prognosis [4].

In the vast majority of patients, esophageal cancer is diagnosed when these affected individuals are already symptomatic with higher levels of dysphagia. At this point of time, the disease has progressed to regional lymph node infiltration and/or distant metastases in most cases, resulting in definitive ineligibility for curative resection. Considering this advanced tumor stage, these patients are provided for palliative treatment settings which aim at an improvement of swallowing, slowing of endoluminal tumor growth, and treatment or prevention of tumor-associated complications. The inability to swallow a regular diet is one of the most severe symptoms reducing quality of life for patients with an incurable disease and short life expectancy [12, 13].

Therefore, the main goal in the palliation of patients with advanced inoperable esophageal carcinoma is a decrease in dysphagia, which always results in an improvement of quality of life. Generally, the prognosis of esophageal carcinoma is poor in palliative settings by obtaining mean survival rates of only several months [14–16].

Concerning this poor prognosis, it is still required to search for further different treatment options which may help improving these disappointing results.

A variety of tumoricidal palliative measures allows at least a limited oral diet, but they require repetitive hospitalization. Many published reports discuss different combinations of tumoricidal procedures, such as endoscopic brachy-radiotherapy and external beam irradiation [17], chemotherapy [18], endoscopic tumor disobliteration with laser technique or cautery [19], stenting [20], photodynamic therapy (PDT) [21], and combinations of these techniques. Therefore, in order to minimize the risk of procedure-induced complications and increase the treatment benefit, different procedures should be combined in the treatment of these terminally ill patients. However, all physicians involved in the management of patients with advanced inoperable cancer of the esophagus know the agony and frustration these patients suffer from. For this reason, PDT increasingly became the focus of attention in these patients.

In the course of time, endoluminal PDT has developed into a promising nonthermal laser technique in the palliative treatment of advanced esophageal cancer. In general, the mechanism of function is based on the illumination of malignant tissue after selective accumulation of photosensitizers in tumor cells resulting in local tumor tissue necrosis [22, 23]. Unlike radiotherapeutic or chemotherapeutic measures, the method is not limited by cumulative doses.

Considering this mode of action, endoscopic PDT is leading to a targeted local destruction of neoplastic under preservation of normal esophageal tissue. As a result, local endoluminal palliation with the aim to improve swallowing, decrement of dysphagia, and therefore improvement of the patient's quality of life are the main therapeutic goals of PDT. After decades of clinical use, PDT is now considered a feasible and efficient minimally invasive endoscopic option for palliation of advanced esophageal cancer [4, 24, 25], applied alone as sole treatment option or as part of a multimodal approach within the palliative treatment regimen tailored to the specific needs of the patient [26–29].

15.2 Basics of PDT for Esophageal Cancer

15.2.1 General Comments on Clinical PDT

The first clinical use of PDT for cancer in modern times dates back to the beginning of the twentieth century when von Tappeiner and colleagues used eosin as topical photosensitizer combined with sunlight to treat facial basal cell carcinoma [30]. Tom Dougherty established the modern era of PDT, when he reported that the systemically injected hematoporphyrin could be activated by red light for the complete eradication of transplanted experimental tumors [31, 32]. He demonstrated the photodynamic effect expressed as necrosis in a variety of cancers which had been conditioned by a chemical photosensitizer and then exposed to an appropriate wavelength of laser light [33].

Since then, much original scientific, experimental, and clinical work has been done, resulting in a considerable number of review articles showing the efficacy of PDT in the treatment of most human cancers. By reason of high inoperability rate of 50–70% at presentation [34–36], patients with advanced esophageal cancer became the focus of attention for PDT initially in the 1980s [29, 37]. Mc Caughan et al. [36] reported a series of patients with large obstructing esophageal cancers treated by hematoporphyrin derivative photodynamic therapy. There was a very good improvement in the patients' ability to swallow and some suggestion of a prolongation of survival. Subsequent evolution expanded its indications to include early inoperable squamous cell carcinoma, as well as severe dysplasia and early adenocarcinoma in Barrett's esophagus [37–43].

The basic requirements for PDT in cancer treatment are a photosensitizing drug, laser light of an adequate wavelength appropriate to the sensitizer, and the presence of molecular oxygen, which has been shown to be crucial for the provoked photochemical reaction.

Generally, in PDT for esophageal cancer, a photosensitizer is first administered intravenously, and then visible light is delivered endoscopically to the cancerous area. Light activation of the photosensitizer triggers a photochemical reaction, resulting in the production of highly reactive oxygen species followed by immediate cell damage and tumor necrosis. The detailed technique of cytotoxic action of PDT has been reviewed extensively [44–52].

Castano [46] explained the function of mechanism as follows: Due to differences in vascular supply and lymphatic clearance from the tumor and the retention of the photosensitizing drug by tumor cells, the photosensitizer is selectively retained in the tumor cells and interstitial tissue of the tumor. The exact mechanism by which photosensitizers preferentially accumulate in tumor cells is unknown. Probably, a combination of multiple factors is important in causing this selective retention. Poorly developed tumor lymphatics may prevent removal of sensitizers from tumor cells. Many sensitizers are relatively hydrophobic and bind to lipoproteins within the plasma. Increased expression of lipoprotein receptors on tumor cell membranes may increase photosensitizer concentration within these cells [53]. Other possible factors include altered pH in tumor cells, tumor neovascularization, and the increased metabolic activity found in tumors [44, 54]. At this point, it is important to know that in the absence of light activation, photosensitizers have no inherent effect on tissue.

In the further course, the photosensitizer binds to the cellular plasma membrane, becomes internalized, and binds to organelle membranes, especially those of mitochondria. These membranes are the initial target of singlet oxygen. However, a greater concentration of the photosensitizer in the tumor than in the adjacent normal tissue can be obtained after 48 hours. The photosensitizer absorbs light energy (photons of appropriate wavelength) and produces singlet oxygen. The photosensitizers work as catalysts when they absorb visible light and then convert molecular oxygen to a range of highly reactive oxygen species (ROS).

The photon is transferred to ground-state triplet oxygen producing the excited singlet oxygen (type II, photo-oxygenation reaction). In the other type of photooxidative process (type I), the excited sensitizer itself initiates a free-radical reaction. Both types of reaction are associated with PDT. Potentially, the ROS that are produced during PDT have been shown to destroy tumors by multifactorial mechanisms [47, 48].

15.2.2 Approved Photosensitizers for PDT

The ideal photosensitizer would have the following properties: high selectivity and concentration in tumor compared to normal tissue, selectable depth of tumor penetration, high-yield production of singlet oxygen, and minimal side effects, especially a low skin photosensitivity.

This type of photosensitizer is not yet available; however, active research on many different types of photosensitizer is ongoing with the goal to define a photosensitizer for early cancer and another one for advanced cancer treatment, each fulfilling the criteria mentioned above.

Within the following paragraphs the three current photosensitizing drugs available for esophageal PDT are described. Considering their different properties and their use in daily routine, porfimer sodium has to be proven to serve as the most appropriate photosensitizer for advanced esophageal cancer.

15.2.2.1 Porfimer Sodium

The compounds are oligomeric mixtures of dihematoporphyrin ethers and esters with very similar photochemical properties. Porfimer sodium represents a purified form of the first clinically used photosensitizer, hematoporphyrin derivative (HpD).

Peak light absorption with porfimer sodium occurs at around 400 nm; however, tissue attenuation at this wavelength is high, limiting the depth of tissue penetration of light and hence of the extent of tumor necrosis. A secondary, lower absorption peak is present around 630 nm, a longer wavelength less attenuated by tissue pigments and hemoglobin. The decreased activation at this wavelength is compensated by increasing the dose of photosensitizer or light. The depth of effective treatment with porfimer sodium at 630 nm varies between 2 and 4 mm. In the case of esophageal cancer, studies have demonstrated that treatment with 630 nm light can produce full-thickness inflammation and tissue injury [55].

Although porfimer sodium is selectively retained in tumor tissue compared to normal tissue at a ratio of at least two to one, the drug does accumulate to significant levels in the skin and in the reticuloendothelial system, especially of the liver and spleen, resulting in photosensitivity lasting 4–6 weeks. The standardized dose of porfimer sodium for PDT of esophageal carcinoma is 2 mg/kg body weight by slow intravenous injection, 48 h prior to irradiation [56].

15.2.2.2 Meta-Tetrahydroxyphenyl-Chlorin (mTHPC)

mTHPC is a second-generation photosensitizer with a shorter period of skin photosensitivity (2 weeks), a longer activation wavelength (652 nm), and therefore increased depth of effect, higher yields of singlet oxygen, and better tumor selectivity. This compound was typically used intravenously at a dose of 0.15–0.3 mg/kg body weight. Early results in regulatory trials, studying palliative treatment of patients with squamous cell carcinoma of the head and neck, were complicated by extensive tissue necrosis, tissue breakdown, and stricture formation probably related to excessive light doses. Phase IIb studies in head and neck cancer have been completed, and mTHPC has also been studied in bronchial and esophageal tumors with encouraging results [57, 58].

15.2.2.3 5-Aminolaevulinic Acid (5-ALA)

5-ALA is an early intermediary in heme biosyntheses. It is formed by the combination of glycine and succinyl CoA. The formation of 5-ALA is the rate-limiting step in the heme pathway and is catalyzed by 5-ALA synthase. 5-ALA is converted to porphobilinogen; porphyrin intermediates via porphobilinogen deaminase and then protoporphyrin IX (PpIX). The final step in this pathway is conversion of PpIX to heme, catalyzed by ferrochelatase. Under normal conditions, 5-ALA production downregulates 5-ALA synthase, limiting heme synthesis; however, administration of large doses of 5-ALA results in intracellular accumulation of PpIX. It has been shown that PpIX accumulates in greater concentrations in adenocarcinoma of the esophagus compared to normal tissue because of a relative increase in activity of porphobilinogen deaminase compared to ferrochelatase [59].

As a photosensitizer, 5-ALA has several advantages compared to other photosensitizers. PpIX is produced in greater levels in the mucosa compared to the submucosa or muscularis propria in the esophagus, making 5-ALA useful for treatment of superficial lesions [54, 60–62].

5-ALA may be administered orally, and the interval between drug administration and activation by laser light is only 6 h.

Skin photosensitivity lasts only for 48 h. However, the oral administration of 5-ALA with necessary doses of 30–75 mg/kg body weight may lead to nausea and vomiting. This is why the application of an antiemetic drug is recommended prior to photosensitization [54, 60].

15.2.3 Light Delivery for PDT

Dosimetry refers to the calculation of light delivery to the tumor. Dosimetry calculations are complex and must take into account multiple factors involving the respective photosensitizer, the method of light delivery, and characteristics of the tumor itself.

One of the first steps in evaluating PDT was to analyze its light dosimetry, or the effects of various light doses on the degree of tumor necrosis. After testing a variety of light doses, a correlation between tissue dose of light (J/cm^2) and the resulting depth of tumor necrosis could be found [63].

The tissue dose of light varies inversely with the luminal diameter of a tumor segment (when the light dose delivered from the fiber tip remains constant) and was calculated by figuring the surface area of an open cylinder. By using this method of analysis and working backward from tissue doses, the light dose of 300 J/cm at the fiber tip was determined to give a safe, yet effective, range of tumor necrosis [27, 28]. This light dose was still lower than that which had been previously used. Furthermore, it was reported that the higher light doses previously used were associated with complications (e.g., pleural effusions, mediastinitis, fistula) [64]. The dose of 300 J/cm at the fiber tip for palliative therapy of malignant esophageal obstruction represents a standardized light dose for PDT.

Several authors have studied the relationship between the extent of tumor necrosis and light dose: They were able to show that the results are consistent with a threshold effect [65]. Bown and coauthors revealed a logarithmic relationship between the extent of tumor necrosis and applied energy fluence rate [66]. However, in all of these studies, only experimental tumors, not human carcinomas, were investigated.

In clinical studies, very high-energy fluence rates up to 4800 J/cm^2 were used in the treatment of colorectal carcinomas [67, 68]. In contrast, Gossner and co-workers showed that colorectal carcinomas react very sensitively to PDT, requiring low energy doses (25–75 J/cm^2) for effective tumor destruction.

They showed that high-energy fluence rates (>300 J/cm^2) were associated with undesired side effects such as hyperthermia [69]. These observations were corroborated by the results of a quantitative study for colorectal cancer [70]. Similarly, high-energy fluence rates were applied to gastric (960 J/cm^2) and esophageal carcinomas (6000 J/cm^2) in other clinical studies [64, 71]. Interestingly, energy fluence higher than 100–150 J/cm^2 did not markedly increase the PDT effect. Thomas and colleagues found an increased risk for complications by using power higher than 1.5 W but were not able to enhance the effectiveness of PDT in tumors by irradiating at the highest doses [64].

These studies demonstrate that the prevailing clinical dosimetry for gastrointestinal carcinomas may result in inadequate treatment by under- or overdosage of the applied energy fluence rate in PDT. However, the dose dependence of light cannot be discussed on the basis of energy fluence alone. On a more fundamental scientific level, one has to keep in mind that the propagation of light in tissue depends on optical parameters: One of them is the absorption coefficient, which varies significantly in different tissues because of the presence of various light-absorbing chromophores (e.g., hemoglobin and melanin) [72]. Therefore, tumors containing a large amount of blood and tissues with great amounts of melanin have a high absorption coefficient.

A further optical parameter is the scattering coefficient, which depends on the structural pattern of tumor tissue [73]. It is easy to understand that different organ tumors might exhibit a variety of structural matrices of differential blood flow. These factors could explain the different reactions of gastrointestinal tumors to the same light dose. The optical parameters of esophageal carcinomas vary with different tumor diameters because of changes in vascular supply and necrosis in its center. The effects these parameters may have on the therapeutic results are expressed by different penetration depth of light at different tumor diameters [74]. Due to the changing optical properties resulting from increased tumor diameter, dosimetry needs to be adapted to achieve proper and comparable results after PDT. In other words, an increasing tumor diameter requires an increasing treatment time of a standardized light dose, depending on its correction factor [74].

15.2.3.1 Light Delivery Systems

Light delivery systems are generally fiber-optic devices adapted for endoscopic use, varying from a quartz light-diffuser fiber placed through the working channel of the endoscope to specially modified diffusers that are put in place with the help of fluoroscopic guidance.

For esophageal carcinoma the cylindrical diffuser fiber is commonly used. It is positioned under endoscopic or fluoroscopic guidance within the lumen of the esophagus. It is necessary to centrally align these applicator systems within the lumen of the esophagus, which in clinical practice is possible only in the presence of high-grade stenosis. Furthermore, they allow light delivery only over a limited distance, and segmental light application is thus needed in longer tumor segments [75].

Additionally, this technique has the disadvantage that it is not possible to flatten the mucosal folds of the esophagus, and thus, some areas may be shielded from the light as the configuration of the lumen continually changes with esophageal peristalsis and respiratory movement [76].

This is thought to produce excessive light dosing in some areas, with the risk of stricture and/or inadequate light dosing to other areas with incomplete mucosal ablation. Modern light systems are optimized to deliver the desired light intensity distribution to the targeted region with minimal losses. In the esophagus this implies the use of a device that is adapted to the shape of the hollow organ, such as an elastic balloon catheter [77].

Superficial light application should therefore always be done by the use of light applicator systems which guarantee a standardized distance of the irradiation source from the surface of the tumor and eliminating the shadowing phenomenon, described above, of a hill-and-valley effect caused by mucosal folds.

Considering the variables for proper calculation of light dosimetry and the irregular endoluminal surface of the tumor, it is impossible to treat the tumor site accurately only by using the light diffuser inserted through the working channel of the endoscope. Based on the findings of optical parameters and considering all the above-mentioned aspects, a very simple and save dosimetry formula was developed and approved in clinical practice of more than 300 esophageal PDTs [74]. The dosimetry formula is based on the use of a standardized light applicator of different length and includes the following variables: the diameter of the tumor stenosis and distance of the irradiation source from the tumor surface, the length of the irradiation source and appropriate light applicator system, the optical parameters of esophageal carcinoma, and finally, the desired light dose (J/cm^2).

For interstitial illumination, the optical fiber can be introduced through the biopsy channel of the endoscope, and the cylindrical diffusing end section is placed into the tumor itself under direct vision [78]. Despite sufficient tumor eradication especially in tumor masses that protrude into the lumen, the risk of perforation and/or penetration might be higher compared to superficial treatment.

15.2.4 Oxygen Supply for Esophageal PDT

Both the photochemical reaction and the efficacy of PDT mainly depend on the presence of molecular oxygen. There seems to be a decreased cell sensitivity to PDT in the presence of low oxygen, which means that hypoxic cells are resistant to PDT.

Furthermore, animal tumor models have demonstrated a decreased effect to PDT under hypoxemic conditions, because hypoxemic tumor cells seem to be less affected by porphyrins and light [79, 80].

PDT-induced cytotoxicity most likely occurs through photooxidative reactions, and there are two major reaction pathways [81]. Type I photooxidation involves a direct reaction of the excited sensitizer with a substrate by a mechanism involving hydrogen or electron transfer to yield transient radicals that further react with oxygen. In the type II photooxidative reaction, energy transfer occurs from the excited triplet state of the sensitizer to molecular oxygen producing singlet oxygen, which can further react with substrates susceptible to oxidation.

Uncommonly, electron transfer from the sensitizer generates a superoxide radical through the activation of hypoxanthine/xanthine oxidase system by catabolism of high-energy phosphate damage of mitochondria and a calcium-dependent protease conversion of xanthine dehydrogenase to xanthine oxidase [82].

Considering these effects, too high fluence rates of exposure light will lead to oxygen depletion: Using transcutaneous oxygen electrodes, Tromberg and co-workers reported that tumor oxygen tension becomes irreversibly low with large PDT fluences (fluence is a measure of time-integrated particle flux, expressed as particles per square centimeter) [83].

Otherwise, too low fluence rates need a long exposure time and lead to vascular shutdown which also cause hypoxia of the tumor tissue [49]. Therefore, the presence of molecular oxygen in tumor tissue is crucial for the effectiveness of PDT.

In clinical practice, different approaches are used to enhance tissue oxygen saturation and thereby the effect of PDT. In the literature, oxygen administration during PDT is described as simple nasal oxygen support of only 3–4 L/min [84], general anesthesia with intubation and FiO_2 of 0.21–1.0 [27, 28, 54, 60, 85], and additional hyperbaric oxygenation with enhanced effectivity [86]. At this time, the general recommendation for clinical practice in PDT application is general anesthesia with orotracheal intubation and oxygen supply using 100% pure oxygen [85].

15.3 Clinical Aspects and Local Antitumor Effect of PDT for Esophageal Carcinoma

15.3.1 Indications for PDT

PDT is an important component of the multimodality treatment in anatomically and/or oncologically non-resectable esophageal cancer with palliative intention. Therefore, PDT is used in a meaningful combination with other endoluminal treatment options, with chemo- and/or radiotherapy, respectively.

15.3.2 Contraindications for PDT

- Porphyria
- Renal insufficiency
- Liver insufficiency
- Preexistent fistula formation
- Allergy against photosensitizing drug
- Tumor involvement of the trachea, the carina, or the main bronchi
- Patient's incompliance regarding required protection from light after photosensitization
- Terminally ill patients

15.3.3 Clinical Examination and Diagnostic Work-Up

Patients' history, recording of dysphagia grade and determination of the Karnofsky performance status, physical examination, routine laboratory parameters including hepatic and renal function.

15.3.4 Clinical Staging

- Esophagogram using water-soluble contrast medium
- Flexible esophagogastroscopy
- Flexible tracheobronchoscopy

- Computed tomographic scan of the chest and abdomen
- Abdominal ultrasonography
- Endoscopic ultrasound
- Electrocardiogram
- Cardiac ultrasonography and spiro-ergomety

Before PDT can be considered as treatment of choice, both flexible esophagogastroscopy and tracheobronchoscopy have to be performed to rule out possible tumor infiltration of the adjacent trachea, the carina, and the main bronchi. If tumorous involvement can be confirmed, PDT must not be carried out due to the considerably increased risk of perforation resulting in a malignant esophagorespiratory fistula.

15.3.5 Patient Information and Education Before PDT

Extensive and complete patient education is of supreme importance in PDT because of the persistence of photosensitivity after drug injection. Patients are instructed to wear opaque clothing that covers most of their skin, including long-sleeve shirt and long pants, gloves, wide-brimmed hat, and sunglasses, when they report for their injection. Because of the persistence of photosensitizer in the skin for several weeks, the patients must wear similar clothing when going outside during this time. The daily application of a potential sunblock with high sun protection factor is mandatory to avoid any event of sunburn. Due to the increasing intensity of the sunlight, we are currently using a special type of sunblock with high sun protection factor (Daylong® extreme SPF 50+, Egerkingen, Suisse).

Inside, the patients are instructed to avoid direct sunlight and bright indoor lighting, although indirect sunlight is acceptable and probably shortens the duration of photosensitivity because of photo-bleaching (removal of the photosensitizer from the skin caused by low-intensity light exposure). Topical sunscreens are generally ineffective in preventing phototoxicity because they block only ultraviolet light.

During endoscopic laser-light treatment, the patient is covered with drapes, taking care to protect also the face. As an alternative, a total sunblock can be applied on the face.

If a patient requires surgery after PDT, drapes must be placed over the skin, liver, and spleen because of the risk of phototoxicity generated by bright operating room lights.

After the average time of skin photosensitivity has elapsed, the patients are instructed to expose the back of one hand to sunlight for 5 min. If no redness or swelling occurs, they can gradually increase their exposure to sunlight over time.

15.3.6 Photosensitization

Intravenous photosensitization of porfimer sodium is recommended at a dose of 2 mg/kg body weight, 48 h prior to irradiation. Before intravenous administration, the cannula should be checked with a test of 20 mL standardized physiologic saline solution because para-venous injection may cause necrosis of the surrounding tissue accompanied by severe pain. Topical application of a sunblock immediately after photosensitization, as well as the above-mentioned general recommendations for light protection, is commonly sufficient to prevent sunburn.

15.3.7 Endoscopic Illumination Procedure During PDT

Forty-eight hours after uneventful photosensitization, esophageal PDT should be done under general anesthesia with endotracheal intubation and appropriate oxygen support. Routine cardiorespiratory monitoring including electrocardiogram, noninvasive continuous blood pressure control, and pulsoxymetry should be provided (Fig. 15.1).

The endoscopic placement of the light delivery system is best done with the help of a flexible endoscope. We are using balloon catheters (PhotoDynamicTherapy®, Vienna, Austria) which are inserted through the biopsy channel of the flexible endoscope (Fig. 15.2).

In case of very advanced esophageal cancer with subsequent extensive tumor stenosis, stepwise endoscopic bouginage using the Savary-Gilliard device may become necessary before PDT can be performed.

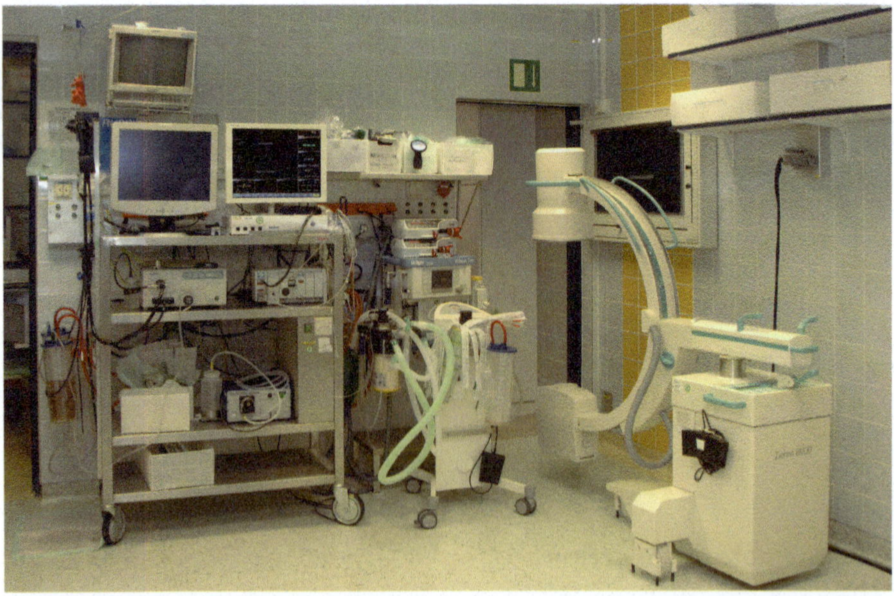

Fig. 15.1 Multifunctional endoscopy unit provided with anesthesiological working station and X-ray fluoroscopic equipment for endoluminal dilatation and stenting

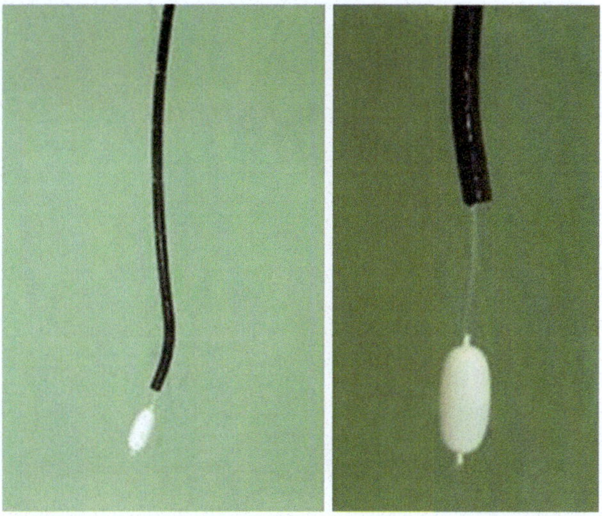

Fig. 15.2 Laser application system (expanded balloon catheter) inserted through the biopsy channel of the flexible endoscope

A flexible guidewire is passed under radiological guidance through the endoscope, and careful and gentle bouginage is done, enabling the passage through the tumor by the flexible gastroscope. In the same intervention, suspicion of perforation can be excluded by esophagogram using a water-soluble contrast medium. Furthermore, it is necessary to pass the entire esophageal carcinoma with the endoscope in order to perform exact measurement of the entire extent of the tumor reaching from the beginning of the upper border to the end of the lower border of the tumor (Fig. 15.3). Meticulous measuring is essential for determination of both the type of balloon catheter and the duration of PDT, respectively.

Afterward, the flexible endoscope is placed above the carcinoma and the laser application system, using a balloon catheter inserted in a folded condition through the biopsy channel. The folded balloon is placed at the distal aspect of the tumor. After instillation of distillated water, the balloon becomes unfolded and is applied very close to the surface of the tumor allowing homogenous light distribution (Fig. 15.4).

Afterward, the expanded balloon is withdrawn proximally in order to treat the entire tumor length. Withdrawal of the expanded balloon occurs step by step, corresponding to the length of the used balloon which usually amounts to 4 cm. The required time for each placement of the balloon is approximately 20 min which is calculated by the help of the dosimetry formula developed at our department as mentioned above [74]. The light dose is calculated as 100 J/cm. Light intensity of 630 nm is applied by a KTP-Nd:YAG laser with DYE-box (Laserscope; Surgical Systems, Gwent, UK). For safety reasons, wavelength and light dose at the tip of the light diffuser should be controlled before and after PDT. Depending on the topography and length of the tumor, single or multiple placements of the balloon may be

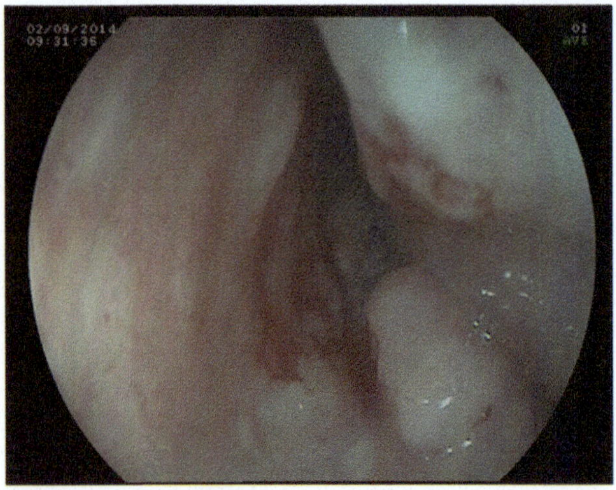

Fig. 15.3 Histologically proven esophageal squamous cell carcinoma causing considerable endoluminal tumor stenosis

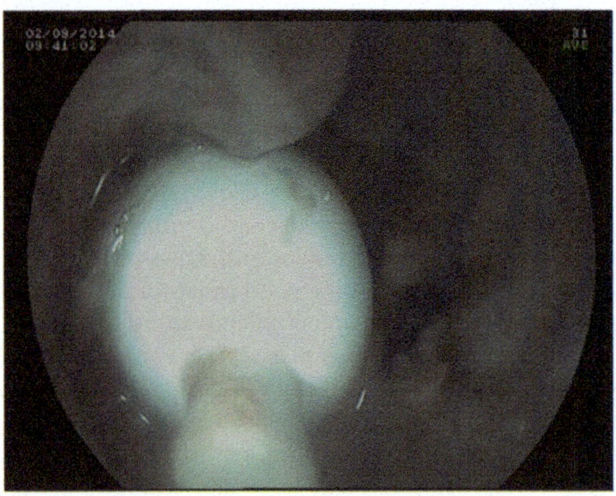

Fig. 15.4 Completely expanded balloon catheter with broad contact to the tumor surface immediately before illumination

necessary in order to illuminate the entire tumor length (Fig. 15.5). Therefore, the required irradiation time may range from at least 20 min up to 90 min. After PDT, a conventional chest X-ray is done to rule out any pneumothorax or pleural effusion.

Five to seven days after successful PDT, flexible esophagogastroscopy under short-term anesthesia is repeated in order to verify the therapeutic response (Fig. 15.6). In the course of this endoscopic intervention, supernumerous necrotic tumor tissue may be removed mechanically by using the forceps or by mere suction

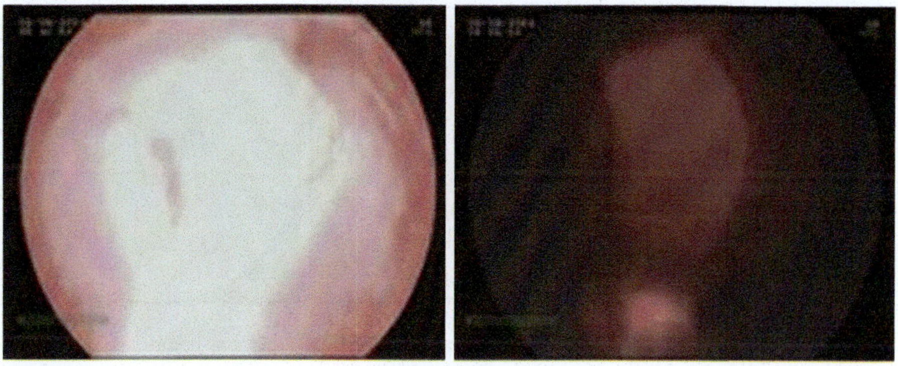

Fig. 15.5 Expanded balloon catheter during illumination procedure using red laser light

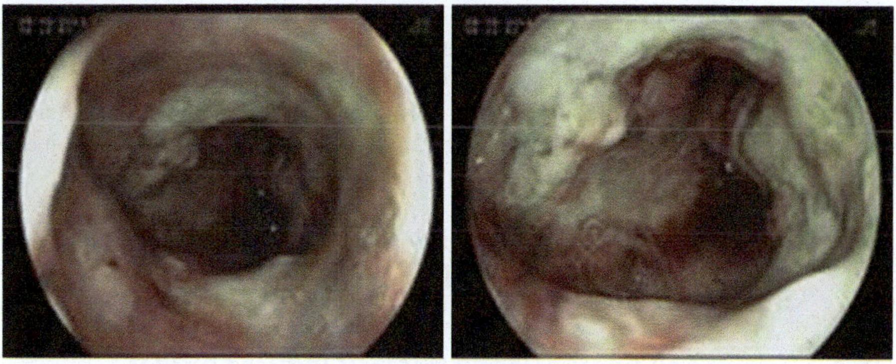

Fig. 15.6 Extensive endoluminal tumor necrosis 1 week after PDT resulting in reopening of the esophageal lumen

when necessary. The depth of tumor necrosis is determined indirectly by the visible increase in luminal diameter measured at the maximal point of constriction or directly by determination using endoscopic ultrasound if available.

Especially in case of large carcinoma with considerable endoluminal tumor load, a single PDT treatment may be not sufficient to induce distinct necrosis of the entire visible tumor. In this case, a second PDT treatment may become necessary and is therefore recommended by various authors [27, 28, 54, 60, 85]. The persistent photosensitization can be used therapeutically by applying the second illumination several days after the first without reapplication of a photosensitizing drug.

15.3.8 Follow-Up After Esophageal PDT

After effective PDT, the patients undergo repetitive endoscopy after 1 month and thereafter every 3 months for the first year (Fig. 15.7). CT scans of the thorax and the abdomen are performed every 6 months [27, 28, 54, 60]. However, increasing

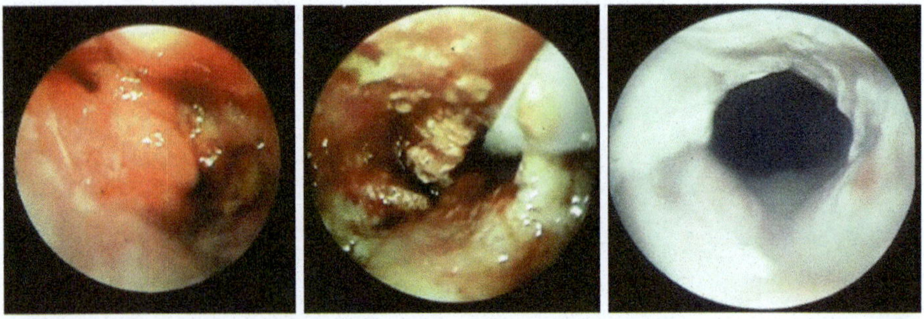

Fig. 15.7 Three-months' follow-up of advanced esophageal carcinoma before, 5 days, and 3 months after PDT: notable tumor necrosis on the fifth post-interventional day and patent esophageal lumen 2 months later

dysphagia at follow-up is still the indication for repetitive PDT treatment. In case of endoscopically verified local tumor progression, the photosensitizer is injected again and PDT can be repeated in the same manner mentioned above. In this context, it has to be mentioned that repeated PDT should never be considered before at least 1 month has passed after first PDT because of the high risk of perforation due to overwhelming tumor necrosis induced by PDT.

15.4 Systemic Inflammatory and Immunological Changes Induced by PDT

15.4.1 Tumor Cell Apoptosis and Necrosis Induced by PDT

As mentioned above, PDT uses nontoxic photosensitizers, visible laser light of appropriate wavelength, and molecular oxygen. None of these three components is individually toxic, but when combined together, they initiate a photochemical reaction that culminates in the generation of highly reactive oxygen species (ROS) that kill malignant cells [87]. This type of ROS produced during PDT causes apoptosis and necrosis by direct action to the cells [49, 50], and shuts down the tumor vasculature, leading to depletion of oxygen and nutrients in the tumor [51, 52].

Although PDT represents an effective local cancer therapy, its effect is not limited to the local site. Furthermore, a considerable inflammatory acute-phase response is induced followed by various immunological consequences as mentioned below.

Moreover, as a result of this traumatic insult to the tumor and its microenvironment, a strong acute inflammatory reaction is provoked at the targeted site. This acute inflammatory response additionally causes infiltration of different host immune cells which remove the damaged cells [88]. This specific PDT-induced inflammation is involved in the development of adaptive antitumor immunity, leading to attraction of leukocytes such as dendritic cells and neutrophils to treated tumors with subsequent antigen uptake and maturation of the local dendritic cells. Finally, PDT also has a significant effect on the immune system, which can be either immune-stimulatory or immune-suppressive. In contrast to chemotherapy, which is mostly immunosuppressive, PDT causes a range of effects that might result in a specific immunological antitumor reaction [89, 90].

On the one hand, PDT directly causes apoptosis [50] and necrosis of illuminated tumor cells: Cell death after PDT may be triggered by influence on signaling pathways [91], mitochondrial events [47, 48, 92], or specific mediators [93].

On the other hand, PDT also affects tumor vasculature where it induces shutdown of vessels, which subsequently induces tumor regression [52, 94]. This effect has also been interpreted as limitation for PDT, the central mechanism of which has been considered to be formation of free radicals. Once oxygen has been consumed, no further formation of radicals will be possible. This is why ways have been sought to evade vascular shutdown.

Some authors have examined the relationship between the mode of tumor cell death and the efficiency of induction of the immune response. It is not clear whether apoptosis or necrosis is more effective in this context [95–97]. There is evidence, however, that PDT causes an unusual mixture of apoptotic and necrotic cell death, whereas most conventional cytotoxic agents only trigger apoptosis [50, 98].

15.4.2 Damage-Associated Molecular Patterns (DAMP) and PDT

Reginato summarized that after the traumatic insult to the tumor induced by PDT, one of the first events occurring at the treatment site is the generation of “danger” signals, so-called damage-associated molecular patterns (DAMPs) that serve as warning signals in innate immunity. DAMPs are endogenous intracellular molecules normally “hidden” within living cells, but upon exposure or secretion from dying and/or damaged cells, they acquire immune-stimulatory properties. DAMPs are thought to be the key mediators of the immunogenicity of tumor cells killed by PDT via necrosis or apoptosis, as the immune system recognizes them signals that “self-altered” antigens have been released from dying cells that trigger a vigorous immunological response [88]. One of these DAMPs released after PDT is the so-called extracellular heat-shock protein 70 (HSP70), a substance induced within the cells following stress and protecting them from death. One of its functions is to chaperone unfolded proteins [99], thus promoting the formation of stable complexes with cytoplasmic tumor antigens. These antigens can either be expressed at the cell surface or may remain intact after cell death. In the latter case, they may interact with antigen-presenting cells (APCs) and stimulate an antitumor immune response [100, 101].

Extracellular HSP70 binds to high-affinity receptors on the surface of the APC, which leads to the activation and maturation of dendritic cells (DC). This enables the cross-presentation of the peptide antigen cargo of HSP70 by the APCs to $CD8^+$ cytotoxic T cells [102]. Interestingly, 15–25% of total cellular HSP70 became exposed at the cell surface almost instantly after a Photofrin-based PDT treatment [101].

PDT also induces the expression of further heat-shock proteins such as HSP47 [103] and HSP60 [104].

15.4.3 Role of Inflammatory Response in Immunological Reaction Following PDT

It has been shown that PDT induces an acute inflammatory response to any tissue where it is applied, and, as a consequence, also a considerable increase of inflammatory cytokines, chemokines, and neutrophils in the circulation can be observed: Elevated levels of IL1, IL6, IL8, and IL10, as well as macrophage inflammatory proteins 1 (MIP1) and MIP2, have been documented [105, 106]. Probably, the inflammatory reaction is triggered by the transcription factors nuclear factor κB and activator protein AP 1, both of which are known to be linked to oxidative stress [107].

Furthermore, C3 complement activation was found after PDT treatment of Lewis lung cancer. Accordingly, the efficacy of the treatment was significantly reduced when C3a or C5a receptors or ICAM1 or IL1, respectively, were blocked [106, 108].

There is evidence that this highly inflammatory reaction is a crucial factor for acute tumor response to PDT. Therefore, innate host immune cells such as monocytes or macrophages, neutrophils, and mainly DC are migrating to the treated site and infiltrate in large numbers the tumor and the peri-tumorous region. Both necrotic and apoptotic cells are incorporated by DC that have accumulated owing to the acute inflammatory response. DC mature after stimulation by cytokines, which are released at the site of inflammation, and home to the regional lymph nodes where they present antigens to T lymphocytes.

Activated T lymphocytes become effector T cells and migrate to the tumor where they kill the tumor cells [109]. It has been shown that these phenomena might induce the development of memory T-cell antitumor response after PDT [46]. The effects of PDT on lymphocytes are controversial: On the one hand, lymphocytes, especially the activated ones [110], suffer lethal damage by PDT [111]. On the other, low doses of PDT can provoke activation of macrophages, which will be induced to secrete TNF-alpha. These PDT-treated macrophages have been shown to exert cytotoxicity to tumor cells [112].

However, the function of these various inflammatory cells, which are recruited to the PDT-treated tumor, is to neutralize the DAMP by engulfing and eliminating the cellular debris, as well as compromised tissue components surrounding the tumor. This cleaning promotes local healing with subsequent production of fibrosis and scar tissue [88].

15.4.4 T-Regulatory Cells and PDT

In addition to directly stimulating antitumor immunity by triggering DC and T-cell activation, PDT may also interfere with immune-suppressive T cells, so-called T-regulatory cells. The main class of T cells suppressing the immune response consists of $CD4^+CD25^+FoxP3^+$ T-regulatory cells. $CD4^+CD25^+$ T-regulatory cells have been shown to suppress the immune response in various instances [113, 114]. If these cells are depleted, cellular immunologic reactions can be enhanced. Following PDT, T-regulatory cells might be specifically influenced by a number of mechanisms: IL6, a cytokine that is abundantly produced after PDT [115], has been shown to inactivate T-regulatory cells.

Since also low-dose cyclophosphamide has been known to cause a $CD4^+$ $CD25^+$ depletion, it was combined with PDT in J774 reticulum cell sarcoma. A high rate of long-term cures and resistance to tumor rechallenge was achieved. In contrast, both cyclophosphamide alone and PDT alone led to therapeutic failure [116].

In a clinical study with human subjects, our research group could confirm that PDT downregulated the immunosuppressive function of peripheral T-regulatory cells in patients with esophageal carcinoma. In contrast to that observed downregulation, the levels of those T-regulatory cells did not change significantly after PDT, whereas a transient reduction of the number of tumor-infiltrating T-regulatory cells around the tissue of the esophageal carcinoma itself could be detected by immunohistochemical staining of biopsies and resected specimens [117].

In a mouse model involving the colon adenocarcinoma CT26 wild-type tumor model, Reginato and co-workers could demonstrate that PDT with low-dose chemotherapy using cyclophosphamide leads to a significant improvement in long-term survival compared with either treatment alone. Further, the development of an immune response to the mouse cancer antigen could be detected [118].

15.4.5 Role of Dendritic Cells in PDT

Reginato and co-workers described that PDT appears to activate DC [88]. In this context, DC are stimulated by the recognition of DAMPs released by dying tumor cells, e.g., HSP70 [119], which forms stable chaperone complexes with cytoplasmic tumor antigens. Thereafter, the HSP-antigen complexes bind to the danger signal receptors, Toll-like receptors 2 and 4 [120] on the surface of DC, which are the most potent APCs. In the absence of inflammation, DC remain in an immature state, but when tissue inflammation and release of DAMPs occur, they mature and migrate in large numbers to the draining lymph nodes. The transition to the mature state of DC involves the upregulation of surface major histocompatibility class I molecules (MHCI) and MHCII and of the costimulatory molecules CD80 and CD86. These changes allow the DC to express peptide-MHC complexes at the cell surface and prime efficiently CD4+ T-helper cells and CD8+ CTLs and hence to initiate an adaptive immune response [88].

Further, DC are known to act as stimulators for natural killer cells and cytotoxic lymphocytes. When DC were injected into experimental tumors treated with PDT, homing to regional and peripheral lymph nodes was achieved. This kind of treatment resulted in good tumor response [121].

This effect does not seem to be linked to a specific type of photosensitizer: In another experimental study in BALB/c mice, PDT using a photosensitizer of the chlorine type was used before intra-tumoral injection of DC. The treatment resulted in a rate of cure much higher than with either of both single procedures [119].

Another group of investigators treated animals with two tumors, in which only one lesion was injected intra-tumorally. Surprisingly it was found that also the second, untreated tumor showed regression accompanied by the presence of tumor-specific lymphocytes [122].

15.4.6 Immunological Response of PDT in the Clinical Setting

Apart from the various numbers of conducted experimental and preclinical studies, only a few clinical trials focusing on the different immunological changes in humans who had undergone PDT have been performed to date. However, some of these recent clinical studies could confirm the supposed relationship between local efficacy and PDT-induced antitumor immunity.

In 2001, enhanced immune cell infiltration after PDT could be detected in a clinical study dealing with intraepithelial neoplasia of the vulva [123]. Those patients who expressed MHCI on the tumor cells were more likely to respond to PDT as compared to patients whose tumors had downregulated MHCI molecules. MHCI recognition is critical for activation of CD8+ T cells, and the downregulation of MHCI molecules is one of the mechanisms used by tumors to evade immune recognition in general and PDT-induced immunity in particular. Those patients who did not respond to PDT had significantly lower CD8+ T-cell infiltration into the treated tumors compared with responders, confirming the important role of CD8+ CTLs in PDT efficacy [88].

In 2003, Yom and co-workers could verify increased postoperative cytokine levels after intraoperative PDT in patients with malignant pleural mesothelioma [124].

In 2007, Thong and co-workers could show that PDT of multifocal angiosarcoma of the head and neck which was located on the right upper limb resulted in an additional spontaneous regression of the untreated distant tumors on the contralateral left upper limb. This phenomenon was accompanied by immune cell infiltration [125].

In 2009, Kabingu and team found out that PDT treatment of superficial basal cell carcinoma enhanced the reactivity of patient's lymphocytes against Hip1, a specific tumor-associated tumor antigen [126].

Conclusion

However, PDT has been shown to serve as an efficient, safe, and feasible treatment option to achieve local control at the site of tumor growth. For this reason, PDT can be combined with other local tumoricidal therapies, e.g., high-dose-rate endoluminal brachytherapy and external beam radiotherapy. Moreover, PDT does not only represent a local anticancer therapy limited to the tumor itself. By inducing a considerable systemic release of various inflammatory cytokines, mediators, and immunological cells culminating in an acute-phase response, PDT is able to interfere with the human immune system. On the one hand, PDT induces the increased synthesis of tumor-specific cytotoxic T cells which are able to eliminate distant untreated tumor cells. On the other hand, PDT leads to the development of antitumor memory immunity that may sustainably prevent tumor recurrence. Further clinical trials are definitively required to investigate and to understand the complex relationship between PDT-induced inflammation and antitumor immune response.

References

1. Enzinger PC, Mayer RJ. Esophageal cancer. N Engl J Med. 2003;349(23):2241–52.
2. Corti L, Skarlatos J, Boso C, Cardin F, Kosma L, Koukourakis MI, Giatromanolaki A, Norberto L, Shaffer M, Beroukas K. Outcome of patients receiving photodynamic therapy for early esophageal cancer. Int J Radiat Oncol Biol Phys. 2000;47(2):419–24.
3. Li Y, Yang H, Cao J. Association between alcohol consumption and cancers in the Chinese population—a systematic review and meta-analysis. PLoS One. 2011;6(4):e18776.
4. Lindenmann J, Maier A, Matzi V, Neuboeck N, Anegg U, Porubsky C, Sankin O, Fell B, Renner H, Swatek P, Smolle-Juettner FM. Photodynamic therapy for esophageal carcinoma. Eur Surg. 2011;43(6):355–65.
5. Brown LM, Hoover R, Silverman D, Baris D, Hayes R, Swanson GM, Schoenberg J, Greenberg R, Liff J, Schwartz A, Dosemeci M, Pottern L, Fraumeni Jr JF. Excess incidence of squamous cell esophageal cancer among US Black men: role of social class and other risk factors. Am J Epidemiol. 2011;153(2):114–22.
6. De Stefani E, Barrios E, Fierro L. Black (air-cured) and blond (flue-cured) tobacco and cancer risk. III: Oesophageal cancer. Eur J Cancer. 1993;29A(5):763–6.
7. Ellis A, Field JK, Field EA, Friedmann PS, Fryer A, Howard P, Leigh IM, Risk J, Shaw JM, Whittaker J. Tylosis associated with carcinoma of the oesophagus and oral leukoplakia in a large Liverpool family—a review of six generations. Eur J Cancer B Oral Oncol. 1994;30B(2):102–12.
8. Risk JM, Mills HS, Garde J, Dunn JR, Evans KE, Hollstein M, Field JK. The tylosis esophageal cancer (TOC) locus: more than just a familial cancer gene. Dis Esophagus. 1999;12(3):173–6.
9. Pera M, Manterola C, Vidal O, Grande L. Epidemiology of esophageal adenocarcinoma. J Surg Oncol. 2005;92(3):151–9.
10. Ellis Jr FH, Watkins Jr E, Krasna MG. Staging of carcinoma of the esophagus and cardia: a comparison of different staging criteria. J Surg Oncol. 1993;52:231–5.
11. Moghissi K. Surgical resection for stage I cancer of the oesophagus and cardia. Br J Surg. 1992;79:935–7.
12. Oliver SE, Robertson CS, Logan RFA. Esophageal cancer: a opulation-based study of survival after treatment. Br J Surg. 1992;79:1321–5.
13. Fok M, Sham JS, Chay D, Chung SWK, Wong J. Postoperative radiotherapy for carcinoma of the esophagus: a prospective, randomized controlled study. Surgery. 1993;113:138–47.
14. Amdal CD, Jacobsen AB, Tausjø JE, Wiig JN, Warloe T, Sandstad B, Bjordal K. Palliative interventions and prognosis in patients with advanced esophageal cancer. Dis Esophagus. 2011;24:502–9.
15. Lord SR, Hall PS, McShane P, Brown J, Seymour MT. Factors predicting outcome for advanced gastroesophageal cancer in elderly patients receiving palliative chemotherapy. Clin Oncol (R Coll Radiol). 2010;22:107–13.
16. Javle M, Ailawadhi S, Yang GY, Nwogu CE, Schiff MD, Nava HR. Palliation of malignant dysphagia in esophageal cancer: a literature-based review. J Support Oncol. 2006;4(8):365–73.
17. Sur RK, Singh DP, Sharma SC. Radiation therapy of esophageal cancer: role of high dose rate brachytherapy. Int J Radiat Oncol Biol Phys. 1992;22:1043–6.
18. Sakamoto T, Kotah H, Shimizu T. Clinical results of treatment of advanced esophageal carcinoma with hyperthermia in combination with chemotherapy. Chest. 1997;112:1487–93.

19. Rutgeerts P, Vantrappen G, Broeckaert L. Palliative Nd:YAG laser therapy for cancer of the esophagus and gastrooesophageal junction: impact on the quality of remaining life. Gastrointest Endosc. 1988;34:87–90.
20. Maier A, Pinter H, Friehs GB, Renner H, Smolle-Juettner FM. Self-expandable coated stent after intraluminal treatment of esophageal cancer: a risky procedure? Ann Thorac Surg. 1999;67:781–4.
21. JS Jr MC. Photodynamic therapy versus Nd:YAG laser treatment of endobronchial or esophageal malignancies. In: Spinelli P, Dal Fante M, Marchesini R, editors. Photodynamic therapy and biomedical lasers. New York: Elsevier; 1992. p. 23–36.
22. Dougherty TJ, Gomer CJ, Henderson BW, Jori G, Kessel D, Korbelik M, Moan J, Peng Q. Photodynamic therapy. J Natl Cancer Inst. 1998;90(12):889–905.
23. Dolmans DE, Fukumura D, Jain RK. Photodynamic therapy for cancer. Nat Rev Cancer. 2003;3(5):380–7.
24. Wu D, Liu Z, Fu Y, Zhang Y, Tang N, Wang Q, Tao L. Efficacy of 2-(1-hexyloxyethyl)-2-devinyl pyropheophorbide-a in photodynamic therapy of human esophageal squamous cancer cells. Oncol Lett. 2013;6(4):1111–9.
25. Yano T, Muto M, Minashi K, Onozawa M, Nihei K, Ishikura S, Kaneko K, Ohtsu A. Long-term results of salvage photodynamic therapy for patients with local failure after chemoradiotherapy for esophageal squamous cell carcinoma. Endoscopy. 2011;43(8):657–63.
26. Lindenmann J, Matzi V, Neuboeck N, Anegg U, Baumgartner E, Maier A, Smolle J, Smolle-Juettner FM. Individualized, multimodal palliative treatment of inoperable esophageal cancer: clinical impact of photodynamic therapy resulting in prolonged survival. Lasers Surg Med. 2012;44(3):189–98.
27. Maier A, Anegg U, Fell B, Tomaselli F, Sankin O, Prettenhofer U, Pinter H, Rehak P, Friehs GB, Smolle-Juttner FM. Effect of photodynamic therapy in a multimodal approach for advanced carcinoma of the gastro-esophageal junction. Lasers Surg Med. 2000;26(5):461–6.
28. Maier A, Tomaselli F, Gebhard F, Rehak P, Smolle J, Smolle-Juttner FM. Palliation of advanced esophageal carcinoma by photodynamic therapy and irradiation. Ann Thorac Surg. 2000;69(4):1006–9.
29. Sibille A, Lambert R, Souquet JC, Sabben G, Descos F. Long-term survival after photodynamic therapy for esophageal cancer. Gastroenterology. 1995;108(2):337–44.
30. von Tappeiner H, Jesionek A Therapeutische Versuche mit fluoreszierenden Stoffen. Munch Med Wochenschr,1903.47:p. 2042–2044.
31. Dougherty TJ. Activated dyes as antitumor agents. J Natl Cancer Inst. 1974;52:1333–13336.
32. Barr H, Dix AJ, Kendall C, Stone N. The potential for photodynamic therapy in the management of upper gastrointestinal disease. Aliment Pharmacol Ther. 2001;15:311–21.
33. Dougherty TJ, Kaufman JE, Goldfarb A, Weishaupt KR, Boule DG, Mitleman A. Photoradiation for the treatment of malignant tumors. Cancer Res. 1978;38:2628–35.
34. Earlam R, Culina-Melo JR. Esophageal squamous cell carcinoma: II. A critical review of radiotherapy. Br J Surg. 1980;67:457–61.
35. Sabanathan S, Shah R, Mearns A, Richardson J, Goulden C, Shakir T. Results of surgical treatment of esophageal cancer. J R Coll Surg Edinb. 1996;41:295–301.
36. McCaughan Jr JS, Hicks W, Laufman L, May E, Roach R. Palliation of esophageal malignancy with photo radiation therapy. Cancer. 1984;54:2905–10.
37. Hayata Y, Kato M, Okitsu M. Photodynamic therapy with haematoporphyrin derivatives in cancer of the upper gastrointestinal tract. Semin Surg Oncol. 1985;1(1):1–11.
38. Overhold BF, Panjenpour M, Haydek JM. Photodynamic therapy for ablation of dysphasia, reduction of specialized mucosa and treatment of superficial esophageal cancer. Gastrointest Endosc. 1995;42:64–70.
39. Overhold BF, Panjenpour M, Haydek JM. Photodynamic therapy of Barrett's oesophagus follow up in 100 patients. Gastrointest Endosc. 1999;49:1–7.

40. Radu A, Wagniere C, Van den Berg H, Monnier P. Photodynamic therapy of early squamous cell cancer of the oesophagus. Gastrointest Endosc Clin N Am. 2000;10:439–60.
41. Barr H, Shepard NA, Dix A, Roberts DJH, Tan WC, Krasner N. Eradication of high grade dysplasia in columnar lined (Barrett's) oesophagus using photodynamic therapy with endogeneously generated protoporhyrin IX. Lancet. 1996;348:584–5.
42. Ackroyd R, Brown NJ, Davis MF, et al. Photodynamic therapy for dysplastic Barrett's oesophagus: a prospective double blind randomized placebo controlled trial. Gut. 2000;47:612–7.
43. Lewis I. The surgical treatment of carcinoma of the oesophagus with special reference to a new operation for growth of the middle third. Br J Surg. 1946;34:18–31.
44. Dougherty TJ, Marcus SL. Photodynamic therapy. Eur J Cancer. 1992;28A:1734–42.
45. Pass HI. Photodynmic therapy in oncology: mechanisms and clinical use. J Natl Cancer Inst. 1993;85:443–56.
46. Castano AP, Mroz P, Hamblin MR. Photodynamic therapy and anti-tumour immunity. Nat Rev Cancer. 2006 Jul;6(7):535–45.
47. Castano AP, Demidova TN, Hamblin MR. Mechanisms in photodynamic therapy: part two-cellular signalling, cell metabolism and modes of cell death. Photodiagnosis Photodyn Ther. 2005;2:1–23.
48. Castano AP, Demidova TN, Hamblin MR. Mechanisms in photodynamic therapy: part three-photosensitizer pharmacokinetics, biodistribution, tumor localization and modes of tumor destruction. Photodiagnosis Photodyn Ther. 2005;2:91–106.
49. Moan J, Peng Q, Sorensen R, Iani V, Nesland JM. The biophysical foundations of photodynmiac therapy. Endoscopy. 1998;30:387–91.
50. Oleinick NL, Morris RL, Belichenko I. The role of apoptosis in response to photodynamic therapy: what, where, why, and how. Photochem Photobiol Sci. 2002;1(1):1–21.
51. Krammer B. Vascular effects of photodynamic therapy. Anticancer Res. 2001;21:4271–7.
52. Dolmans DE. Vascular accumulation of a novel photosensitizer, MV6401, causes selective thrombosis in tumor vessels after photodynamic therapy. Cancer Res. 2002;62:2151–6.
53. Lindenmann J, Matzi V, Neubock N, Maier A, Smolle-Juttner FM. The clinical impact of photodynamic therapy in thoracic surgery. Eur Surg. 2010;42(5):220–8.
54. Maier A, Tomaselli F, Matzi V, Rehak P, Pinter H, Smolle-Juettner FM. Does new photosensitizer improve photodynamic therapy in advanced esophageal carcinoma? Lasers Surg Med. 2001;29(4):323–7.
55. Grosjean P, Wagnieres G, Fontolliet C. Clinical photodynamic therapy for superficial cancer in the esophagus and the bronchi: 514 nm compared with 630 nm light irradiation after sensitization with Photofrin II. Br J Cancer. 1998;77:1989–95.
56. Lightdale CJ. Role of photodynamic therapy in the management of advanced esophageal cancer. Gastrointest Endosc Clin N Am. 2000;10:397–408.
57. Savary JF, Grosjean P, Monnier P. Photodynamic therapy for early squamous cell carcinomas of the esophagus: a review of 31 cases. Endoscopy. 1998;30:258–65.
58. Grosjean P, Savary JF, Wagnieres G. Tetra(m-hydroxyphenyl)-chlorin clinical photodynamic therapy of early bronchial and oesophageal cancers. Lasers Med Sci. 1996;11:227–35.
59. Hinnen P, de Rooij FW, van Velthuysen ML. Biochemical basis of 5-aminolaevulinic acid induced protoporphyrin IX accumulation: a study in patients with (pre)malignant lesions of the oesophagus. Br J Cancer. 1998;78:679–82.
60. Maier A, Tomaselli F, Matzi V, Rehak P, Pinter H, Smolle-Juettner FM. Photosensitization with hematoporphyrin derivative are more effective compared to 5-aminolaevulinic acid for photodynamic therapy of advanced esophageal carcinoma. Ann Thorac Surg. 2001;72:1136–40.
61. Loh CS, Vernon D, MacRobert AJ. Endogeneous porphyrin distribution induced by 5-aminolaevulinic acid in the tissue layers of the gastrointestinal tract. J Photochem Photobiol. 1993;20:47–54.
62. Webber J, Kessel D, Fromm D. Side effects and photosensitization of human tissues after aminolaevulinic acid. J Surg Res. 1997;68:31–7.

63. Heier SK, Rothmann K, Heier LM, Rosenthal WS. Randomized trial and light dosimetry of palliative photodynamic therapy (abstract). Gastrointest Endosc. 1992;38:279.
64. Thomas RJ, Abott M, Bhathal PS, St John DJB, Morstyn G. High dose photoirradiation of esophageal cancer. Ann Surg. 1987;206:193–9.
65. Fingar VH, Henderson BW. Drug and light dose dependence of photodynamic therapy: a study of tumor and normal tissue response. Photochem Photobiol. 1987;46:837–41.
66. Bown SG, Tralu CJ, Coleridge-Smith PD, Akdemir D, Wieman TJ. Photodynamic therapy with porphyrin and phtalocyanine senstization: quantitative studies in normal rat liver. Br J Cancer. 1986;54:43–52.
67. Karanov S, Shopova M, Getov H. Photodynamic therapy in gastrointestinal cancers. Lasers Surg Med. 1991;11:395–8.
68. Song SZ, Li JH, Zou J, Shu MY, Zhao FY, Jin ML, Guo ZH. Hematoporphyrin derivative and laser photodynamic reaction in the diagnosis and treatment of malignant tumors. Lasers Surg Med. 1985;5:61–6.
69. Gossner L, Wittke H, Warzecha A, Sroka R, Ernst H, Meier M, Ell C. Dose-dependent destruction of human gastrointestinal neoplasms by photodynamic therapy: a quantitative pilot study in athymic nude mice. Eur J Gastroenterol Hepatol. 1994;6:159–65.
70. Barr H, Krasner N, Boulus PB, Chatlani P, Bown SG. Photodynamic therapy for colorectal cancer: a quantitative pilot study. Br J Surg. 1990;77:93–6.
71. Kato H, Kawaguchi M, Konaka C, Hayata Y, Okitsu H. Evaluation of photodynamic therapy in gastric cancer. Lasers Med Sci. 1986;1:67–74.
72. Cheong WF, Prahl SA, Welch AJ. A review of the optical properties of biological tissues. IEEE J Quantum Electron. 1990;26:2166–85.
73. Wilson BC, Patterson MS. The physics of photodynamic therapy. Phys Med Biol. 1986;31:327–60.
74. Maier A, Sullmann D, Anegg U, Tomaselli F, Rehak P, Hutten H, Pinter H, Smolle-Jüttner FM. In vivo determination of tumor optical parameters in esophageal carcinoma. Lasers Surg Med. 2000;27:350–7.
75. Gossner L, May A, Sroka R. A new long-range through-the-scope balloon applicator for photodynamic therapy in the esophagus and cardia. Endoscopy. 1999;31:370–6.
76. Wang KK. Current status of photodynamic therapy of Barrett's esophagus. Gastrointest Endosc. 1999;49:S20–3.
77. van den Bergh H. On the evolution of some endoscopic light delivery systems for photodynamic therapy. Endoscopy. 1998;30:392–407.
78. Moghissi K, Dixon K, Thorpe JAC, Stringer M, Moore PJ. The role of photodynamic therapy (PDT) in inoperable oesophageal cancer. Eur J Cardiothorac Surg. 2000;17:95–100.
79. Gomer C, Razum N. Acute skin response in albino mice following porphyrin photsensitization under oxic and anoxic conditions. Photochem Photobiol. 1984;40:435–9.
80. Jirsa Jr M, Pouckiva P, Dolezal J, Pospísil J, Jirsa M. Hyperbaric oxygen and photodynmaic therapy in tumor bearing nude mice. Eur J Cancer. 1991;27:109.
81. Foote CS. Mechanisms of photooxygenation. In: Doiron DR, Gomer CJ, editors. Porphyrin localization and treatment of tumors. New York: Alan R. Liss; 1984. p. 3.
82. Athar M, Elmets CA, Bickers DR. A novel mechanism for the generation of superoxide anions in hematoporphyrin derivative mediated cutaneous photosensitization. Activation of the xanthine oxidase pathway. J Clin Invest. 1989;83:1137–43.
83. Tromberg BJ, Orenstein A, Kimel S. In vivo tumor oxygen tension measurements fort the evaluation of the efficiency of photodynamic therapy. Photochem Photobiol. 1990;52(2):375–85.
84. McCaughan JS, Williams TE, Bethel BH. Palliation of esophageal malignancy with photodynamic therapy. Ann Thorac Surg. 1985;40(2):113–20.
85. Moghissi K, Dixon K. Photodynamic therapy (PDT) in esophageal cancer: A surgical view of its indications based on 14 years experience. Technol Cancer Res Treat. 2003;2(4):319–26.

86. Maier A, Tomaselli F, Anegg U, Rehak P, Fell B, Luznik S, Pinter H, Smolle-Jüttner FM. Combined photodynamic therapy and hyperbaric oxygenation in carcinoma of the esophagus and the esophago-gastric junction. Eur J Cardiothorac Surg. 2000;18:649–55.
87. Moan J, Peng Q. An outline of the hundred-year history of PDT. Anticancer Res. 2003;23(5A):3591–600.
88. Reginato E, Wolf P, Hamblin MR. Immune response after photodynamic therapy increases anti-cancer and anti-bacterial effects. World J Immunol. 2014;4(1):1–11.
89. van Duijnhoven FH, Aalbers RI, Rovers JP, Terpstra OT, Kuppen PJ. The immunological consequences of photodynamic treatment of cancer, a literature review. Immunobiology. 2003;207:105–13.
90. Canti G, De Simone A, Korbelik M. Photodynamic therapy and the immune system in experimental oncology. Photochem Photobiol Sci. 2002;1:79–80.
91. Agostinis P, Buytaert E, Breyssens H, Hendrickx N. Regulatory pathways in photodynamic therapy induced apoptosis. Photochem Photobiol Sci. 2004;3:721–9.
92. Castano AP, Hamblin MR. Anti-tumor immunity generated by photodynamic therapy in a metastatic murine tumor model. Proc SPIE. 2005;5695:7–16.
93. Plaetzer K, Kiesslich T, Oberdanner CB, Krammer B. Apoptosis following photodynamic tumor therapy: induction, mechanisms and detection. Curr Pharm Des. 2005;11:1151–65.
94. Chen B, Roskams T, Xu Y, Agostinis P, de Witte PA. Photodynamic therapy with hypericin induces vascular damage and apoptosis in the RIF-1 mouse tumor model. Int J Cancer. 2002;98:284–90.
95. Scheffer SR, Nave H, Korangy F, Schlote K, Pabst R, Jaffee EM, Manns MP, Greten TF. Apoptotic, but not necrotic, tumor cell vaccines induce a potent immune response in vivo. Int J Cancer. 2003;103:205–11.
96. Magner WJ, Tomasi TB. Apoptotic and necrotic cells induced by different agents vary in their expression of MHC and costimulatory genes. Mol Immunol. 2005;42:1033–42.
97. Bartholomae WC, Rininsland FH, Eisenberg JC, Boehm BO, Lehmann PV, Tary-Lehmann M. T cell immunity induced by live, necrotic, and apoptotic tumor cells. J Immunol. 2004;173:1012–22.
98. Lilge L, Portnoy M, Wilson BC. Apoptosis induced in vivo by photodynamic therapy in normal brain and intracranial tumour tissue. Br J Cancer. 2000;83:1110–7.
99. Yenari MA, Liu J, Zheng Z, Vexler ZS, Lee JE, Giffard RG. Antiapoptotic and anti-inflammatory mechanisms of heat-shock protein protection. Ann N Y Acad Sci. 2005;1053:74–83.
100. Korbelik M, Sun J. Photodynamic therapy-generated vaccine for cancer therapy. Cancer Immunol Immunother. 2005;55:900–9.
101. Korbelik M, Sun J, Cecic I. Photodynamic therapy-induced cell surface expression and release of heat shock proteins: relevance for tumor response. Cancer Res. 2005;65:1018–26.
102. Todryk S, Melcher AA, Hardwick N, Linardakis E, Bateman A, Colombo MP, Stoppacciaro A, Vile RG. Heat shock protein 70 induced during tumor cell killing induces Th1 cytokines and targets immature dendritic cell precursors to enhance antigen uptake. J Immunol. 1999;163:1398–408.
103. Verrico AK, Haylett AK, Moore JV. In vivo expression of the collagen-related heat shock protein HSP47, following hyperthermia or photodynamic therapy. Lasers Med Sci. 2001;16:192–8.
104. Hanlon JG, Adams K, Rainbow AJ, Gupta RS, Singh G. Induction of Hsp60 by Photofrin-mediated photodynamic therapy. J Photochem Photobiol B. 2001;64:55–61.
105. Gollnick SO, Evans SS, Baumann H, Owczarczak B, Maier P, Vaughan L, Wang WC, Unger E, Henderson BW. Role of cytokines in photodynamic therapy-induced local and systemic inflammation. Br J Cancer. 2003;88:1772–9.
106. Cecic I, Korbelik M. Mediators of peripheral blood neutrophilia induced by photodynamic therapy of solid tumors. Cancer Lett. 2002;183:43–51.

107. Balkwill F, Charles KA, Mantovani A. Smoldering and polarized inflammation in the initiation and promotion of malignant disease. Cancer Cell. 2005;7:211–7.
108. Cecic I, Serrano K, Gyongyossy-Issa M, Korbelik M. Characteristics of complement activation in mice bearing Lewis lung carcinomas treated by photodynamic therapy. Cancer Lett. 2005;225:215–23.
109. Krosl G, Korbelik M, Dougherty GJ. Induction of immune cell infiltration into murine SCCVII tumour by photofrin-based photodynamic therapy. Br J Cancer. 1995;71:549–55.
110. Jiang H, Granville DJ, North JR, Richter AM, Hunt DW. Selective action of the photosensitizer QLT0074 on activated human T lymphocytes. Photochem Photobiol. 2002;76:224–31.
111. Jiang H, Granville DJ, McManus BM, Levy JG, Hunt DW. Selective depletion of a thymocyte subset in vitro with an immunomodulatory photosensitizer. Clin Immunol. 1999;91:178–87.
112. Korbelik M, Krosl G. Enhanced macrophage cytotoxicity against tumor cells treated with photodynamic therapy. Photochem Photobiol. 1994;60:497–502.
113. Takahashi T, Tagami T, Yamazaki S, Uede T, Shimizu J, Sakaguchi N, Mak TW, Sakaguchi S. Immunologic self-tolerance maintained by CD25(+)CD4(+) regulatory T cells constitutively expressing cytotoxic T lymphocyte-associated antigen 4. J Exp Med. 2000;192:303–10.
114. Von Boehmer H. Mechanisms of suppression by suppressor T cells. Nat Immunol. 2005;6:338–44.
115. Gollnick SO, Liu X, Owczarczak B, Musser DA, Henderson BW. Altered expression of interleukin 6 and interleukin 10 as a result of photodynamic therapy in vivo. Cancer Res. 1997;57:3904–9.
116. Castano AP, Mroz P, Wu MX, Hamblin MR. Photodynamic therapy plus low-dose cyclophosphamide generates antitumor immunity in a mouse model. Proc Natl Acad Sci U S A. 2008 8;105(14):5495–500.
117. Reginato E, Lindenmann J, Langner C, Schweintzger N, Bambach I, Smolle-Jüttner F, Wolf P. Photodynamic therapy downregulates the function of regulatory T cells in patients with esophageal squamous cell carcinoma. Photochem Photobiol Sci. 2014;13(9):1281–9.
118. Reginato E, Mroz P, Chung H, Kawakubo M, Wolf P, Hamblin MR. Photodynamic therapy plus regulatory T-cell depletion produces immunity against a mouse tumour that expresses a self-antigen. Br J Cancer. 2013;109(8):2167–74.
119. Jalili A, et al. Effective photoimmunotherapy of murine colon carcinoma induced by the combination of photodynamic therapy and dendritic cells. Clin Cancer Res. 2004;10(13):4498–508.
120. Vabulas RM, Wagner H, Schild H. Heat shock proteins as ligands of toll-like receptors. Curr Top Microbiol Immunol. 2002;270:169–84.
121. Wahl SM, Swisher J, McCartney-Francis N, Chen W. TGF-β: the perpetrator of immune suppression by regulatory T cells and suicidal T cells. J Leukoc Biol. 2004;76:15–24.
122. Saji H, Song W, Furumoto K, Kato H, Engleman EG. Systemic antitumor effect of intratumoral injection of dendritic cells in combination with local photodynamic therapy. Clin Cancer Res. 2006;12:2568–74.
123. Abdel-Hady ES, Martin-Hirsch P, Duggan-Keen M, Stern PL, Moore JV, Corbitt G, Kitchener HC, Hampson IN. Immunological and viral factors associated with the response of vulval intraepithelial neoplasia to photodynamic therapy. Cancer Res. 2001;61:192–6.
124. Yom SS, Busch TM, Friedberg JS, Wileyto EP, Smith D, Glatstein E, Hahn SM. Elevated serum cytokine levels in mesothelioma patients who have undergone pleurectomy or extrapleural pneumonectomy and adjuvant intraoperative photodynamic therapy. Photochem Photobiol. 2003;78:75–81.
125. Thong PS, Ong KW, Goh NS, Kho KW, Manivasager V, Bhuvaneswari R, Olivo M, Soo KC. Photodynamic-therapy-activated immune response against distant untreated tumours in recurrent angiosarcoma. Lancet Oncol. 2007;8(10):950–2.
126. Kabingu E, Oseroff AR, Wilding GE, Gollnick SO. Enhanced systemic immune reactivity to a Basal cell carcinoma associated antigen following photodynamic therapy. Clin Cancer Res. 2009;15(13):4460–6.

Biomarkers for Predicting Neoplastic Progression in Barrett's Esophagus

16

F. Durchschein and G. Absenger

Contents

F. Durchschein (✉)
Division of Gastroenterology and Hepatology, Department of Internal Medicine, Medical University of Graz, Graz, Austria
e-mail: franziska.durchschein@medunigraz.at

G. Absenger
Division of Clinical Oncology, Department of Internal Medicine, Medical University of Graz, Graz, Austria

J. Haybaeck (ed.), *Mechanisms of Molecular Carcinogenesis – Volume 1*,
DOI 10.1007/978-3-319-53659-0_16

Abstract

Barrett's esophagus (BE), a complication of chronic gastroesophageal reflux disease (GORD), represents the strongest risk of esophageal adenocarcinoma (EAC). The low risk of progression together with the economic costs for surveillance argue for biomarkers predicting the likelihood of BE progression. In the last decades several promising biomarkers have been developed to estimate the risk of malignant transformation. In this review we summarize the current knowledge regarding these biomarkers for an individualized risk prediction and therapeutic outcome.

Abbreviations

APC	Adenomatous polyposis coli
BE	Barrett's esophagus
COX	Cyclooxygenase
DNA	Deoxyribonucleic acid
EAC	Esophageal adenocarcinoma
EGF	Epidermal growth factor
EGFR	Epidermal growth factor receptor
FISH	DNA fluorescent in situ hybridization
GERD	Gastroesophageal reflux
HGD	High-grade dysplasia
LGD	Low-grade dysplasia
LOH	Loss of heterozygosity
Mcm protein	Minichromosome maintenance protein
miRNA	MicroRNA
NBI	Narrowband imaging
ND	Nondysplastic
NF-kB	Nuclear factor "kappa-light-chain-enhancer" of activated B cells
PCNA	Proliferating cell nuclear antigen
TFF3	Trefoil factor
TGF-α	Transforming growth factor alpha

16.1 Introduction: Need for Biomarkers

The importance of Barrett's esophagus (BE) lies in its increasing prevalence and strong association to esophageal adenocarcinoma (EAC) [1]. While the risk and incidence of distal gastric cancer are decreasing worldwide, EAC has the most rapidly rising incidence in the Western world [2–4]. BE is characterized by the replacement of the normal stratified squamous epithelium of the distal esophagus by columnar epithelium with specialized intestinal metaplasia (IM) containing goblet cells [5] (Fig. 16.1). It is a premalignant condition, and patients with BE have a 30–60 times greater risk of developing adenocarcinoma of the esophagus than the general population [3, 5, 6]. The risk of developing cancer is higher among men, older patients, and patients with long segments of Barrett's mucosa or dysplasia [7].

Despite the increased risk of cancer development, the natural history of BE is incompletely understood [1]. The progression of BE from a columnar-lined esophagus to EAC is an established, gradual process from nondysplastic (ND) BE to low-grade dysplasia (LGD) and high-grade dysplasia (HGD) before the development of invasive cancer [8]. However, the individual risk of cancer progression is difficult to ascertain as only a small number of patients with BE will progress to EAC [2, 9]. Approximately 0.2–0.5% of patients with ND BE will develop EAC annually [9, 10], and only 5% of patients with EAC are known with a prior diagnosis of BE [5, 11]. Besides, some patients with dysplastic BE will also regress, with no further dysplasia detectable [12]. At present, there are no clinical or histological features to stratify the risk of progression or regression of patients with BE [8], and these patients are evaluated by the histological grade of dysplasia [5]. Based on this finding, the interval of endoscopic surveillance is determined individually [5, 12, 13]. Furthermore, despite advanced techniques, including narrowband imaging (NBI) and chromoendoscopy, endoscopic detection of BE is difficult [4, 10], and dysplastic areas in BE can be missed because of biopsy sampling errors [9]. Dysplasia is often patchy in extension and severity, and several biopsies are necessary to detect BE reliably [14]. Besides, histological diagnosis and grading of dysplasia are also potential limitations [9]. Interobserver variability

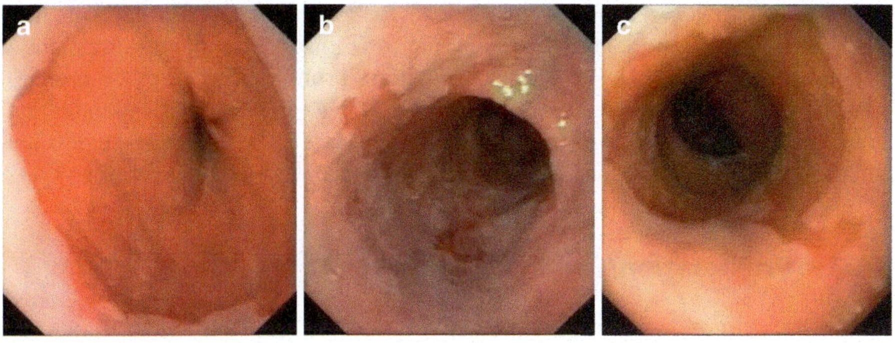

Fig. 16.1 Endoscopic findings of (**a**) short-segment Barrett's esophagus and (**b**, **c**) long-segment Barrett's esophagus

is a known problem especially for discrimination between ND and LGD. Most studies comparing diagnosis of dysplasia among different pathologist have concluded that there is a significant intra- and interobserver variability [15, 16]. Consequently, recent international guidelines for the management of LGD in BE recommend that the diagnosis of LGD should be confirmed by a second pathologist with specialized expertise in gastrointestinal (GI) pathology [13]. Furthermore, the difficulty to discriminate inflammatory and reactive changes from true dysplasia complicates the diagnosis of dysplasia [17].

These limitations, the low risk of progression, together with the economic costs for surveillance, argue for biomarkers predicting the likelihood of BE progression [14] and allowing targeting of screening for those most at risk [8, 9, 12]. The increasing number of publications, seen in the past few years, reflects the ongoing research for effective biomarkers, as well as the lack of clinically validated prognostic tools [9]. Several clinical, endoscopic, and genetic markers have been studied to risk stratify patients with BE in terms of their risk of progression [18].

16.2 Biomarkers

To validate and integrate biomarkers for the early detection of cancer and for clinical use, the National Cancer Institute Early Detection Research Network (EDRN) has proposed five phases, which are analogue to the process in therapeutic drug studies [4, 19]. Phase I consists of preclinical exploratory studies to identify potential biomarkers; phase II comprises clinical assay developments to determine sensitivity and specificity of markers in patients with the disease compared to healthy controls; phase III is composed of retrospective studies on specimens from subjects prior to their diagnosis; phase IV consists of prospective screening studies; phase V constitutes cancer control studies to detect whether screening with biomarkers reduces cancer incidence [4]. In BE, the majority of biomarkers have never been studied beyond phase I or II, and most studies used complex technologies not useful for clinical practice [4, 12, 14, 20].

Similar to other malignant tumors, carcinogenesis of EAC is characterized by several genetic and epigenetic aberrations [4, 9]. At least 5–10 genetic alterations are necessary to generate a malignant phenotype [3]. In the last few decades, multiple genes have been identified which seem to be involved in the development of neoplastic lesions in BE [4]. These markers are proliferation/cell cycle proteins, tumor suppressor genes, adhesion molecules, DNA content, and inflammation-associated markers [5, 12]. Some of these changes are early events in the development of cancer and might serve as biomarkers for risk stratification [9].

16.3 Proliferation/Cell Cycle Proteins

Hyperproliferation of endothelial cells is detectable in BE with an increase during progression from metaplasia to dysplasia [4]. In general, proliferative stimuli to cells in chronic GERD and BE are gastric acid and bile acids [21]. Some studies have demonstrated that pulsatile exposure to low pH leads to hyperproliferation of endothelial cells compared to growth at neutral pH [22]. To replace injured tissue after gastroesophageal reflux, cells need to progress from the G1 to the S phase in cell cycle [5], which is controlled by several key proteins. Mutations of any of these proteins regulating cell cycle may result in BE progression and may be useful to predict progression.

16.4 Ki67

The proliferation marker Ki67 (usually stained with the monoclonal antibody MIBI), which is upregulated in all active phase of cell cycle, may be a reliable biomarker [2]. The determination of KI67 expression has become routine in various malignant tumors, for example, it is a major biomarker for treatment decisions in breast cancer [23]. Ki67 expression in BE showed a stepwise increase with neoplastic progression [5, 24] and differences in expression levels between ND, LGD, and HGD BE [25]. The number and localization of Ki67-positive nuclei were significantly altered between ND, LGD, or HGD BE and EAC [26–28].

16.5 PCNA

The proliferation marker PCNA (proliferating cell nuclear antigen) is an indicator of cell cycle progression at the G1/S transition [2]. Studies have shown an increased proportion of cells stained with this antibody parallel to progression from metaplasia to dysplasia [2]. PCNA immunostaining was mainly seen in the basal cells of the epithelial compartment of glands in ND BE [20]. However, in mucosa of HGD BE, the proliferative compartment extended upward into the superficial layers of glands [28, 29].

16.6 Cyclins

Cyclins are potential biomarkers to predict BE progression. These proteins play a key role in cell cycle regulation [30]. Cyclin D1 is a proto-oncogene controlling the G1-S transition [12]. Studies have postulated that BE showing cyclin D1 overexpression is 6–7 times more likely to develop EAC [31]. However, other studies were not able to confirm this finding [5, 32–34]. At present, abnormalities of cyclin D1 expression cannot be used as routine biomarkers to predict progression risk [12]. The proto-oncogene cyclin A is expressed in 76% of patients with BE in the proliferative compartment [12]. With increasing grades of dysplasia, cyclin A expression shifts toward the mucosal surface [12]. In ND BE, 24% of patients express cyclin A at the surface epithelium compared with 59% of patients with LGD, 87% of patients with HDG, and 100% of patients with EAC [35].

16.7 Mcm

Mcm proteins (minichromosome maintenance proteins) are essential for DNA replication [4] and are expressed in all proliferating cells throughout the cell cycle [4]. Overexpression of the minichromosome maintenance deficient 2 (Mcm2) in BE biopsies was significantly associated with higher risk of EAC [9, 36]. BE biopsies of patients who progressed to EAC had Mcm2 overexpression in 28.4% of the luminal cells compared with 3.4% in nonprogressors [36].

In summary, abnormalities of proteins controlling cell cycle may be biomarkers to predict neoplastic progression [5]. However, further larger prospective studies with standardized techniques and definitions to measure proliferation are needed [18].

16.8 Tumor Suppressor Genes

Tumor suppressor genes regulate cell proliferation, apoptosis, cell adhesion, and gene expression [3]. Various studies have evaluated their ability to predict progression in BE.

16.9 P53

P53 is expressed by the TP53 gene (chromosome 17p) and is one of the most commonly mutated tumor suppressor genes in human cancers [37]. P53 is responsible for the activation of DNA repair mechanisms, activates cell cycle arrest at the G1/S cell cycle checkpoint, and initiates apoptosis if DNA damage cannot be repaired [8, 38, 39]. Alterations of p53 in EAC and its precursor lesions have been detected in several studies [4, 12].

The p53 protein has a short half-life and is, in general, not detectable immunohistochemically at all or only at low levels [12, 40]. In dysplastic BE, p53 function is often lost due to point mutations in the DNA binding domain of the gene [8]. This results in an increased half-life of p53 protein, and its accumulation in the cell nucleus generates levels that can be detected by immunohistochemistry [5, 8]. A stepwise overexpression of p53 with increasing grades of dysplasia in BE has been shown in several studies [24, 41–43]: immunohistochemical analysis has shown a low percentage of p53 overexpression in ND BE (5%), increasing to 10–20% in LGD and to more than 60% in HGD [44, 45]. Patients with LGD show an increased risk of progression to HGD and cancer in case of p53 overexpression [46]. Besides, TP53 point mutations in EAC can be detected in up to 70% [7, 47, 48] and are associated with poor tumor differentiation, as well as reduced overall survival following surgical resection [20].

Next to point mutations, loss of heterozygosity (LOH) is a frequent alteration of p53 in BE. LOH refers to the loss of the normal, functional allele at a heterozygous locus in which the other allele has already been inactivated [9, 49]. Studies have revealed that LOH of p53 (17pOH) could be a biomarker to predict cancer progression in BE. 17pOH has been shown to occur in 0–6% of BE without dysplasia, in 20–27% with LGD, in 57% with HGD, and in 54–92% with EAC [5]. In BE biopsies containing different grades of dysplasia, the 3-year cumulative incidence of cancer was 38% (95% CI, 26.0–54.0) in those with proven 17pLOH compared to

3.3% (95% CI, 1.4–8.0) in biopsy samples without 17pLOH [50]. Reid et al. reported that 17pLOH is associated with a 16-fold increased risk of progression to cancer [50]. In that study, 17pLOH was a significant predictor of progression to HGD in patients with initial ND, indefinite dysplasia, or LGD [50].

In conclusion, p53 gene alterations (mutations and LOH) are early and frequent events in EAC and seem to be associated with malignant transformation of BE [3]. However, the sensitivity of this marker alone to predict cancer risk seems to be of limited value [5]. Immunohistochemistry of the mutated p53 was shown to be 88–100% sensitive and 75–93% specific for predicting progression from LGD BE to HGD [46, 48, 51] but only 32% sensitive to predict progression from ND BE to HGD [34]. Besides, some mutations result in a truncated p53 protein, which is undetectable by immunohistochemistry [40]. There was no detectable accumulation by immunohistochemistry in 31% of patients with proved p53 mutation [52]. In addition, not all p53 protein accumulations are caused by mutations, as inflammation or cellular stress can upregulate p53, too [5, 18, 40, 53].

Consequently, 17pLOH and p53 immunostaining seem to represent useful biomarkers to predict BE progression, especially in combination with other high-risk markers [3, 4]. However, they have to be proved in large-scale, multicenter trials [18], and newer genotyping technologies may overcome some of the current limitations surrounding p53 [12].

16.10 P16

The tumor suppressor gene p16 is located at chromosome 9p21 and encodes a cell cycle regulator protein. Its inactivation results in uncontrolled cell proliferation [5]. Acid and bile exposure of the esophageal mucosa may mediate inactivation of p16, resulting in BE progression to dysplasia and EAC [54]. Alterations of p16 can be detected in all grades of dysplasia [12] and in up to 85% of EAC [9]. It occurs as a result of hypermethylation, mutation, LOH, or methylation of the promotor regions [55].

Hypermethylation of the p16 promoter is a common mechanism of p16 inactivation during neoplastic progression in BE and is already present in ND premalignant BE [56]. In a retrospective study of 53 patients, it was associated with an increased risk of progression from ND to HGD BE or invasive cancer (OR 1.74: 95% CI 1.33–2.20) [57]. Another genetic event leading to loss of p16 is LOH, detectable in approximately 75% of EAC tissue samples [58]. P16 LOH seems to be associated with subsequent clonal expression along the Barrett segment, favoring further mutations and disease progression [59]. The combination of p16 mutations and LOH in 9p21 seems to occur early, prior to the development of aneuploidy or cancer [5, 58], and may be a predictive biomarker panel. Furthermore, allelic loss of p16 seems to predict lack of response to photodynamic therapy in patients with HGD BE and cancer [9, 18]. However, larger studies evaluating the efficiency of p16 as biomarker for tumor progression have to be performed.

16.11 Further Promising Tumor Suppressor Genes/Proto-oncogenes to Predict BE Progression

The tumor suppressor p27 inhibits cyclin E/Cdk2 complexes, preventing cells from entering cell cycle into S phase [12]. P27 knockout mice showed an increased risk of EAC development compared to wild-type mice [60]. In BE and EAC, loss of p27 expression is associated with malignant transformation and a poorer prognosis [12, 60].

The tumor suppressor gene adenomatous polyposis coli (APC), a regulator of the WNT pathway, seems to be altered in BE by methylation and LOH [61, 62]. However, further studies have to determine its predictive ability [18].

A strong association has been found between 17p13 LOH and an abnormal flow cytometric DNA content in BE [63]. Reid et al. showed that 37% of patients with LOH at 17p13 progressed from ND BE to EAC, compared to 3% of patients without LOH at this allele [64].

The bcl-2 proto-oncogene, which blocks apoptosis, seems to be overexpressed early in the dysplasia-to-carcinoma sequence of BE [3] and may be a potential biomarker for predicting progression.

However, all these genes have to be evaluated in further studies to assess their role in predicting BE progression to EAC.

16.12 Chromosomal Abnormalities

A further possibility of predicting BE progression to EAC lies in chromosomal abnormalities. DNA content abnormalities refer to numerical and structural changes in chromosomes, including aneuploidy and tetraploidy [9]. Aneuploidy is the presence of an abnormal number of chromosomes in a cell, unlike the normal content of 46 chromosomes [18]. Tetraploidy refers to the instance when the chromosomal number of a cell is twice as high as that of normal cells [18].

Abnormalities in DNA ploidy correlate well with conventional histologic diagnoses of dysplasia and carcinoma, and several studies suggest that this marker might represent a valuable adjunctive tool in the evaluation of patients with Barrett's esophagus [49, 55]. In biopsies with ND or LGD BE without aneuploidy or increased tetraploidy, the 5-year cancer incidence was found to be 0% [49]. However, with biopsies containing the same grades of dysplasia demonstrating either aneuploidy or increased tetraploidy, the 5-year risk of cancer progression was 28% [49, 64, 65]. Over 90% of HGD BE and EAC show DNA aneuploidy, and there is a significant relation between the presence of DNA aneuploid population and the progression form ND BE to dysplasia and EAC [2, 66].

In summary, DNA content abnormalities seem to be an accurate marker of progression in subjects with BE, but have not been widely used due to technical challenges with flow cytometry [18].

16.13 FISH

DNA fluorescent in situ hybridization (FISH) is a technique in which small fluorescently labeled DNA probes are used for detection of chromosomal and gene aberrations [9]. This method can detect various types of cytogenetic alterations, including aneusomy, duplication, amplification, deletion, and translocation [18]. In the past, several studies used FISH probes directed against different tumor suppressor or proto-oncogenes like p53 (17q13.1), p16 (9p21), or HER-2/neu (17q11.2) to find biomarkers predicting progression of BE [67–70]. Amplification of at least one of these loci occurred in 14% of HGD and increased to 50% in EAC [71]. A prospective follow-up study showed promising results in identifying high-risk BE patients with a FISH assay, including the tumor suppressor genes p53 and p16 and centromeric probes of chromosomes 7 and 17 to detect aneuploidy [9]. Aberrations of chromosomes 7 and 17 were detected in 13% of ND BE, increased with dysplastic stage, and detected HGD/EAC with a sensitivity and specificity of 85% and 84% [9, 72]. Besides, a multicolored FISH assay has been developed for detection of dysplasia in BE [73]. This probe set showed a sensitivity of 84–93% and specificity of 93% to identify HGD and EAC [18, 73]. Furthermore, FISH-based biomarkers may also be used to predict response to ablation therapy and help to guide therapy decisions [74]. In summary, genetic abnormalities detected by FISH appear to be a promising method for BE progression. However, further validation in larger studies is needed [18].

16.14 Methylation

DNA hypermethylation is an early event in tumorigenesis and causes inactivation of tumor suppressor genes, as well as chromosomal instability [4]. Methylation-induced inactivation of genes, which is involved in cell cycle and cell differentiation during BE pathogenesis, was shown in several studies [4], and patients with a dense methylation pattern in EAC showed a worse survival after surgery [57].

Methylation of the p16 tumor suppressor gene is a common genetic abnormality found in BE [18] and can be detected in 34–66% [57, 75, 76]. Methylation of the tumor suppressor gene CDKN2A, which inhibits cell cycle progression and abrogates expression of p16, seems to be associated with the progression from BE to EAC [8]. It occurs early in the metaplasia-dysplasia-carcinoma sequence [77] and can be detected in 3–77% of BE patients [77]. Besides, it was shown to be related to 17pLOH and chromosomal abnormalities like tetraploidy and aneuploidy [8]. Based on the methylation of some genes (p16, HPP1, RUNX3) and clinical parameters (gender, BE segment length, and histopathology), a model was developed to stratify patients with BE into low-, intermediate- and high-risk groups [78]. This may represent a useful biomarker panel to predict BE progression. Hypermethylation of other genes like APC and T1MP1 has been detected in patients with BE [18]. However, convincing studies on their predictive ability are lacking [18]. Moreover, DNA methylation is a reversible event [4]. Consequently, modulation of the epigenetically involved pathways by using small molecules might become a therapeutically option for patients with BE [4].

16.15 Biomarker Panels

Combinations of biomarkers in panels may be better in predicting the risk of neoplastic progression in patients with BE than individual biomarkers alone [68, 69, 79]. Biopsies demonstrating high diversity seem to be more likely to progress to EAC [8, 70]. Due to technical progression, several molecular aberrations can be analyzed simultaneously with the aid of panels of biomarkers [8]. Using aneuploidy/increased tetraploidy, 17pLOH, and 9pLOH in combination, the presence of all three abnormalities predicted an 80% risk of cancer progression in BE at 6 years [9, 79, 80]. Moreover, a study demonstrated that the combination of LGD, abnormal DNA ploidy, and *Aspergillus oryzae* lectin can predict progression from BE to HGD and EAC [81]. Besides, multicolored FISH might be an option to analyze several biomarkers in a single assay [69].

16.16 Further Potential Biomarkers

16.16.1 HER2/neu

The proto-oncogene HER2/neu (c-erbB2) encodes a transmembrane glycoprotein with intrinsic tyrosine kinase activity [3]. Alteration of HER2/neu can be detected in approximately 10–70% of EAC [3, 82]. HER2/neu overexpression in EAC correlates significantly with tumor invasion, distant metastasis, lymph node involvement, and status of residual tumor after resection [3, 83], but it offers therapeutic options in the combination of chemotherapy and trastuzumab [84]. As HER2/neu overexpression is not detectable in dysplastic BE, it seems to be a late event in the dysplasia to carcinoma sequence [85]. Further studies evaluating the potential of HER2/neu to predict BE progression are necessary.

16.17 Several Growth Factors

Epidermal growth factor (EGF), epidermal growth factor receptor (EGFR), and transforming growth factor alpha (TGF-α) are important members of the family of growth factors [3]. Some studies show the correlation of EGF, EGFR, and TGF-α overexpression with the degree of mucosal dysplasia and the occurrence of EAC [3, 82, 86–88].

Neovascularization seems to be an early event in the pathogenesis of BE [4]. An increased number of small vessels can be detected in dysplastic BE, and an increasing microvessel density can be seen from BE to HGD or intramucosal carcinoma [4, 89]. Overexpression of VEGF and VEGFR can be detected in dysplastic BE and EAC [90]. Furthermore, COX-2 expression is associated with neovascularization, suggesting that bile and gastric acid may induce angiogenesis via COX-2 expression [89]. Besides, in other tissues, COX-2 inhibitors can suppress vessel growth [89]. However, trails of a selective COX-2 inhibitor, celecoxib, did not show any protective effect against BE progression to EAC [91].

16.18 NF-kB

The transcription factor NF-kB (nuclear factor "kappa-light-chain-enhancer" of activated B cells) regulates proinflammatory genes, differentiation, and growth [12]. Cytokines, free radicals, and acid stimulate translocation of NF-kB to the nucleus, where it binds specific DNA sites and upregulates the expression of genes involved in inflammatory process [92]. NF-kB expression is stepwise increased in patients with BE adjacent to EAC [93, 94]. In patients with ND BE, NF-kB overexpression was detected in 50%, with LGD BE in 63% and with HGD BE in 100% [93]. NF-kB can be activated by deoxycholic acid, a bile acid and a common component of reflux, or acid pH [89]. However, further studies are needed to determine the role of this molecule in the metaplasia-carcinoma sequence [12].

16.19 Cyclooxygenase 2 (COX-2)

Cyclooxygenase (COX) catalyzes the rate-limiting step in prostaglandin synthesis [3]. COX-1 is constitutively expressed, whereas COX-2 is undetectable in most cells. However, it can be activated by cytokines, gastric acid, and bile acids. Studies revealed that COX-2 is involved in cell proliferation, reducing apoptosis and promoting angiogenesis [12, 95]. Unconjugated bile acids, one of the major components of gastroesophageal reflux, can stimulate COX-2 expression through a reactive oxygen species-mediated signaling pathway [96]. COX-2 expression cannot be measured in normal esophageal mucosa [4], but a progressive increase of COX-2 expression along the metaplasia-dysplasia sequence was described [97]. Additionally, COX-2 is expressed in 70–80% of patients with EAC [3]. Besides, nonsteroidal anti-inflammatory drug (NSAID) intake was shown to have a protective effect and reduces the risk of EAC, especially in patients with several molecular high-risk abnormalities [80]. However, at the moment, there are not enough data that support the role of COX-2 as a useful biomarker [12].

16.20 MicroRNAs

MicroRNAs (miRNAs) are small segments of noncoding RNA of 20–24 nucleotides regulating the translation of mRNA. They play a role in cell proliferation and function as tumor suppressor genes or oncogenes [4]. MicroRNAs may be useful biomarkers, as they are present in circulating blood plasma in a highly stable, cell-free form included in lipid or lipoprotein complexes [98]. Several studies have examined the role of miRNAs in progression from BE to EAC [98–100] and detected alterations in miRNA expression profiles between ND BE, HGD BE, and EAC [98]. Alterations in miR-25, miR-93, and miR-106b have been reported in BE and EAC compared to normal esophageal tissue [100]. Furthermore, in samples of EAC, an upregulation of mi-21 and mi-192 has been detected [101].

There are several miRNAs that have been found to be up- or downregulated in different stages in the progression from BE to EAC [98]. Identifying specific miRNA patterns in BE might help to detect dysplasia with more progressive potential and might help to distinguish low-risk from high-risk patients [4]. Further work is required in order to use miRNAs for risk stratification in the progression from BE to EAC [98].

16.21 Endoscopic Measurements

Next to reliable biomarkers, methods to detect areas of concern for biopsies are needed. The direct application of molecular markers during endoscopy to allow visualization of dysplasia without the need for histopathology is a further promising field of BE research [8]. The use of fluorescent probes to molecules involved in the dysplasia sequence of BE may allow for targeting areas of concern [8]. The majority of these studies rely on the use of confocal imaging [8]. The development of a polyclonal antiperiostin antibody against periostin, which is expressed differentially in ND and dysplastic BE, is an example for this new method [102]. However, periostin is also expressed in inflamed tissue [8] limiting the sensitivity and specificity of this marker. The peptide probe sequence ASYNYDA has been fluorescently labeled to be visible in vivo by using fluorescence microscopy [8]. The sensitivity and specificity for the detection of dysplastic BE was 82% and 85%, respectively [8]. However, at present, confocal imaging is not a standard endoscopic technique, and a more clinically applicable fluorescence dye visible with a standard endoscope is needed [8].

16.22 Non-endoscopic Methods to Detect Dysplasia

The costs, as well as discomfort of the numerous surveillance endoscopies of patients with BE, have argued for non-endoscopic alternatives to detect BE. The Cytosponge is a capsule on a string that is swallowed by the patient. When the capsule reaches the stomach, the capsule dissolves and releases a spherical sponge which is retrieved [8]. During the retrieval through the esophagus, cells adhere to the sponge and can be immunohistochemical analyzed for the presence of TFF3 (trefoil factor) [8]. TFF3 is a marker of columnar epithelium and is expressed in a variety of tissues, including goblet cells of the intestines and colon. It promotes mucosal healing and epithelial restitutions in vivo in the gastrointestinal mucosa. Detecting TFF3-positive glandular cells in the Cytosponge indicates the presence of BE. In a study with 500 patients, the sensitivity and specificity of this method for detection of BE were 73 and 93% for short-segment BE and increased to 90 and 93% for long-segment BE [103]. Furthermore, in 19 of 22 sponge samples taken from patients with known high-grade dysplasia, mutations in the TP53 gene could be detected. By contrast, no TP53 mutations were found in the sponge samples of healthy controls or patients with BE without dysplasia [47]. However, due to false positivity, the clinical utility of TFF3 may be limited in the cardia [104].

Serum biomarkers for the detection of patients at an increased risk of EAC are under intensive investigation [8]. Telomere length in blood samples of patients with BE without dysplasia was assessed and followed for 5.8 years. Patients with shorter telomere length at baseline were at increased risk of developing EAC [105].

Conclusion

The major risk of patients with BE of developing EAC has generated interest in defining subgroups of high-risk patients who can be surveilled effectively [106]. However, the natural history of BE is still very difficult to predict for one individual patient [106]. Several promising candidate biomarkers and biomarker panels have been described: proliferation markers, chromosomal abnormalities, tumor suppressor genes, DNA hypermethylation, as well as FISH or microRNAs might be able to predict Barrett's progression. The development of a Barrett's risk score incorporating clinical variables and biomarker panels may be an option to stratify patients into low-risk and high-risk subsets [18]. However, there are several problems to translate the use of these biomarkers into practice like the need for special media for biopsies, interlaboratory variation in methodology, and lack of standardization [18]. The majority of these studies were performed retrospectively and included only a small number of patients [106]. Consequently, the majority of these markers need to be evaluated in large-scale prospective clinical trials. Prolonged follow-up of patients ranging between 5 and 10 years is required leading to logistical problems [18]. Besides, in order to develop useful biomarkers, we need to further understand molecular and genetic abnormalities associated with BE [12], and it still needs to be proven that these biomarkers will reduce cancer incidence [106].

In the next years, we can expect more studies attempting to find new methods that effectively predict BE progression.

References

1. Findlay JM, Middleton MR, Tomlinson I. Genetic biomarkers of Barrett's esophagus susceptibility and progression to dysplasia and cancer: a systematic review and meta-analysis. Dig Dis Sci. 2016;61(1):25–38.
2. Flejou JF. Barrett's oesophagus: from metaplasia to dysplasia and cancer. Gut. 2005;54(Suppl 1):i6–12.
3. Wijnhoven BP, Tilanus HW, Dinjens WN. Molecular biology of Barrett's adenocarcinoma. Ann Surg. 2001;233(3):322–37.
4. Tischoff I, Tannapfel A. Barrett's esophagus: can biomarkers predict progression to malignancy? Expert Rev Gastroenterol Hepatol. 2008;2(5):653–63.
5. Kerkhof M, Kusters JG, van Dekken H, Kuipers EJ, Siersema PD. Biomarkers for risk stratification of neoplastic progression in Barrett esophagus. Cell Oncol. 2007;29(6):507–17.
6. Zagorowicz E, Jankowski J. Molecular changes in the progression of Barrett's oesophagus. Postgrad Med J. 2007;83(982):529–35.
7. Hvid-Jensen F, Pedersen L, Drewes AM, Sorensen HT, Funch-Jensen P. Incidence of adenocarcinoma among patients with Barrett's esophagus. N Engl J Med. 2011;365(15):1375–83.
8. Zeki S, Fitzgerald RC. The use of molecular markers in predicting dysplasia and guiding treatment. Best Pract Res Clin Gastroenterol. 2015;29(1):113–24.
9. Timmer MR, Sun G, Gorospe EC, Leggett CL, Lutzke L, Krishnadath KK, et al. Predictive biomarkers for Barrett's esophagus: so near and yet so far. Dis Esophagus. 2013;26(6):574–81.
10. Overholt BF, Lightdale CJ, Wang KK, Canto MI, Burdick S, Haggitt RC, et al. Photodynamic therapy with porfimer sodium for ablation of high-grade dysplasia in Barrett's esophagus: international, partially blinded, randomized phase III trial. Gastrointest Endosc. 2005;62(4):488–98.
11. Corley DA, Levin TR, Habel LA, Weiss NS, Buffler PA. Surveillance and survival in Barrett's adenocarcinomas: a population-based study. Gastroenterology. 2002;122(3):633–40.
12. Moyes LH, Going JJ. Still waiting for predictive biomarkers in Barrett's oesophagus. J Clin Pathol. 2011;64(9):742–50.
13. Wang KK, Sampliner RE. Practice Parameters Committee of the American College of Gastroenterology. Updated guidelines 2008 for the diagnosis, surveillance and therapy of Barrett's esophagus. Am J Gastroenterol. 2008;103(3):788–97.
14. Abela JE, Going JJ, Mackenzie JF, McKernan M, O'Mahoney S, Stuart RC. Systematic four-quadrant biopsy detects Barrett's dysplasia in more patients than nonsystematic biopsy. Am J Gastroenterol. 2008;103(4):850–5.
15. Montgomery E, Bronner MP, Goldblum JR, Greenson JK, Haber MM, Hart J, et al. Reproducibility of the diagnosis of dysplasia in Barrett esophagus: a reaffirmation. Hum Pathol. 2001;32(4):368–78.
16. Reid BJ, Haggitt RC, Rubin CE, Roth G, Surawicz CM, Van Belle G, et al. Observer variation in the diagnosis of dysplasia in Barrett's esophagus. Hum Pathol. 1988;19(2):166–78.
17. Spechler SJ. Dysplasia in Barrett's esophagus: limitations of current management strategies. Am J Gastroenterol. 2005;100(4):927–35.
18. Prasad GA, Bansal A, Sharma P, Wang KK. Predictors of progression in Barrett's esophagus: current knowledge and future directions. Am J Gastroenterol. 2010;105(7):1490–502.
19. Pepe MS, Etzioni R, Feng Z, Potter JD, Thompson ML, Thornquist M, et al. Phases of biomarker development for early detection of cancer. J Natl Cancer Inst. 2001;93(14):1054–61.
20. Williams LJ, Guernsey DL, Casson AG. Biomarkers in the molecular pathogenesis of esophageal (Barrett) adenocarcinoma. Curr Oncol. 2006;13(1):33–43.
21. Ouatu-Lascar R, Fitzgerald RC, Triadafilopoulos G. Differentiation and proliferation in Barrett's esophagus and the effects of acid suppression. Gastroenterology. 1999;117(2):327–35.
22. Fitzgerald RC, Omary MB, Triadafilopoulos G. Dynamic effects of acid on Barrett's esophagus. An ex vivo proliferation and differentiation model. J Clin Invest. 1996;98(9):2120–8.

23. Senkus E, Kyriakides S, Ohno S, Penault-Llorca F, Poortmans P, Rutgers E, et al. Primary breast cancer: ESMO Clinical Practice Guidelines for diagnosis, treatment and follow-up. Ann Oncol. 2015;26(Suppl 5):v8–30.
24. Ohbu M, Kobayashi N, Okayasu I. Expression of cell cycle regulatory proteins in the multistep process of oesophageal carcinogenesis: stepwise over-expression of cyclin E and p53, reduction of p21(WAF1/CIP1) and dysregulation of cyclin D1 and p27(KIP1). Histopathology. 2001;39(6):589–96.
25. Hong MK, Laskin WB, Herman BE, Johnston MH, Vargo JJ, Steinberg SM, et al. Expansion of the Ki-67 proliferative compartment correlates with degree of dysplasia in Barrett's esophagus. Cancer. 1995;75(2):423–9.
26. Polkowski W, van Lanschot JJ, Ten Kate FJ, Baak JP, Tytgat GN, Obertop H, et al. The value of p53 and Ki67 as markers for tumour progression in the Barrett's dysplasia-carcinoma sequence. Surg Oncol. 1995;4(3):163–71.
27. Rioux-Leclercq N, Turlin B, Sutherland F, Heresbach N, Launois B, Campion JP, et al. Analysis of Ki-67, p53 and Bcl-2 expression in the dysplasia-carcinoma sequence of Barrett's esophagus. Oncol Rep. 1999;6(4):877–82.
28. Lauwers GY, Kandemir O, Kubilis PS, Scott GV. Cellular kinetics in Barrett's epithelium carcinogenic sequence: roles of apoptosis, bcl-2 protein, and cellular proliferation. Mod Pathol. 1997;10(12):1201–8.
29. Jankowski J, McMenemin R, Yu C, Hopwood D, Wormsley KG. Proliferating cell nuclear antigen in oesophageal diseases; correlation with transforming growth factor alpha expression. Gut. 1992;33(5):587–91.
30. Casson AG, Zheng Z, Evans SC, Geldenhuys L, van Zanten SV, Veugelers PJ, et al. Cyclin D1 polymorphism (G870A) and risk for esophageal adenocarcinoma. Cancer. 2005;104(4):730–9.
31. Bani-Hani K, Martin IG, Hardie LJ, Mapstone N, Briggs JA, Forman D, et al. Prospective study of cyclin D1 overexpression in Barrett's esophagus: association with increased risk of adenocarcinoma. J Natl Cancer Inst. 2000;92(16):1316–21.
32. Jaskiewicz K, Louw J, Anichkov N. Barrett's oesophagus: mucin composition, neuroendocrine cells, p53 protein, cellular proliferation and differentiation. Anticancer Res. 1994;14(5A):1907–12.
33. Lao-Sirieix P, Brais R, Lovat L, Coleman N, Fitzgerald RC. Cell cycle phase abnormalities do not account for disordered proliferation in Barrett's carcinogenesis. Neoplasia. 2004;6(6):751–60.
34. Murray L, Sedo A, Scott M, McManus D, Sloan JM, Hardie LJ, et al. TP53 and progression from Barrett's metaplasia to oesophageal adenocarcinoma in a UK population cohort. Gut. 2006;55(10):1390–7.
35. Lao-Sirieix P, Lovat L, Fitzgerald RC. Cyclin A immunocytology as a risk stratification tool for Barrett's esophagus surveillance. Clin Cancer Res. 2007;13(2 Pt 1):659–65.
36. Sirieix PS, O'Donovan M, Brown J, Save V, Coleman N, Fitzgerald RC. Surface expression of minichromosome maintenance proteins provides a novel method for detecting patients at risk for developing adenocarcinoma in Barrett's esophagus. Clin Cancer Res. 2003;9(7):2560–6.
37. Hollstein M, Sidransky D, Vogelstein B, Harris CC. p53 mutations in human cancers. Science. 1991;253(5015):49–53.
38. Zhang XP, Liu F, Wang W. Two-phase dynamics of p53 in the DNA damage response. Proc Natl Acad Sci U S A. 2011;108(22):8990–5.
39. Vousden KH, Lu X. Live or let die: the cell's response to p53. Nat Rev Cancer. 2002;2(8):594–604.
40. Keswani RN, Noffsinger A, Waxman I, Bissonnette M. Clinical use of p53 in Barrett's esophagus. Cancer Epidemiol Biomarkers Prev. 2006;15(7):1243–9.
41. Symmans PJ, Linehan JM, Brito MJ, Filipe MI. p53 expression in Barrett's oesophagus, dysplasia, and adenocarcinoma using antibody DO-7. J Pathol. 1994;173(3):221–6.
42. Ramel S, Reid BJ, Sanchez CA, Blount PL, Levine DS, Neshat K, et al. Evaluation of p53 protein expression in Barrett's esophagus by two-parameter flow cytometry. Gastroenterology. 1992;102(4 Pt 1):1220–8.

43. Krishnadath KK, Tilanus HW, van Blankenstein M, Hop WC, Teijgeman R, Mulder AH, et al. Accumulation of genetic abnormalities during neoplastic progression in Barrett's esophagus. Cancer Res. 1995;55(9):1971–6.
44. Chatelain D, Flejou JF. High-grade dysplasia and superficial adenocarcinoma in Barrett's esophagus: histological mapping and expression of p53, p21 and Bcl-2 oncoproteins. Virchows Arch. 2003;442(1):18–24.
45. Younes M, Lebovitz RM, Lechago LV, Lechago J. p53 protein accumulation in Barrett's metaplasia, dysplasia, and carcinoma: a follow-up study. Gastroenterology. 1993;105(6):1637–42.
46. Weston AP, Banerjee SK, Sharma P, Tran TM, Richards R, Cherian R. p53 protein overexpression in low grade dysplasia (LGD) in Barrett's esophagus: immunohistochemical marker predictive of progression. Am J Gastroenterol. 2001;96(5):1355–62.
47. Weaver JM, Ross-Innes CS, Shannon N, Lynch AG, Forshew T, Barbera M, et al. Ordering of mutations in preinvasive disease stages of esophageal carcinogenesis. Nat Genet. 2014;46(8):837–43.
48. Younes M, Ertan A, Lechago LV, Somoano JR, Lechago J. p53 Protein accumulation is a specific marker of malignant potential in Barrett's metaplasia. Dig Dis Sci. 1997;42(4):697–701.
49. Souza RF. Biomarkers in Barrett's esophagus. Tech Gastrointest Endosc. 2010;12(2):116–1212.
50. Reid BJ, Prevo LJ, Galipeau PC, Sanchez CA, Longton G, Levine DS, et al. Predictors of progression in Barrett's esophagus II: baseline 17p (p53) loss of heterozygosity identifies a patient subset at increased risk for neoplastic progression. Am J Gastroenterol. 2001;96(10):2839–48.
51. Skacel M, Petras RE, Rybicki LA, Gramlich TL, Richter JE, Falk GW, et al. p53 expression in low grade dysplasia in Barrett's esophagus: correlation with interobserver agreement and disease progression. Am J Gastroenterol. 2002;97(10):2508–13.
52. Coggi G, Bosari S, Roncalli M, Graziani D, Bossi P, Viale G, et al. p53 protein accumulation and p53 gene mutation in esophageal carcinoma. A molecular and immunohistochemical study with clinicopathologic correlations. Cancer. 1997;79(3):425–32.
53. Dolan K, Walker SJ, Gosney J, Field JK, Sutton R. TP53 mutations in malignant and premalignant Barrett's esophagus. Dis Esophagus. 2003;16(2):83–9.
54. Paulson TG, Galipeau PC, Xu L, Kissel HD, Li X, Blount PL, et al. p16 mutation spectrum in the premalignant condition Barrett's esophagus. PLoS One. 2008;3(11):e3809.
55. Noffsinger AE. Defining cancer risk in Barrett's esophagus: a pathologist's perspective. Gastrointest Cancer Res. 2008;2(6):308–10.
56. Klump B, Hsieh CJ, Holzmann K, Gregor M, Porschen R. Hypermethylation of the CDKN2/p16 promoter during neoplastic progression in Barrett's esophagus. Gastroenterology. 1998;115(6):1381–6.
57. Schulmann K, Sterian A, Berki A, Yin J, Sato F, Xu Y, et al. Inactivation of p16, RUNX3, and HPP1 occurs early in Barrett's-associated neoplastic progression and predicts progression risk. Oncogene. 2005;24(25):4138–48.
58. Barrett MT, Sanchez CA, Galipeau PC, Neshat K, Emond M, Reid BJ. Allelic loss of 9p21 and mutation of the CDKN2/p16 gene develop as early lesions during neoplastic progression in Barrett's esophagus. Oncogene. 1996;13(9):1867–73.
59. Maley CC, Galipeau PC, Li X, Sanchez CA, Paulson TG, Reid BJ. Selectively advantageous mutations and hitchhikers in neoplasms: p16 lesions are selected in Barrett's esophagus. Cancer Res. 2004;64(10):3414–27.
60. Ellis Jr FH, Xu X, Kulke MH, LoCicero 3rd J, Loda M. Malignant transformation of the esophageal mucosa is enhanced in p27 knockout mice. J Thorac Cardiovasc Surg. 2001;122(4):809–14.
61. Clement G, Braunschweig R, Pasquier N, Bosman FT, Benhattar J. Alterations of the Wnt signaling pathway during the neoplastic progression of Barrett's esophagus. Oncogene. 2006;25(21):3084–92.

62. Zhuang Z, Vortmeyer AO, Mark EJ, Odze R, Emmert-Buck MR, Merino MJ, et al. Barrett's esophagus: metaplastic cells with loss of heterozygosity at the APC gene locus are clonal precursors to invasive adenocarcinoma. Cancer Res. 1996;56(9):1961–4.
63. Galipeau PC, Prevo LJ, Sanchez CA, Longton GM, Reid BJ. Clonal expansion and loss of heterozygosity at chromosomes 9p and 17p in premalignant esophageal (Barrett's) tissue. J Natl Cancer Inst. 1999;91(24):2087–95.
64. Reid BJ, Levine DS, Longton G, Blount PL, Rabinovitch PS. Predictors of progression to cancer in Barrett's esophagus: baseline histology and flow cytometry identify low- and high-risk patient subsets. Am J Gastroenterol. 2000;95(7):1669–76.
65. Rabinovitch PS, Longton G, Blount PL, Levine DS, Reid BJ. Predictors of progression in Barrett's esophagus III: baseline flow cytometric variables. Am J Gastroenterol. 2001;96(11):3071–83.
66. Reid BJ, Blount PL, Rubin CE, Levine DS, Haggitt RC, Rabinovitch PS. Flow-cytometric and histological progression to malignancy in Barrett's esophagus: prospective endoscopic surveillance of a cohort. Gastroenterology. 1992;102(4 Pt 1):1212–9.
67. Rygiel AM, van Baal JW, Milano F, Wang KK, ten Kate FJ, Fockens P, et al. Efficient automated assessment of genetic abnormalities detected by fluorescence in situ hybridization on brush cytology in a Barrett esophagus surveillance population. Cancer. 2007;109(10):1980–8.
68. Fels Elliott DR, Fitzgerald RC. Molecular markers for Barrett's esophagus and its progression to cancer. Curr Opin Gastroenterol. 2013;29(4):437–45.
69. Geppert CI, Rummele P, Sarbia M, Langer R, Feith M, Morrison L, et al. Multi-colour FISH in oesophageal adenocarcinoma-predictors of prognosis independent of stage and grade. Br J Cancer. 2014;110(12):2985–95.
70. Maley CC, Galipeau PC, Finley JC, Wongsurawat VJ, Li X, Sanchez CA, et al. Genetic clonal diversity predicts progression to esophageal adenocarcinoma. Nat Genet. 2006;38(4):468–73.
71. Rygiel AM, Milano F, Ten Kate FJ, Schaap A, Wang KK, Peppelenbosch MP, et al. Gains and amplifications of c-myc, EGFR, and 20.q13 loci in the no dysplasia-dysplasia-adenocarcinoma sequence of Barrett's esophagus. Cancer Epidemiol Biomarkers Prev. 2008;17(6):1380–5.
72. Rygiel AM, Milano F, Ten Kate FJ, de Groot JG, Peppelenbosch MP, Bergman JJ, et al. Assessment of chromosomal gains as compared to DNA content changes is more useful to detect dysplasia in Barrett's esophagus brush cytology specimens. Genes Chromosomes Cancer. 2008;47(5):396–404.
73. Brankley SM, Wang KK, Harwood AR, Miller DV, Legator MS, Lutzke LS, et al. The development of a fluorescence in situ hybridization assay for the detection of dysplasia and adenocarcinoma in Barrett's esophagus. J Mol Diagn. 2006;8(2):260–7.
74. Timmer MR, Brankley SM, Gorospe EC, Sun G, Lutzke LS, Iyer PG, et al. Prediction of response to endoscopic therapy of Barrett's dysplasia by using genetic biomarkers. Gastrointest Endosc. 2014;80(6):984–91.
75. Bian YS, Osterheld MC, Fontolliet C, Bosman FT, Benhattar J. p16 inactivation by methylation of the CDKN2A promoter occurs early during neoplastic progression in Barrett's esophagus. Gastroenterology. 2002;122(4):1113–21.
76. Eads CA, Lord RV, Wickramasinghe K, Long TI, Kurumboor SK, Bernstein L, et al. Epigenetic patterns in the progression of esophageal adenocarcinoma. Cancer Res. 2001;61(8):3410–8.
77. Kaz AM, Grady WM. Epigenetic biomarkers in esophageal cancer. Cancer Lett. 2014;342(2):193–9.
78. Sato F, Jin Z, Schulmann K, Wang J, Greenwald BD, Ito T, et al. Three-tiered risk stratification model to predict progression in Barrett's esophagus using epigenetic and clinical features. PLoS One. 2008;3(4):e1890.

79. Souza RF. The molecular basis of carcinogenesis in Barrett's esophagus. J Gastrointest Surg. 2010;14(6):937–40.
80. Galipeau PC, Li X, Blount PL, Maley CC, Sanchez CA, Odze RD, et al. NSAIDs modulate CDKN2A, TP53, and DNA content risk for progression to esophageal adenocarcinoma. PLoS Med. 2007;4(2):e67.
81. Bird-Lieberman EL, Dunn JM, Coleman HG, Lao-Sirieix P, Oukrif D, Moore CE, et al. Population-based study reveals new risk-stratification biomarker panel for Barrett's esophagus. Gastroenterology. 2012;143(4):927–35. e3
82. Jankowski J, Hopwood D, Wormsley KG. Expression of epidermal growth factor, transforming growth factor alpha and their receptor in gastro-oesophageal diseases. Dig Dis. 1993;11(1):1–11.
83. Polkowski W, van Sandick JW, Offerhaus GJ, ten Kate FJ, Mulder J, Obertop H, et al. Prognostic value of Lauren classification and c-erbB-2 oncogene overexpression in adenocarcinoma of the esophagus and gastroesophageal junction. Ann Surg Oncol. 1999;6(3):290–7.
84. Bang YJ, Van Cutsem E, Feyereislova A, Chung HC, Shen L, Sawaki A, et al. Trastuzumab in combination with chemotherapy versus chemotherapy alone for treatment of HER2-positive advanced gastric or gastro-oesophageal junction cancer (ToGA): a phase 3, open-label, randomised controlled trial. Lancet. 2010;376(9742):687–97.
85. Hardwick RH, Barham CP, Ozua P, Newcomb PV, Savage P, Powell R, et al. Immunohistochemical detection of p53 and c-erbB-2 in oesophageal carcinoma; no correlation with prognosis. Eur J Surg Oncol. 1997;23(1):30–5.
86. Jankowski J, Hopwood D, Wormsley KG. Flow-cytometric analysis of growth-regulatory peptides and their receptors in Barrett's oesophagus and oesophageal adenocarcinoma. Scand J Gastroenterol. 1992;27(2):147–54.
87. Al-Kasspooles M, Moore JH, Orringer MB, Beer DG. Amplification and over-expression of the EGFR and erbB-2 genes in human esophageal adenocarcinomas. Int J Cancer. 1993;54(2):213–9.
88. Jankowski J. Gene expression in Barrett's mucosa: acute and chronic adaptive responses in the oesophagus. Gut. 1993;34(12):1649–50.
89. Picardo SL, Maher SG, O'Sullivan JN, Reynolds JV. Barrett's to oesophageal cancer sequence: a model of inflammatory-driven upper gastrointestinal cancer. Dig Surg. 2012;29(3):251–60.
90. Auvinen MI, Sihvo EI, Ruohtula T, Salminen JT, Koivistoinen A, Siivola P, et al. Incipient angiogenesis in Barrett's epithelium and lymphangiogenesis in Barrett's adenocarcinoma. J Clin Oncol Off J Am Soc Clin Oncol. 2002;20(13):2971–9.
91. Heath EI, Canto MI, Piantadosi S, Montgomery E, Weinstein WM, Herman JG, et al. Secondary chemoprevention of Barrett's esophagus with celecoxib: results of a randomized trial. J Natl Cancer Inst. 2007;99(7):545–57.
92. Yamamoto Y, Gaynor RB. Therapeutic potential of inhibition of the NF-kappaB pathway in the treatment of inflammation and cancer. J Clin Invest. 2001;107(2):135–42.
93. O'Riordan JM, Abdel-latif MM, Ravi N, McNamara D, Byrne PJ, McDonald GS, et al. Proinflammatory cytokine and nuclear factor kappa-B expression along the inflammation-metaplasia-dysplasia-adenocarcinoma sequence in the esophagus. Am J Gastroenterol. 2005;100(6):1257–64.
94. Abdel-Latif MM, O'Riordan J, Windle HJ, Carton E, Ravi N, Kelleher D, et al. NF-kappaB activation in esophageal adenocarcinoma: relationship to Barrett's metaplasia, survival, and response to neoadjuvant chemoradiotherapy. Ann Surg. 2004;239(4):491–500.
95. Tsujii M, Kawano S, Tsuji S, Sawaoka H, Hori M, DuBois RN. Cyclooxygenase regulates angiogenesis induced by colon cancer cells. Cell. 1998;93(5):705–16.
96. Song S, Guha S, Liu K, Buttar NS, Bresalier RS. COX-2 induction by unconjugated bile acids involves reactive oxygen species-mediated signalling pathways in Barrett's oesophagus and oesophageal adenocarcinoma. Gut. 2007;56(11):1512–21.

97. Morris CD, Armstrong GR, Bigley G, Green H, Attwood SE. Cyclooxygenase-2 expression in the Barrett's metaplasia-dysplasia-adenocarcinoma sequence. Am J Gastroenterol. 2001;96(4):990–6.
98. Sakai NS, Samia-Aly E, Barbera M, Fitzgerald RC. A review of the current understanding and clinical utility of miRNAs in esophageal cancer. Semin Cancer Biol. 2013;23(6 Pt B):512–21.
99. Wu X, Ajani JA, Gu J, Chang DW, Tan W, Hildebrandt MA, et al. MicroRNA expression signatures during malignant progression from Barrett's esophagus to esophageal adenocarcinoma. Cancer Prev Res. 2013;6(3):196–205.
100. Kan T, Meltzer SJ. MicroRNAs in Barrett's esophagus and esophageal adenocarcinoma. Curr Opin Pharmacol. 2009;9(6):727–32.
101. Feber A, Xi L, Luketich JD, Pennathur A, Landreneau RJ, Wu M, et al. MicroRNA expression profiles of esophageal cancer. J Thorac Cardiovasc Surg. 2008;135(2):255–60. ; discussion 60
102. Wong GS, Habibollahi P, Heidari P, Lee JS, Klein-Szanto AJ, Waldron TJ, et al. Optical imaging of periostin enables early endoscopic detection and characterization of esophageal cancer in mice. Gastroenterology. 2013;144(2):294–7.
103. Kadri SR, Lao-Sirieix P, O'Donovan M, Debiram I, Das M, Blazeby JM, et al. Acceptability and accuracy of a non-endoscopic screening test for Barrett's oesophagus in primary care: cohort study. BMJ. 2010;341:c4372.
104. Peitz U, Kouznetsova I, Wex T, Gebert I, Vieth M, Roessner A, et al. TFF3 expression at the esophagogastric junction is increased in gastro-esophageal reflux disease (GERD). Peptides. 2004;25(5):771–7.
105. Risques RA, Vaughan TL, Li X, Odze RD, Blount PL, Ayub K, et al. Leukocyte telomere length predicts cancer risk in Barrett's esophagus. Cancer Epidemiol Biomarkers Prev. 2007;16(12):2649–55.
106. Flejou JF, Svrcek M. Barrett's oesophagus—a pathologist's view. Histopathology. 2007;50(1):3–14.

The Role of Macrophages Within Microenvironment in a Lung Cancer Development and Progression

17

Luka Brčić, Luka Opašić, and Helmut Popper

Contents

Abstract

Lung cancer, including carcinogenesis and therapy, is one of the main research focuses today. One of the main reasons for that is the very high mortality rate of patients with lung cancer. Cancer tissue is very heterogeneous, consisting of malignant tumor cells with many different cell types, proteins, and signaling molecules, all together forming the tumor microenvironment. The concept that tumor development is primarily based on mutations has been reapproached from the side of interaction between immune cells of the host, tumor cells, and tumor microenvironment. All components of the cancer microenvironment interact with each other and with tumor cells in a complex manner, both promoting tumor cell growth and development, as well as suppressing it. This interplay is very complicated and today still not completely understood. The most prevalent cells

L. Brčić (✉) • H. Popper
Institute of Pathology, Medical University of Graz, Auenbruggerplatz 25, Graz 8036, Austria
e-mail: luka.brcic@medunigraz.at

L. Opašić
Max Planck Institute for Evolutionary Biology,
August-Thienemann-Str. 2, Plön 24306, Germany

J. Haybaeck (ed.), *Mechanisms of Molecular Carcinogenesis – Volume 1*,
DOI 10.1007/978-3-319-53659-0_17

among leukocytes in the cancer microenvironment are macrophages. These are called tumor-associated macrophages and are still very difficult to differentiate and identify by single markers. However, it is clear that they have a very important role in tumor development and progression in lung cancer, as in many other cancers. In patients with lung carcinoma, there is a correlation between tumor-associated macrophages and prognosis, although not uniform.

17.1 Introduction

Today, lung cancer is still the number one cause of cancer-related death throughout the world. The discovery of targetable/drugable mutations, primarily in lung adenocarcinomas, had a huge impact on the quality of life, however, only to a small subset of patients, and has not much changed the overall survival of patients with lung carcinomas. In addition, many of these patients treated with these new drugs will develop resistance to this therapy. And while enthusiasm about these targeted therapies is slowly melting away, more and more concentration, research, and therapeutical attempts are being redirected toward tumor microenvironment. Microenvironment, with its plethora of different cells, extracellular matrix, and very complicated interactions and impact on tumors, will be shortly presented, with emphasis on tumor-associated macrophages and their role in lung cancer.

17.2 Microenvironment of Lung Cancer

Cancer tissue is very heterogeneous, and within it, we can always differentiate many components. Apart from tumor cells, there are many other different cell types, proteins, and signal molecules, which all together form the tumor microenvironment (TME). Usually, the most abundant component in tumor mass consists of cancer cells, including cancer stem cells. Other cell types found are tumor-activated fibroblasts with altered phenotype, called cancer-associated fibroblasts (CAF; in reality myofibroblasts); endothelial cells, forming the vasculature within tumor; and infiltrative immune cells (Fig. 17.1). They all together are embedded in the extracellular matrix of the tumor stroma. Extracellular matrix is composed of structural molecules like collagen, fibronectin, laminin, tenascin, and other glycoproteins and proteoglycans which are produced by cancer-associated fibroblasts [1]. In the last 10 years, research has focused on other cells within the tumor. All those components of cancer microenvironment closely interact with each other in a complex manner, both inducing or promoting tumor cell growth and development and suppressing it. Lung tissue is physiologically highly oxygenized and vascularized, and is in close contact with the outer environment. It can easily recruit immune inflammatory cells to manage injuries. On the other hand, additional factors, such as tobacco smoke, increase injury incidence and promote chronic inflammation, which increases the probability of malignant alteration of epithelial cells [1, 2]. When such alteration occurs, malignant cells start to recruit and alter the phenotype of the surrounding stroma cells. They do that by secreting many different factors, such as interleukins,

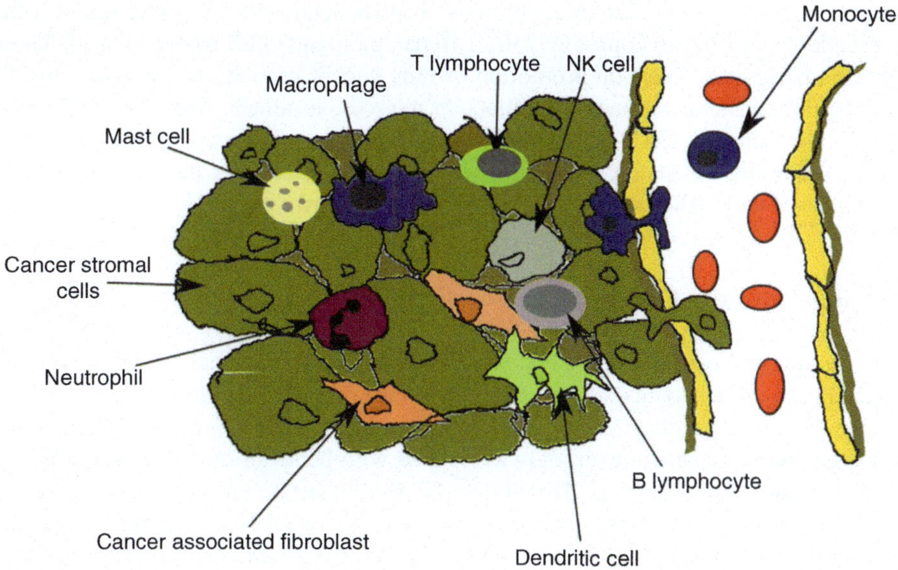

Fig. 17.1 Cellular components of tumor microenvironment

transforming growth factor β (TGFβ), vascular endothelial growth factor (VEGF), platelet-derived growth factor (PDGF), and basic fibroblast growth factor (bFGF), inducing in this way stromal reactions [3].

17.2.1 Cancer-Associated Fibroblasts (CAF) in Lung Carcinogenesis at the Border of Supply

Fibroblasts are cells responsible for the production of structural components of extracellular matrix and, in healthy tissues, are usually dormant until homeostasis is compromised. Once activated, they secrete mediators of inflammation, growth factors, and pro-migratory extracellular matrix components, all of which can contribute to carcinogenesis [4]. Studies on other tumors and on mouse models have shown that most CAF are myofibroblasts and develop from locally present precursor cells. They can also differentiate from bone marrow-derived precursor cells, however, in much smaller numbers. Interestingly, epithelial cells can also be the source of CAF after epithelial-mesenchymal transition [5]. In some pulmonary carcinomas, the desmoplastic stroma is in part formed by the tumor cells themselves using EMT. It is known that desmoplasia, defined as increased fibrocytic component within the tumor, is strongly associated with NSCLC. Vincent [6] has shown that both normal fibroblast and CAF promote growth of NSCLC and that CAF are derived from locally present normal myofibroblast. Normal fibroblasts under long exposure to cancer cells start to express the same cytokines that are expressed by CAF. Cardiotrophin-like cytokine factor 1 (*CLCF1*) and interleukin 6 (*IL6*) are genes that are upregulated in CAF in comparison with normal fibroblast. CLCF1 and IL6 are members of the interleukin 6 family that activate JAK-STAT and MAPK cascade via gp130 (glycoprotein 130), LIF receptor, and OSM receptor [7–9]. They exhibit paracrine effects on promotion of lung cancer growth and thus are considered an important component of molecular microenvironment. Alongside signaling molecule secretion, CAF also produce serine proteases and matrix metalloproteases which remodel the matrix and therefore facilitate migration of both immune cells into cancer and malignant cells out of the primary tumor mass [1].

17.2.2 Immune Contexture

Interestingly, Virchow was the first to describe leukocyte infiltration within cancers back in [10]. It is known today, primarily from the research on colorectal cancer (and some other tumors), that differences in type, number, and level of mutual interactions between immune cells are strongly associated with behavior of cancer, response to therapy, and patient survival. That link is so strong that, according to some authors, type and density of T cells within certain tumors are regarded as a better prognostic factor than standard pathological criteria [11]. Yet, in lung tumors, immune cells are not evenly distributed. Instead, those cells are organized into tertiary lymphoid structures composed out of dendritic cells (DC), T-cells clusters, and B-cell follicles. Those structures are not present in healthy adult lung, but are found in diseased fetal and infant lung [12]. Interestingly, the number of those tertiary lymphoid structures and density of mature dendritic cells within are in positive correlation with patient survival [13]. A subset of dendritic cells known as killer dendritic cells also promote elimination of cancer cells via apoptosis and necrosis in some cancers [14]. That is the reason why immunotherapy for lung cancer containing dendritic cells is currently undergoing clinical trials [15]. Dendritic cells, together with mast cells, macrophages, neutrophils, eosinophils, and basophils, comprise the innate immune microenvironment. They are the first line of defense against altered/malignant cells. NK cells have direct cytotoxic effects on cancer cells and positive role in immunosurveillance. They kill cells that fail to present MHC class I molecules, "missing self-cells" [16]. NK cells play an ambiguous role in cancer. Type 1 NK cells promote immunosurveillance, whereas type 2 downregulates immunosurveillance and suppresses immune response against cancer. Other cell types that promote cancer growth and development are myeloid-derived suppressor cells [17]. They are a heterogeneous population of immature myeloid cells that inhibit the function of T and NK cells. One suppressing mechanism is overexpression of arginase 1, which depletes L-arginine required for T-cell development [18]. Furthermore, they promote angiogenesis by secreting proangiogenic mediators, produce transforming growth factor β (TGFβ) and reactive oxygen species, reduce antigen presentation capability of dendritic cells, disrupt polarization of M1 macrophages, and are involved in other processes promoting cancer growth [17, 11]. Lymphocytes usually comprise up to two thirds of all nonmalignant cells inside tumor mass [19], with T cells being the most abundant (around 80% of all lymphocytes).

Al-Shibli and his colleagues [20] have shown in a series of resected NSCLC that the number of $CD4^+$ and $CD8^+$ lymphocytes in stroma is positively correlated with prolonged survival of patients, yet Wakabayashi and colleagues, few years before [21], demonstrated just the opposite, correlating the number of intratumoral $CD8^+$ lymphocytes with shorter overall survival (OS) in NSCLC. Some research suggests that $CD8^+$ lymphocytes which infiltrate cancer do not secrete IFNγ and are thus not able to deal with malignant cells [22]. Regulatory T cells are known to suppress immune reactions, which potentially benefit tumor development. This suppression is mediated by cyclooxygenase-2 (COX-2), and by its inhibition, tumor burden can be decreased [23, 24].

17.3 Role of Macrophages in Cancer

The concept that tumor development is primarily based on mutations has been reapproached from the side of interplay between immune cells of the host, tumor cells, and tumor microenvironment [25, 26]. Very important in these interactions are macrophages, which are involved in innate immunity and are important for immunological reactions, as well as for tissue healing processes. Macrophages play an important role in maintaining tissue homeostasis. Macrophages within tumor, tumor-associated macrophages (TAMs), have also important roles in cancer development. Meta-analysis by Zhang et al. [27, 28] showed statistically significant correlation between the $CD68^+$ macrophage densities in solid tumors and decreased overall survival.

Tumors use factors produced by macrophages for their progression and metastasis, in other words to avoid potentially harmful reaction of the host's immune system. TAM is a very interesting cell population, because TAMs can present with various phenotypes and comprise heterogeneous population [29, 30]. They are close to tumor cells, and they produce and release many different cytokines, growth factors, and chemokines, such as VEGF and IL-10. Many of these are important for tumor progression (local growth and metastatic potential) [31]. They are activated by different molecules, such as IFN-gamma, TNF, IL-4, and IL-10, and are profiled as M1 or M2 macrophages [31–35]. In which direction they differentiate depends mainly on the signals they receive from microenvironment. Macrophages are attracted to tumor microenvironment by hypoxia [36], but also by chemokine secretion, like CCL2 [37, 38].

M1 macrophages are involved in inflammation and infection, mainly as antigen-presenting cells and activators of inflammation [39], but are also directed against tumor cells [35, 40, 41]. M1 macrophages are activated by interferon γ (IFNγ) and lipopolysaccharide to act bactericidally and promote inflammation and T helper 1 (TH1) responses. TAM with phenotype similar to M1 can be found in early stages of cancer development and chronic inflammation that precedes it. Macrophages are generally not tumoricidal. Only when activated, they can destroy cancer cells directly or indirectly by stimulation of other cells. Direct cytotoxicity can be macrophage-mediated tumor cytotoxicity (MTC) in which toxic factors, such as TNFα, are secreted onto cancer cells, causing lysis or antibody-dependent cellular cytotoxicity (ADCC) [42]. In contrast, M2 macrophages, activated through IL-10 and TGFβ, influence angiogenesis through VEGF [32], and indirectly increase the expression of angiogenin via IL-1b and TNFα [42, 43] in vitro, as well as basic fibroblast growth factor (bFGF, also known as FGF2), chemokine (C-X-C motif) ligand 8 (CXCL8; also known as IL-8), cyclooxygenase 2 (COX2, also known as PTGS2), plasminogen activator, urokinase (uPA, also known as PLAU), and platelet-derived growth factor β (PDGFβ) [44]. Matrix metalloproteinases (MMP) excreted by TAM increase cancer growth also by promotion of angiogenesis (MMP7, MMP9, and MMP12) and, on the other hand, along with serine proteinases such as urokinase, metalloproteases as collagenase (MMP-1), gelatinase A and B (MMP-2, MMP-9), stromelysin (MMP-3), and macrophage elastase (MMP-12) degrade basal membrane and connective tissue, facilitating in this way tumor growth and migration of tumor cells [42, 45]. They also influence tissue remodeling [35, 40]. M2 macrophages improve tumor cell growth in in vitro conditions [46], and in vivo they act anti-inflammatory, preventing T-cell proliferation and antigen presentation and secreting IL-10 as an anti-inflammatory cytokine [35, 47, 48]. M2 macrophages are immunosuppressive and promote T helper 2 (TH2) responses. As tumor develops, TAM starts to express genes typical of M2 phenotype to become M2-skewed TAM, a predominant type of TAM (Fig. 17.2) [44].

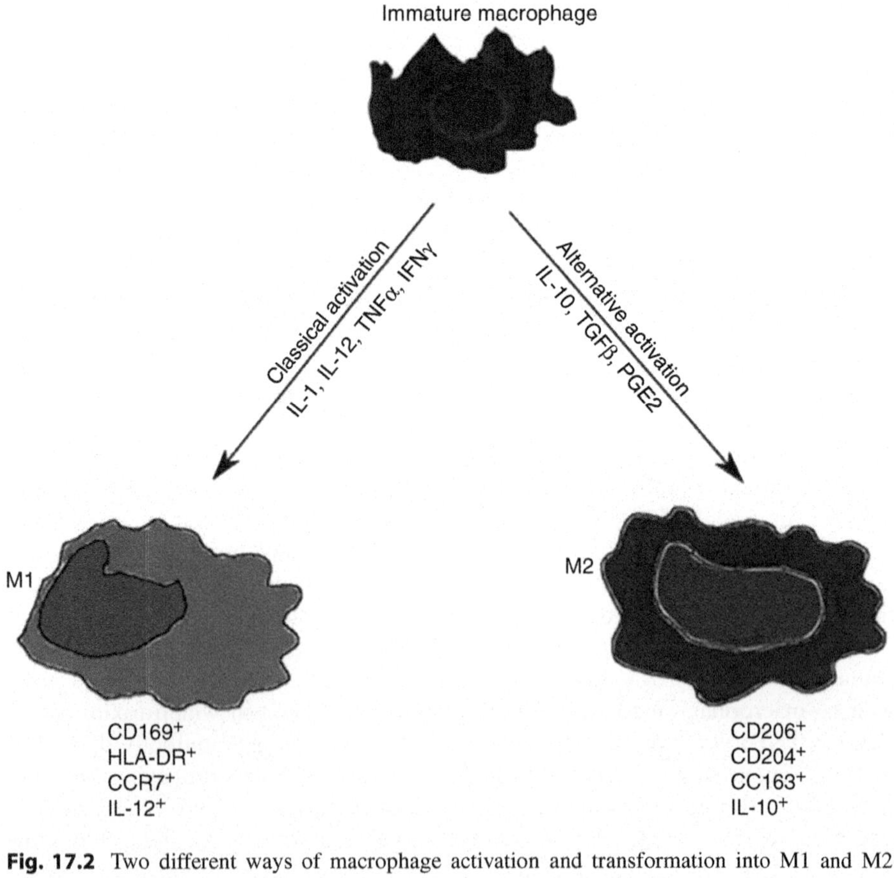

Fig. 17.2 Two different ways of macrophage activation and transformation into M1 and M2 subtypes

Although there is an overlap, generally two subpopulations of macrophages can be distinguished by immunohistochemistry. CD68 is a general macrophage marker, and CD163 and iNOS can be used as markers for M2 and M1, respectively. M1 macrophages express IL-1, IL-12, and iNOS [31, 32]. M2 are characterized by CD204 and CD206 (Fig. 17.3) [35, 40].

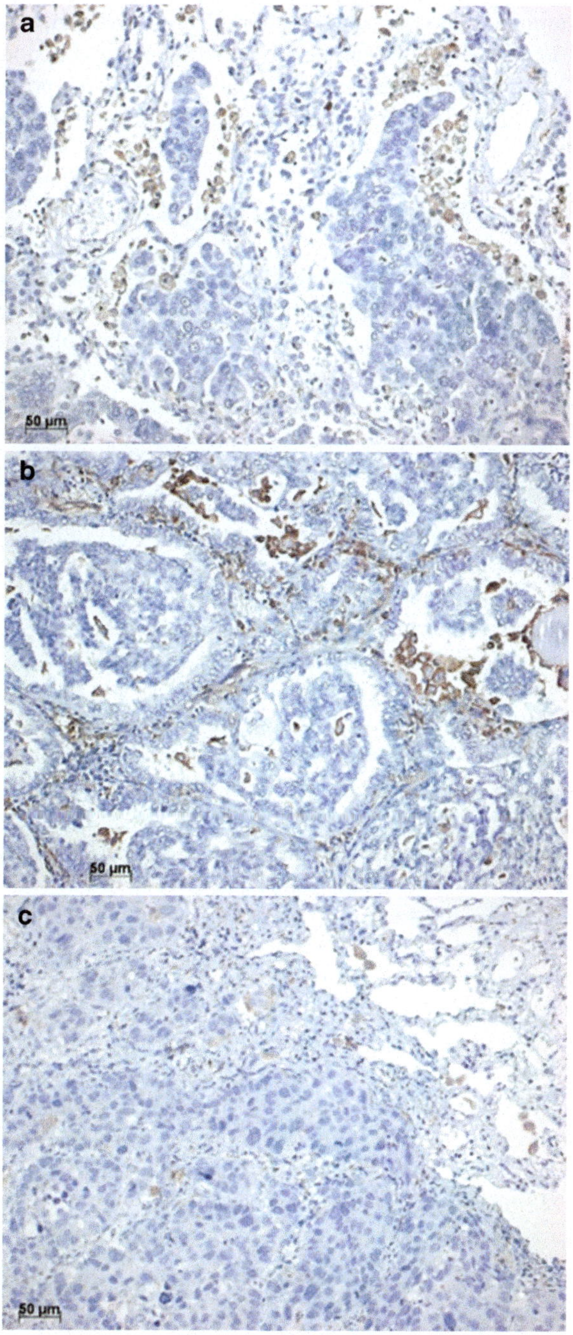

Fig. 17.3 Immunohistochemical presentation of squamous cell carcinoma (**a**) and adenocarcinoma (**b**) with many CD206-positive macrophages and solid area of another lung adenocarcinoma (**c**) with only few CD206-positive macrophages

In a study by van Overmeire et al. [49], the authors clearly demonstrated the importance of hypoxia for macrophage differentiation toward M1 or M2 line. Masumoto et al. [50] showed that increased metabolic activity in tumor, as well as not adequate vascularization, resulted in hypoxia (chronic or transient). And hypoxia is important for invasion of tumor cells, as well as for the resistance to therapy. Macrophages, especially M2 subpopulations, are present in hypoxic areas of the tumors. However, some authors [51] showed that improving hypoxic conditions, by regulation of blood vessel formation in tumor, does not decrease the number of M2 in the hypoxic areas, but only regulates the production of factors like VEGF A, GLUT 1, GLUT3, and iNOS, stimulating angiogenesis induced by TAMs [51].

17.4 Macrophages in Lung Cancer

Data on the importance of TAM in lung cancer has emerged, especially in the last 15–20 years, since it has been found that they have an interesting, but not easily understandable role in lung cancer. For example, some studies have clearly shown that patients with higher numbers of macrophages in primary lung carcinomas have shorter survival, while, on the other hand, others showed total lack of significance in correlation between TAM and survival. Studies concentrating on the localization of the macrophages, with regard to tumor cells and tumor stroma, showed favorable outcomes for the patients who had higher density of macrophages between tumor cells and an unfavorable one if the density was higher in the stroma [52, 53]. However, macrophage density was higher in adenocarcinomas than in other lung carcinomas analyzed (squamous and large cell) [52]. In one study comparing survival after platinum-based first-line therapy and macrophage infiltration in between tumor cells, the author found no association with the number of macrophages and survival [54], but also showed significantly better survival in patients with lower number of stromal macrophages. Only one of the previously mentioned studies [55] did not find any significance of stromal macrophages density in relation to the patient survival. Although studies used different approaches and methods, and in spite of studies demonstrating no relation of macrophages and survival, we can conclude (based on the all available data) that a higher number of macrophages between tumor cells is prognostically better, while the opposite is true for the macrophage numbers in tumor stroma. It is evident that looking only into the number of macrophages infiltrating lung carcinomas is not enough, and that macrophage differentiation toward M1 and M2 has to be evaluated. A single marker which can reliably and specifically detect M1 macrophages does not exist [30, 56], and in further studies, a panel of antibodies has to be used for distinction and characterization of macrophages in lung carcinomas. In a very nice and a clear-cut study by Ohri and his colleagues [32] using patients with long survival (mean 92.7 ± 7.2 months) and short survival (7.7 ± 0.7 months), the authors showed that in the prognostically better group, there are more M1 than M2 macrophages, and that patients whose M1 density was under the median had a 5-year survival rate of $<5\%$, in comparison with the ones with M1 density over the median whose 5-year survival rate was >75% [32]. Some studies showed that higher number of TAM ($CD68^+$) is correlated with survival in NSCLC patients [53, 57], while others were not able to find this connection [58, 59]. M2 macrophages ($CD204^+$), in a study by Ohtaki et al. [60], where only adenocarcinomas were included, demonstrated significant correlation with survival, but also with vascular and pleural invasion and stage.

On the other hand, Zhang and his colleagues [61] demonstrated that the number of M2 macrophages is to be an independent prognostic factor for overall survival. The confusion does not end here: while Ohri et al. [32] clearly showed that M1 macrophages (CD68 iNOS+) are significantly increased between tumor cells of the patients with longer survival, making them good prognostic markers, Almatroodi et al. [62] compared the same M1 population in tumor and non-tumor tissue, demonstrating decreased iNOS expression in squamous cell lung carcinomas and adenocarcinomas, but not in large cell carcinomas (in comparison with the matched non-tumorous tissue). Decreased iNOS expression is associated with deregulation of NF-KB signaling pathway having as a consequence non-adequate immunological response [63]. M2 macrophages (CD163+) stimulate proliferation of tumor, mainly through angiogenesis activation [31]. Higher number of M2 inside tumor cell aggregates correlates with metastatic tumor potential [61] and is very high in progressive disease [64]. Almatroodi et al. [62] showed that in all NSCLC types (adenocarcinomas, squamous cell lung carcinomas, large cell lung carcinomas), CD163 macrophages are present in a greater number than in non-tumor tissue. M2 have been in positive correlation with poor prognosis, TNM staging, and metastases to lymph nodes [60, 61]. This great variability of presented results is probably the consequence of a nonexistent specific marker for TAM (M2). They can be characterized by cell surface proteins, transcription factors, enzyme, and cytokine production. However, these markers change their expression over time, depending on cell activity, depending on the tumor type in which they are analyzed, and even depending on the smoking status of patient.

It is very important to stress once more that TAMs, like macrophages in general, are a very heterogeneous population and that there is a wide continuum of possible phenotypes between M1 and M2 macrophages. Many potential factors have influence on this differentiation process, among which tumor type and stage, as well as microenvironment, play an important role [30, 31].

M2 accelerates proliferation and migration of Lewis lung carcinoma (LLC1) cells [65] and activates lymphangiogenesis [61]. Zhang showed accelerated proliferation and invasiveness of LLC cells after cultivation with mouse macrophages (cell line Raw264.7), equivalent to M2 macrophages [27, 28]. A possible mechanism of M2 macrophage (CD206+ macrophages) activation according to the Unver experimental model [66] is also through chemokine ligand 7 (CXCL7) (which is a member of ELR+ CXCL chemokines promoting tumor progression mainly through angiogenesis [67]), indicating once again the importance of chemokine interplay in microenvironment that is crucial for early tumor development and progression. Furthermore, CCR2 and CX3CR1 are two receptors present on the macrophages and, when bonded to their ligand CCL2 and CX3CL1, influence signaling pathways such as JAK-STAT, PI3-K, and MAPK [68–71]. In a very nice and comprehensive study by Schmall et al. [72], the authors showed importance of CCR and CX3CR1 signaling for lung cancer progression. In mice without host CCR2 and CX3CR1, LLC1 tumor and progression decreased, and M2 macrophages were repolarized in M1 direction, influencing also angiogenesis and resulting in better survival. In the same study, they showed significant correlation between CCR2 expression with tumor stage and metastasis. They reconfirmed M2 as major player for tumor progression. They showed for the first time that through IL-10 secreted from macrophages, upregulation of CCR2 and CX3CR1 occurs in lung cancer cells. In this way, lung cancer cells behave like the macrophages, and with CCL2 and CX3CL1, secretion attracts more macrophages creating amplification loop, resulting in cancer cell proliferation and migration, as well as metastasis and creation of new blood supply network for microenvironment.

17.5 Tumor-Associated Macrophages and Antitumor Therapy

Another emerging and interesting interaction exists between TAMs and applied antitumor therapy, since it is now evident that TAMs influence response to chemotherapy, both positively and negatively, depending on the cytotoxic agent. Doxorubicin promotes M1 population, having a positive antitumoral effect, while on the other hand, different therapeutical protocols might induce PD-L1 expression on macrophages, followed by CD8+ T lymphocyte inhibition and unsuccessful therapy outcomes [73, 74]. On the other hand, there are some recent promising studies [75] in mouse models of breast carcinoma where TAMs were used as gene delivery vehicles for interferon alpha, activating immunity and inhibiting progression of the breast carcinoma. Another possible approach is reversing polarization of M2 toward M1, as in the study by Chen et al. [76], where M1 macrophages were induced by neuropeptide methionine enkephalin, resulting in antitumor activity or macrophage depletion as in study by Fritz et al. [77]. Latter macrophage depletion in mouse models of lung carcinoma was induced by clodronate-encapsulated liposomes, resulting in lower tumor burden and lower tumor cell proliferation. Especially interesting are immunotherapies using monoclonal antibodies against immune checkpoints [78, 79], where repolarization toward M1 might also improve efficacy [56].

Conclusion

After the above, only shortly, presented overview of macrophages and their role in cancer, it is clear that macrophages have a great plasticity and impressive variety of secreted molecules, such as growth factors, enzymes, cytokines, and chemokines. Today, it is known that not only the majority of cancer types contain macrophages but that they also form a symbiotic relationship in which cancer cells recruit macrophages and support their growth in exchange for factors that promote angiogenesis and tumor growth produced by macrophages. Most of this complicated interplay is yet to be revealed and then transferred to lung carcinoma diagnosis and treatment. It is clear that macrophage polarization into M2 is crucial for angiogenesis in tumor, tumor growth, and metastasis. The majority of the studies have demonstrated negative correlation with M2 macrophages and prognosis of patients with lung carcinoma. What makes the integration of presented as well as future studies' results and cross-integration very hard is still nonexistent single marker for M2 population, resulting in many different combinations of selected markers for presumably the same cells. However, even with the knowledge we have accumulated so far, new promising therapeutical options and treatment approaches are emerging.

References

1. El-Nikhely N, Larzabal L, Seeger W, Calvo A, Savai R. Tumor-stromal interactions in lung cancer: novel candidate targets for therapeutic intervention. Expert Opin Investig Drugs. 2012;21(8):1107–22.
2. Mahale J, Smagurauskaite G, Brown K, Thomas A, Howells LM. The role of stromal fibroblasts in lung carcinogenesis: a target for chemoprevention? Int J Cancer. 2016;138(1): 30–44.
3. Mueller MM, Fusenig NE. Friends or foes—biopolar effects of the tumour stroma in cancer. Nat Rev Cancer. 2004;4(11):839–49.
4. Pickup MW, Mouw JK, Weaver VM. The extracellular matrix modulates the hallmarks of cancer. EMBO Rep. 2014;15:1243–53.
5. Ostman A, Augsten M. Cancer-associated fibroblasts and tumor growth—bystanders turning into key players. Curr Opin Genet Dev. 2009;19:67–73.
6. Vicent S, Sayles LC, Vaka D, Khatri P, Gevaert O, Chen R, Zheng Y, et al. Cross-species functional analysis of cancer-associated fibroblasts identifies a critical role for CLCF1 and IL6 in non-small cell lung cancer in vivo. Cancer Res. 2012;72:5744–56.
7. Dagoneau N, Bellais S, Blanchet P, Sarda P, Al-Gazali LI, Di Rocco M, Huber C, Djouadi F, Le Goff C, Munnich A, Cormier-Daire V. Mutations in cytokine receptor-like factor 1 (CRLF1) account for both crisponi and cold-induced sweating syndromes. Am J Hum Genet. 2007;80:966–70.
8. Heinrich PC, Behrmann I, Haan S, Hermanns HM, Müller-Newen G, Schaper F. Principles of interleukin (IL)-6-type cytokine signalling and its regulation. Biochem J. 2003;374(Pt 1): 1–20.
9. Shiga K, Hara M, Nagasaki T, Sato T, Takahashi H, Takeyama H. Cancer-associated fibroblasts: their characteristics and their roles in tumor growth. Cancer. 2015;7:2443–58.
10. Virchow R. Aetologie der neoplastichen Geschwulste/Pathogenie der neoplastischen Geschwulste. Verlag von August Hirschwald: Die Krankhaften Geschwulste Berlin; 1863. p. 57–101.
11. Remark R, Becker C, Gomez JE, Damotte D, Dieu-Nosjean MC, Sautès-Fridman C, Fridman WH, Powell CA, Altorki NK, Merad M, Gnjatic S. The non-small cell lung cancer immune contexture: a major determinant of tumor characteristics and patient outcome. Am J Respir Crit Care Med. 2014;191(4):377–90.
12. Bremnes RM, Al-Shibli K, Donnem T, Sirera R, Al-Saad S, Andersen S, Stenvold H, Camps C, Busund LT. The role of tumor-infiltrating immune cells and chronic inflammation at the tumor site on cancer development, progression, and prognosis emphasis on non-small cell lung cancer. J Thorac Oncol. 2011;6:824–33.
13. Dieu-Nosjean M-C, Antoine M, Danel C, Heudes D, et al. Long-term survival for patients with non–small-cell lung cancer with intratumoral lymphoid structures. J Clin Oncol. 2008;26:4410–7.

14. Wesa AK, Storkus WJ. Killer dendritic cells: mechanisms of action and therapeutic implications for cancer. Cell Death Differ. 2008;15:51–7.
15. Wang Y, Xu Z, Zhou F, Sun Y, Chen J, Li L, Jin H, Qian Q. The combination of dendritic cells-cytotoxic T lymphocytes/cytokine-induced killer (DC-CTL/CIK) therapy exerts immune and clinical responses in patients with malignant tumors. Exp Hematol Oncol. 2015;4:32.
16. Vivier E, Tomasello E, Baratin M, Walzer T, Ugolini S. Functions of natural killer cells. Nat Immunol. 2008;9:503–10.
17. Quail DF, Joyce JA. Microenvironmental regulation of tumor progression and metastasis. Nat Med. 2013;19(11):1423–37.
18. Gabrilovich DI, Nagaraj S. Myeloid-derived suppressor cells as regulators of the immune system. Nat Rev Immunol. 2009;9:162–74.
19. Kataki A, Scheid P, Piet M, et al. Tumor infiltrating lymphocytes and macrophages have a potential dual role in lung cancer by supporting both host-defense and tumor progression. J Lab Clin Med. 2002;140:320–8.
20. Al-Shibli KI, Donnem T, Al-Saad S, Persson M, Bremnes RM, Busund LT. Prognostic effect of epithelial and stromal lymphocyte infiltration in non-small cell lung cancer. Clin Cancer Res. 2008;14(16):5220–7.
21. Wakabayashi O, Yamazaki K, Oizumi S, Hommura F, Kinoshita I, Ogura S, Dosaka-Akita H, Nishimura M. CD4+ T cells in cancer stroma, not CD8+ T cells in cancer cell nests, are associated with favorable prognosis in human non-small cell lung cancers. Cancer Sci. 2003;94(11):1003–9.
22. Trojan A, Urosevic M, Dummer R, Giger R, Weder W, Stahel RA. Immune activation status of CD8+ T cells infiltrating non-small cell lung cancer. Lung Cancer. 2004;44(2):143–7.
23. Sharma S, Yang SC, Zhu L, Reckamp K, Gardner B, Baratelli F, Huang M, Batra RK, Dubinett SM. Tumor cyclooxygenase-2/prostaglandin E2-dependent promotion of FOXP3 expression and CD4+ CD25+ T regulatory cell activities in lung cancer. Cancer Res. 2005;65(12):5211–20.
24. Liu CY, Wang YM, Wang CL, Feng PH, Ko HW, Liu YH, Wu YC, Chu Y, Chung FT, Kuo CH, Lee KY, Lin SM, Lin HC, Wang CH, Yu CT, Kuo HP. Population alterations of L-arginase- and inducible nitric oxide synthaseexpressed CD11b+/CD14⁻/CD15+/CD33+ myeloid-derived suppressor cells and CD8+ T lymphocytes in patients with advanced-stage non-small cell lung cancer. J Cancer Res Clin Oncol. 2010;136(1):35–45.
25. Coussens LM, Werb Z. Inflammation and cancer. Nature. 2002;420:860–7.
26. Hanahan D, Weinberg RA. The hallmarks of cancer. Cell. 2000;100:57–70.3.
27. Zhang JB, Zhang Y, Yao G, Gao J, Yang B, Zhao Y, et al. M2-polarized macrophages promote metastatic behavior of Lewis lung carcinoma cells by inducing vascular endothelial growth factor-C expression. Clinics (Sao Paulo). 2012;67:901–6.
28. Zhang QW, Liu L, Gong CY, Shi HS, Zeng YH, Wang XZ, Zhao YW, Wei YQ. Prognostic significance of tumor-associated macrophages in solid tumor: a meta-analysis of the literature. PLoS One. 2012;7:e50946. doi:10.1371/journal.pone.0050946.
29. Edin S, Wikberg ML, Dahlin AM, Rutegard J, Oberg A, Oldenborg PA, Palmqvist R. The distribution of macrophages with a M1 or M2 phenotype in relation to prognosis and the molecular characteristics of colorectal cancer. PLoS One. 2012;7(10):e47045. doi:10.1371/journal.pone.0047045.
30. Quatromoni JG, Eruslanov E. Tumor-associated macrophages: function, phenotype, and link to prognosis in human lung cancer. Am J Transl Res. 2012;4(4):376–89.
31. Mantovani A, Sozzani S, Locati M, Allavena P, Sica A. Macrophage polarization: tumor-associated macrophages as a paradigm for polarized M2 mononuclear phagocytes. Trends Immunol. 2002;23(11):549–55.
32. Ohri CM, Shikotra A, Green RH, Waller DA, Bradding P. Macrophages within NSCLC tumour islets are predominantly of a cytotoxic M1 phenotype associated with extended survival. Eur Respir J. 2009;33(1):118–26.
33. Lewis CE, Pollard JW. Distinct role of macrophages in different tumor microenvironments. Cancer Res. 2006;66(2):605–12. doi:10.1158/0008-5472.CAN-05-4005.

34. Lopez-Gonzalez JS, Avila-Moreno F, Prado-Garcia H, AguilarCazares D, Mandoki JJ, Meneses-Flores M. Lung carcinomas decrease the number of monocytes/macrophages (CD14+ cells) that produce TNF-[alpha]. Clin Immunol. 2007;122(3):323–9.
35. Mantovani A, Sica A, Sozzani S, Allavena P, Vecchi A, Locati M. The chemokine system in diverse forms of macrophage activation and polarization. Trends Immunol. 2004;25(12):677–86.
36. Murdoch C, Giannoudis A, Lewis CE. Mechanisms regulating the recruitment of macrophages into hypoxic areas of tumors and other ischemic tissues. Blood. 2004;104:2224–34.
37. Qian BZ, Pollard JW. New tricks for metastasis-associated macrophages. Breast Cancer Res. 2012;14:316.
38. Zhang J, Patel L, Pienta KJ. CC chemokine ligand 2 (CCL2) promotes prostate cancer tumorigenesis and metastasis. Cytokine Growth Factor Rev. 2010;21:41–8.
39. Mantovani A, Sica A. Macrophages, innate immunity and cancer: balance, tolerance, and diversity. Curr Opin Immunol. 2010;22:231–7.
40. Sica A, Larghi P, Mancino A, Rubino L, Porta C, Totaro MG, et al. Macrophage polarization in tumour progression. Semin Cancer Biol. 2008;18:349–55.
41. van Ravenswaay Claasen HH, Kluin PM, Fleuren GJ. Tumour infiltrating cells in human cancer. On the possible role of CD16+ macrophages in anti-tumour cytotoxicity. Lab Investig. 1992;67:166–74.
42. Bingle L, Brown NJ, Lewis CE. The of tumour-associated macrophages in tumour progression: implications for new anticancer therapies. J Pathol. 2002;196:254–65.
43. Etoh T, Shibuta K, Barnard GF, Kitano S, Mori M. Angiogenin expression in human colorectal cancer: the role of focal macrophage infiltration. Clin Cancer Res. 2000;6:3543–51.
44. Murdoch C, Muthana M, Coffelt SB, Lewis CE. The role of myeloid cells in the promotion of tumour angiogenesis. Nat Rev Cancer. 2008;8:618–31.
45. Komohara Y, Fujiwara Y, Ohnishi K, Takeya M. Tumor-associated macrophages: potential therapeutic targets for anti-cancer therapy. Adv Drug Deliv Rev. 2015; doi:10.1016/j.addr.2015.11.009.
46. Chang CI, Liao JC, Kuo L. Macrophage arginase promotes tumor cell growth and suppresses nitric oxide-mediated tumor cytotoxicity. Cancer Res. 2001;61(3):1100–6.
47. Redente EF, Dwyer-Nield LD, Merrick DT, Raina K, Agarwal R, Pao W, Rice PL, Shroyer KR, Malkinson AM. Tumor progression stage and anatomical site regulate tumor-associated macrophage and bone marrow-derived monocyte polarization. Am J Pathol. 2010;176(6):2972–85.
48. Zea AH, Rodriguez PC, Atkins MB, Hernandez C, Signoretti S, Zabaleta J, McDermott D, Quiceno D, Youmans A, O'Neill A, Mier J, Ochoa AC. Arginase-producing myeloid suppressor cells in renal cell carcinoma patients: a mechanism of tumor evasion. Cancer Res. 2005;65(8):3044–8.
49. Van Overmeire E, Laoui D, Keirsse J, Van Ginderachter J, Sarukhan A. Mechanisms driving macrophage diversity and specialization in distinct tumor microenvironments and parallelisms with other tissues. Front Immunol. 2014;26:5–127.
50. Matsumoto S, Yasui H, Mitchell JB, Krishna MC. Imaging cycling tumor hypoxia. Cancer Res. 2010;70:10019–23.
51. Laoui D, Van Overmeire E, Di Conza G, Aldeni C, Keirsse J, Morias Y, Movahedi K, Houbracken I, Schouppe E, Elkrim Y, et al. Tumor hypoxia does not drive differentiation of tumor-associated macrophages but rather fine-tunes the M2-like macrophage population. Cancer Res. 2013;74(1):24–30. doi:10.1158/0008-5472.CAN-13-1196.
52. Dai F, Liu L, Che G, Yu N, Pu Q, Zhang S, Ma J, Ma L, You Z. The number and microlocalization of tumor-associated immune cells are associated with patient's survival time in non-small cell lung cancer. BMC Cancer. 2010;10:220.
53. Welsh TJ, Green RH, Richardson D, Waller DA, O'Byrne KJ, Bradding P. Macrophage and mastcell invasion of tumor cell islets confers a marked survival advantage in non-small-cell lung cancer. J Clin Oncol. 2005;23:8959–67.
54. Kawai O, Ishii G, Kubota K, Murata Y, Naito Y, Mizuno T, Aokage K, et al. Predominant infiltration of macrophages and cd8(+) t cells in cancer nests is a significant predictor of survival in stage iv nonsmall cell lung cancer. Cancer. 2008;113:1387–95.

55. Kim DW, Min HS, Lee KH, Kim YJ, Oh DY, Jeon YK, Lee SH, Im SA, Chung DH, Kim YT, Kim TY, Bang YJ, Sung SW, Kim JH, Heo DS. High tumour islet macrophage infiltration correlates with improved patient survival but not with egfr mutations, gene copy number or protein expression in resected non-small cell lung cancer. Br J Cancer. 2008;98:1118–24.
56. Heusinkveld M, van der Burg SH. Identification and manipulation of tumor associated macrophages in human cancers. J Transl Med. 2011;9:216.
57. Ohri CM, Shikotra A, Green RH, Waller DA, Bradding P. The tissue microlocalisation and cellular expression of CD163, VEGF, HLA-DR, iNOS, and MRP 8/14 is correlated to clinical outcome in NSCLC. PLoS One. 2011;6(7):e21874. doi:10.1371/journal.pone.0021874.
58. Tataroğlu C, Kargi A, Ozkal S, Eşrefoğlu N, Akkoçlu A. Association of macrophages, mast cells and eosinophil leukocytes with angiogenesis and tumor stage in non-small cell lung carcinomas (NSCLC). Lung Cancer. 2004;43(1):47–54.
59. Toomey D, Smyth G, Condron C, Kelly J, Byrne AM, Kay E, et al. Infiltrating immune cells, but not tumour cells, express FasL in non-small cell lung cancer: no association with prognosis identified in 3-year follow-up. Int J Cancer. 2003;103(3):408–12.
60. Ohtaki Y, Ishii G, Nagai K, Ashimine S, Kuwata T, Hishida T, Nishimura M, Yoshida J, Takeyoshi I, Ochiai A. Stromal macrophage expressing CD204 is associated with tumor aggressiveness in lung adenocarcinoma. J Thorac Oncol. 2010;5(10):1507–15.
61. Zhang B, Yao G, Zhang Y, Gao J, Yang B, Rao Z, Gao J. M2-polarized tumor-associated macrophages are associated with poor prognoses resulting from accelerated lymphangiogenesis in lung adenocarcinoma. Clinics. 2011;66(11):1879–86.
62. Almatroodi SA, McDonald CF, Darby IA, Pouniotis DS. Characterization of M1/M2 tumour-associated macrophages (TAMs) and Th1/Th2 cytokine profiles in patients with NSCLC. Cancer Microenviron. 2016;9(1):1–11.
63. Biswas SK, Gangi L, Paul S, Schioppa T, Saccani A, Sironi M, et al. A distinct and unique transcriptional program expressed by tumor-associated macrophages (defective NF-kappaB and enhanced IRF-3/STAT1 activation). Blood. 2006;107(5):2112–22.
64. Chung FT, Lee KY, Wang CW, Heh CC, Chan YF, et al. Tumor-associated macrophages correlate with response to epidermal growth factor receptor-tyrosine kinase inhibitors in advanced non-small cell lung cancer. Int J Cancer. 2012;131(3):E227–35.
65. Lamagna C, Aurrand-Lions M, Imhof BA. Dual role of macrophages in tumor growth and angiogenesis. J Leukoc Biol. 2006;80:705–13.
66. Unver N, Esendagli G, Yilmaz G, Guc D. CXCL7-induced macrophage infiltration in lung tumor is independent of CXCR2 expression: CXCL7-induced macrophage chemotaxis in LLC tumors. Cytokine. 2015;75:330–7.
67. Belperio JA, Keane MP, Arenberg DA, Addison CL, Ehlert JE, Burdick MD, et al. CXC chemokines in angiogenesis. J Leukoc Biol. 2000;68:1–8.
68. Franklin RA, Liao W, Sarkar A, Kim MV, Bivona MR, Liu K, Pamer EG, Li MO. The cellular and molecular origin of tumor-associated macrophages. Science. 2014;344:921–5.
69. Mellado M, Rodriguez-Frade JM, Aragay A, del Real G, Martin AM, Vila-Coro AJ, Serrano A, Mayor Jr F, Martinez-A C. The chemokine monocyte chemotactic protein 1 triggers Janus kinase 2 activation and tyrosine phosphorylation of the CCR2B receptor. J Immunol. 1998;161:805–13.
70. Qian BZ, Li J, Zhang H, Kitamura T, Zhang J, Campion LR, Kaiser EA, Snyder LA, Pollard JW. CCL2 recruits inflammatory monocytes to facilitate breast-tumour metastasis. Nature. 2011;475:222–5.
71. Wolf MJ, Hoos A, Bauer J, Boettcher S, Knust M, Weber A, Simonavicius N, Schneider C, Lang M, Sturzl M, et al. Endothelial CCR2 signaling induced by colon carcinoma cells enables extravasation via the JAK2-Stat5 and p38MAPK pathway. Cancer Cell. 2012;22:91–105.
72. Schmall A, Al-Tamari HM, Herold S, Kampschulte M, Weigert A, Wietelmann A, Vipotnik N, Grimminger F, Seeger W, Pullamsetti SS, Savai R. Macrophage and cancer cell cross-talk via CCR2 and CX3CR1 is a fundamental mechanism driving lung cancer. Am J Respir Crit Care Med. 2015;191(4):437–47.

73. De Palma M, Lewis CE. Macrophage regulation of tumor responses to anticancer therapies. Cancer Cell. 2013;23(3):277–86.
74. Hasan A, Ghebeh H, Lehe C, Ahmad R, Dermime S. Therapeutic targeting of B7-H1 in breast cancer. Expert Opin Ther Targets. 2011;15(10):1211–25.
75. Escobar G, Gentner B, Naldini L, Mazzieri R. Engineered tumor-infiltrating macrophages as gene delivery vehicles for interferon-α activates immunity and inhibits breast cancer progression. OncoImmunology. 2014;3:e28696.
76. Chen W, Liu J, Meng J, Lu C, Li X, Wang E, et al. Macrophage polarization induced by neuropeptide methionine enkephalin (MENK) promotes tumoricidal responses. Cancer Immunol Immunother. 2012;61(10):1755–68.
77. Fritz JM, Tennis MA, Orlicky DJ, Lin H, Ju C, Redente EF, et al. Depletion of tumor-associated macrophages slows the growth of chemically induced mouse lung adenocarcinomas. Front Immunol. 2014;5:587.
78. Brahmer JR, Tykodi SS, Chow LQ, Hwu WJ, Topalian SL, Hwu P, et al. Safety and activity of anti-PD-L1 antibody in patients with advanced cancer. N Engl J Med. 2012;366(26): 2455–65.
79. Tomasini P, Khobta N, Greillier L, Barlesi F. Ipilimumab: its potential in non-small cell lung cancer. Ther Adv Med Oncol. 2012;4(2):43–50.
80. Hanahan D, Weinberg RA. Hallmarks of cancer: the next generation. Cell. 2011;144(5):646–74. doi:10.1016/j.cell.2011.02.013.
81. Movahedi K, Laoui D, Gysemans C, Baeten M, Stangé G, Van den Bossche J, Mack M, Pipeleers D, In't Veld P, De Baetselier P, et al. Different tumor microenvironments contain functionally distinct subsets of macrophages derived from Ly6C(high) monocytes. Cancer Res. 2010;70:5728–39. doi:10.1158/0008-5472.CAN-09-4672.

Progression of Lung Cancer: Role of Hypoxia and the Metabolic Tumor Microenvironment

18

Katharina Leithner and Horst Olschewski

Contents

Abstract

Hypoxia and nutrient deprivation are frequently present in the microenvironment of solid tumors, like lung cancer. Poor perfusion due to aberrant tumor vessels and large diffusion distances, as well as high consumption (e.g., of glucose), are the underlying causes. In addition, lactate accumulates, creating an acidic tumor microenvironment. The cancer-promoting role of hypoxia and the underlying molecular mechanisms are quite well characterized: activation of angiogenesis via upregulation of vascular endothelial growth factor (VEGF), induction of apoptosis resistance, selection of resistant clones under severe hypoxia, and others. In contrast, the impact of nutrient deprivation and lactate accumulation on cancer progression and cancer cell metabolism are less well understood. In the present chapter, we summarize recent clinical and preclinical data on hypoxia

K. Leithner (✉) • H. Olschewski
Division of Pulmonology, Department of Internal Medicine,
Medical University of Graz, Auenbruggerplatz 15, Graz, A-8036, Austria
e-mail: katharina.leithner@medunigraz.at

J. Haybaeck (ed.), *Mechanisms of Molecular Carcinogenesis – Volume 1*,
DOI 10.1007/978-3-319-53659-0_18

and nutrient deprivation in cancer with special emphasis on lung cancer. The contribution of cofactors, like anemia, and the consequences for carcinogenesis and cell metabolism are discussed.

18.1 Lung Cancer

Lung cancer remains the leading cause of cancer deaths worldwide [1]. Lung cancer is often advanced at diagnosis, and 5-year survival among lung cancer patients is poor [1]. In advanced-stage lung cancer, platinum-based chemotherapy is the backbone of treatment [2, 3]. However, chemotherapy resistance, primary or acquired, is frequent [2, 3].

Histologic classification divides lung cancer into two major categories, non-small cell lung cancer (NSCLC) and small-cell lung cancer (SCLC). NSCLCs include three major subtypes, adenocarcinoma, squamous cell carcinoma, and large-cell carcinoma. Further sub-differentiation is performed according to oncogenic driver mutations, guiding targeted therapy, e.g., activating mutations of the epidermal growth factor receptor (EGFR) or chromosomal rearrangements, leading to echinoderm microtubule-associated protein-like 4 (EML)-anaplastic lymphoma kinase (ALK) fusion proteins [3].

18.2 The Metabolic Cancer Microenvironment

Nutrient and oxygen (O_2) deprivation is frequent in solid cancers, like lung cancer [4, 5]. Although angiogenesis is activated early in cancer growth, the newly formed vascular network is frequently aberrant, with leaky vessels and irregular blood flow [6]. In addition, cancers "outgrow" their supply by continuing proliferation and consumption of glucose and O_2 [6]. Cancer cells, therefore, face the challenge of limited and unreliable supply of O_2 and nutrients [7]. Hypoxia, nutrient limitation, and lactate accumulation all put environmental pressure on cancer cells. In the present chapter, recent clinical and preclinical data on hypoxia and nutrient deprivation in cancer with special emphasis on lung cancer are summarized, and the consequences of these microenvironmental factors for carcinogenesis and cell metabolism are discussed.

18.2.1 Necrosis: Crisis at the Border of Supply

With increasing distance from blood vessels, O_2 and glucose concentrations rapidly decline, often associated with the development of necrosis [4, 5, 8]. In lung cancer, necrosis resulting from nutrient and O_2 limitation is frequently present, especially in squamous cell carcinoma, but also in giant-cell carcinoma and small-cell lung cancer, and to a lesser extent also in adenocarcinoma [9]. In contrast, preinvasive lesions, like squamous dysplasia/carcinoma in situ (CIS) or atypical adenomatous hyperplasia, do not contain necrosis [9]. Likewise, in bronchioloalveolar carcinoma, a form of adenocarcinoma in situ, necrosis is absent [9]. Since the limit of O_2 diffusion and the viable zone around microvessels have been shown to overlap in early studies, necrosis has been attributed primarily to critical hypoxia (anoxia), which does not permit cell survival (for review see [6]).

18.2.2 Hypoxia

18.2.2.1 Cause and Incidence of Hypoxia in Cancer

Hypoxia is caused by poor perfusion ("perfusion-limited hypoxia") and by the diffusion limit for O_2 ("diffusion-limited hypoxia") in solid cancers [6, 10]. Perfusion-limited hypoxia results from fluctuations in tumor microvessel oxygenation and/or perfusion, e.g. if vessels are temporarily closed [11]. Also reverse flow can occur. This fluctuating flow can result in transient hypoxia. If closed vessels are reperfused after re-opening, this may result in hypoxia-reperfusion injury of affected tumor tissue and microvascular endothelial cells [6]. In contrast, diffusion-limited hypoxia occurs in tumor areas located near the diffusion limit for O_2, which has been shown to be less than 200 μm [12]. Hypoxia is typically present in the vicinity of a necrosis; however, from a macroscopic point of view, hypoxic areas are heterogeneously distributed within a tumor [10]. Tumor hypoxia can be further enhanced by reduced O_2 transport in the blood, as in anemia (found in roughly 30% of patients at diagnosis) [10], or reduced blood oxygenation due to lung diseases [13] (Fig. 18.1).

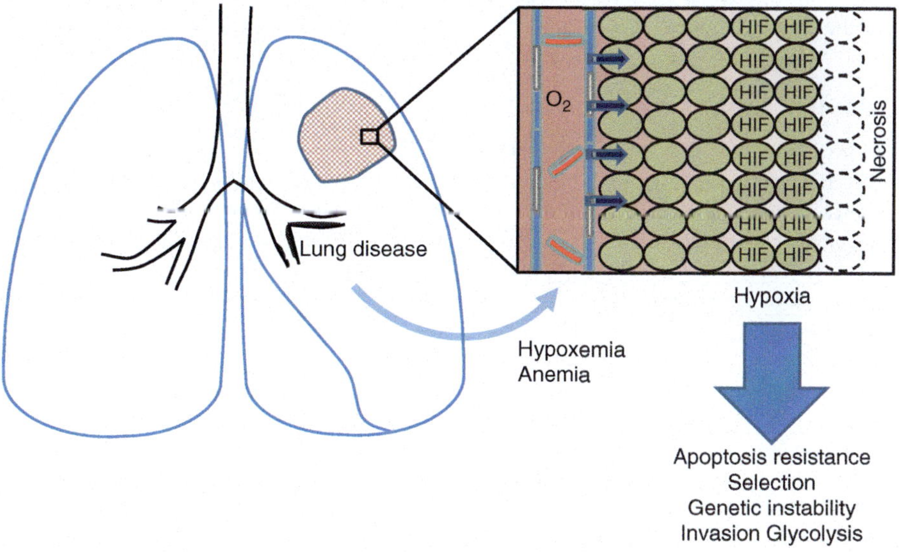

Fig. 18.1 Causes and consequences of hypoxia in lung cancer. The oxygen (O_2) supply of cancer cells is limited by diffusion, by irregular blood flow, and by an abnormal vascular architecture. Furthermore, anemia and lung diseases may reduce O_2 availability. *HIF* hypoxia-inducible factor

Direct O_2 measurements in cancers revealed that the mean O_2 levels are lower in cancers compared to corresponding normal tissues [8, 10]. The normal O_2 concentration is 80–100 mmHg in the blood and between approximately 25 and 70 mmHg in different normal tissues [8, 10]. Direct intraoperative O_2 measurements using polarographic electrodes in two studies on NSCLC patients revealed median tumor pO_2 values of 16.6 mmHg and 13.5 mmHg, respectively [14, 15]. The median tumor pO_2 was consistently lower than values from adjacent normal lung tissues [14].

Additional information on lung cancer oxygenation was obtained from studies using hypoxic radiotracers, e.g., [^{18}F]fluoromisonidazole ([^{18}F]FMISO) or [^{18}F]fluoroazomycin arabinoside, which allow the detection of hypoxia in vivo. [^{18}F]FMISO enters the cells by diffusion and is reduced by nitroreductase enzymes to form reduction products that bind to intracellular macromolecules when the oxygen tension is less than 10 mmHg and is then trapped intracellularly, allowing the determination of the hypoxic fraction in tumors in vivo (for review see [16]). In NSCLC, the mean hypoxic fraction (defined as proportion of pixels with elevated [^{18}F] FMISO signal, i.e., a tumor-to-blood ratio of >1.2 or >1.4.) was found to be variable, ranging from 1.3 to 94.7%. The median values were 48% and 58% in two different studies, respectively (for review see [16]). Overall, the oxygenation status in lung cancer varies from hypoxic to nearly normal; in general, hypoxia appears to be less pronounced than in, e.g., head and neck squamous cell carcinoma [15].

18.2.2.2 Role of Hypoxia in Cancer Progression and Therapy Resistance

Hypoxia exerts multiple effects in cancers, listed in Table 18.1. Importantly, it activates angiogenesis, enhances invasion and metastasis, and leads to radio- and chemotherapy resistance. Radio- and chemotherapy resistance caused by hypoxia has

Table 18.1 Effects of hypoxia in tumor biology

Effect	Mechanism
Selection of hypoxia-resistant clones	Cell death under severe hypoxia or hypoxia reoxygenation and selection of genotypes favoring survival
Suppression of apoptosis	Changes in gene expression, e.g., downregulation of proapoptotic Bid and Bax
Activation of autophagy	Changes in gene expression
Activation of glycolysis	HIF-induced overexpression of GLUT1, HK2, ENO1, and other glycolytic genes
Activation of angiogenesis	HIF-induced overexpression of VEGF, FLT1, ANG1, ANG2, and TIE2
Increased invasion and metastasis	HIF-induced overexpression of c-Met, CXCR4, RIOK3, and LOX
Attraction of tumor-associated macrophages	Increased expression of monocytic chemotactic proteins
Loss of genomic stability	Increased generation of reactive oxygen species
Decreased DNA repair	Downregulation of DNA repair pathways

HIF hypoxia-inducible factor. For reference see [17, 20]

different underlying mechanisms, but induction of apoptosis resistance is the best characterized one. The present view is that apoptosis resistance is either caused by selection of apoptosis-resistant clones or by suppression of apoptosis by alterations in gene expression [4, 17]. The latter seems to occur at rather mild hypoxia, while cell death and thus selection occur under severe hypoxia [4]. Apoptosis resistance under mild hypoxia (1% O_2) was shown to be reversible after 24–48 hours of reoxygenation and was associated with downregulation of Bcl-2-associated X protein (Bax) in NSCLC cell lines [18].

Patients with reduced blood oxygenation due to airway obstruction, e.g., patients with chronic obstructive pulmonary disease (COPD), a common smoking-related disorder, are at elevated risk of developing lung cancer [19]. When mice were subjected to intermittent hypobaric hypoxia (10% O_2) after lung cancer initiation with two different chemical carcinogens, a significantly increased tumor volume, but no increase in tumor frequency, was found [13]. Tumors from hypoxic mice showed increased proliferation and angiogenesis, and the pro-angiogenic growth factors VEGF and FGF were enhanced both in the lungs and tumors of hypoxic mice [13]. However, the exact role of lung diseases, such as COPD in lung carcinogenesis, is yet to be elucidated.

18.2.2.3 Hypoxia-Inducible Transcription Factors

Many of the responses to hypoxia are mediated by the transcription factors hypoxia-inducible factor 1α (HIF1α) and HIF2α, which dimerize with HIF1β and bind to hypoxia-response elements to induce expression of many genes [6]. HIF1α and HIF2α are constitutively expressed. Under normoxic conditions, they are constantly degraded by prolyl hydroxylases, which require O_2 as cofactor. Under hypoxia, HIF1α and HIF2α are stabilized. HIF1α and HIF2α can also be stabilized in a hypoxia-independent manner by growth factor-activated signaling cascades [6, 20].

HIFs play a role in the progression of lung cancer, but also other lung diseases, like pulmonary arterial hypertension and acute lung injury (for review see [21]). Both SCLC and NSCLC exhibit high levels of HIF1α and HIF2α, both of which are associated with poor prognosis [21]. A small molecule inhibitor of HIF1α, PX-478, was effective against tumor growth in an orthotopic mouse model of human lung cancer; however, silencing HIF1α in A549 lung adenocarcinoma cells impaired tumor vascularization and increased the necrotic area when grown as subcutaneous tumors, but did not reduce tumor cell proliferation and only slightly reduced tumor growth [21]. In contrast, reduction of HIF1α levels markedly impaired metastasis in murine models of human lung and mammary cancer [22].

18.2.3 Glucose Deprivation

18.2.3.1 Cause and Incidence of Glucose Deprivation in Cancer

Similar to oxygen, glucose levels decrease in underperfused tumor areas [5] (Fig. 18.2). In normal individuals, the average plasma glucose concentration is approximately 5.5 mM (100 mg/dL), ranging from approximately 3.2 mM (60 mg/dL) to approximately 7.8 mM (140 mg/dL) after meals [23]. In lung cancer,

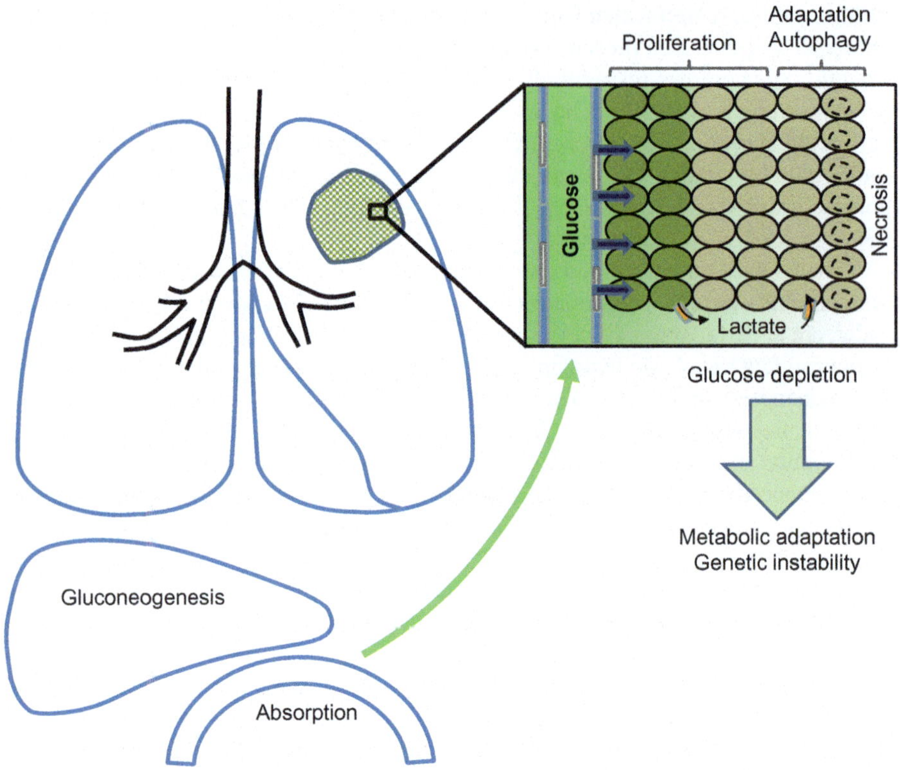

Fig. 18.2 Causes and consequences of glucose deprivation in lung cancer. The blood glucose is kept relatively constant by absorption from the intestine and by gluconeogenesis, which takes place mainly in the liver. Glucose is avidly consumed by cancer cells. High glucose consumption and reduced blood flow cause steep glucose gradients in solid cancers. In the low glucose micro-environment, alternative carbon sources and energy fuels, like lactate, are used by cancer cells, or autophagy is initiated. *Open circles* symbolize autophagic vacuoles

similar to other solid cancers, the glucose concentration is consistently lower than in corresponding normal lung tissue, as shown, e.g., by in vitro magnetic nuclear spectroscopy of excised tissues [24–26]. Similar results have been obtained in other solid human cancers [27–29]. The concentration of glucose was estimated to be 3–10 times lower in tumors compared to corresponding normal tissues [30].

Using imaging bioluminescence, which allows the histographical mapping of glucose and lactate concentrations in tissue sections at a high spatial resolution, glucose levels were shown to approach zero in the viable tumor area of some experimental tumors [31, 32]. At present, it is poorly understood how changes in blood glucose affect tumor glucose levels. However, when given intravenously, glucose is trapped intracellularly as glucose 6-phosphate in many cancers. This is the underlying mechanism of tumor imaging by ^{18}F-deoxyglucose positron-emission tomography (FDG-PET), a routinely used diagnostic imaging technique in the clinical staging of cancer patients [33].

Glucose deprivation in cancers is regarded as a consequence of high glucose consumption. In 1968, ascites fluid in the peritoneal cavity of mice inoculated with Ehrlich ascites tumor cells was shown to virtually lack glucose, in contrast to peritoneal fluid from non-inoculated mice [34]. This had been attributed to the high glycolytic activity of tumor cells [34]. When two different human cervical cancer cell lines were grown as subcutaneous tumors in SCID mice, xenografts formed from the cell line with higher glycolytic activity and OC316 displayed significantly reduced glucose levels compared to xenografts from the less glycolytic IGROV-1 cell line [35]. Similar to other aggressive cancers, invasive non-small cell lung cancers and small-cell lung cancers typically show a high maximum standardized uptake value (SUVmax) in FDG-PET, indicating high glucose uptake, while the noninvasive bronchioloalveolar carcinoma and seldom metastasizing typical carcinoid tumors generally show low uptake of FDG [9, 36].

In cancer cells, glucose is metabolized primarily by glycolysis [37–40]. This "aerobic glycolysis," described by the Nobel laureate Otto Heinrich Warburg as early as 1924, is observed in cancer cells but also in other highly proliferative cells and ensures the generation of building blocks for cell growth. It enhances flux of glucose to glycolytic metabolites and further along to the oxidative and non-oxidative branches of the pentose phosphate pathway (PPP), thus providing NADPH (reduced nicotinamide adenine dinucleotide phosphate) and ribose [37–40]. Furthermore, glucose is diverted to the synthesis of glycerol, serine, and hexosamines [41, 42].

Low availability of glucose dramatically reduces the flux via glycolysis. The fact that glucose consumption decreases with reduced glucose availability in tumor tissue was already observed by Warburg [43]. However, since glucose was regarded as a major fuel for cancer cells, the mechanisms of adaptation of cancer cells to glucose deprivation and a possible contribution of a glucose-deprived microenvironment to carcinogenesis have long been neglected.

18.2.3.2 Impact of Glucose Deprivation on Carcinogenesis

It has been suggested that glucose deprivation, on the one hand, and hypoxia, on the other hand, select for particular genetic abnormalities in cancer cells [44]. In a study by Yun et al. [44], colon cancer cell lines that survived glucose limitation (0.5 mM) have been shown to possess activating mutations in the gene encoding KRAS (4.4% of the clones) or BRAF (0.8% of the clones). In contrast, no KRAS or BRAF mutations were identified in clones generated in the presence of high (25 mM) concentrations of glucose. When cells with mutant KRAS or BRAF alleles were mixed with an excess of cells containing wild-type KRAS or BRAF alleles, respectively, and were incubated in either low (0.5 mM) or high (25 mM) concentrations of glucose, cells with mutant KRAS or BRAF alleles overtook the population in low-glucose conditions, but not in high-glucose conditions [44].

In a study published by Schlappack et al. [45], murine cancer cells formed a 5 to more than 20 times higher number of lung metastases after injection into mouse veins after exposure to glucose starvation (0 mM) for 48 h. Exposure of cells to low pH also had a metastasis promoting effect in that study. This suggests that glucose deprivation may enhance metastasis formation in some cancers. However, further studies are warranted to clarify the role of glucose depletion in cancer progression.

Only a limited number of studies assessed the correlation between glucose deprivation and tumor aggressiveness. Higher grade breast cancers exhibited lower glucose levels compared to lower grade breast cancer [46]. In contrast, the glucose levels in human cervical cancer xenografts from two different cell lines either forming spontaneous metastases or not forming spontaneous metastases were not significantly different [47]. In this study, however, metastasis formation was associated with increased lactate levels in the primary tumor, which correlated with the hypoxic fraction [47]. To the best of our knowledge, no studies on the relation between glucose levels in lung cancer tissue and survival have been published.

18.2.3.3 Survival of Cancer Cells Under Low Glucose

How cancer cells, which are reprogrammed to utilize high amounts of glucose, adapt to a decline in extracellular glucose levels is poorly understood. Recent studies show that cancer cell metabolism is more complex and intricate than previously thought and that metabolic flexibility allows cells to survive conditions of nutrient shortage [48]. Using a small interfering RNA (siRNA) screen, it has been found that cancer cell lines are dependent on respiratory chain proteins for survival and growth under chronically reduced glucose conditions (0.75 mM glucose) [30]. On the other hand, storage of glucose in the form of glycogen, which is activated by hypoxia, was shown to protect cancer cells from acute glucose deprivation [49].

Alternative fuels are utilized by cancer cells for biomass and energy production under glucose starvation, e.g., acetate, fatty acids, and amino acids. This is accomplished by altered expression of central metabolic enzymes but also by enhanced expression of membrane transporters (for review see [48]). Metabolic adaptation in tumors also involves metabolic symbiosis between cancer cells in different tumor compartments or between cancer cells and stroma cells, which excrete metabolites used by cancer cells (for review see [50, 51]). However, the use of alternative carbon sources for the generation of glycolysis-derived metabolites, like ribose (for DNA synthesis) and glycerol, would require the action of a gluconeogenesis enzyme, phosphoenolpyruvate carboxykinase (PEPCK) [52, 53]. We have recently shown that the mitochondrial isoform, of PEPCK and PCK2, is expressed and active in lung cancers, mediating the conversion of lactate into the glycolytic/gluconeogenic intermediate phosphoenolpyruvate under glucose deprivation [54]. Subsequently, PCK2 has also been found to play a role in the survival of other cancer cell lines and to be activated by endoplasmic reticulum stress and glutamine deprivation [55]. Silencing of PCK2 using shRNA led to significantly decreased growth of lung cancer cell xenografts in vivo [56].

Autophagy is a tightly regulated pro-survival pathway that captures, degrades, and recycles intracellular proteins and organelles in lysosomes [57, 59]. It involves the action of specific autophagy-executing proteins, e.g., microtubule-associated protein 1 light chain 3 (MAP1LC3, best known as LC3). Nutrient depletion is the most potent known physiological inducer of autophagy. Acute autophagy induction is critical for mammalian cells but also for yeast cells to survive nutrient depletion, which is attributed to the recycling of intracellular components into metabolic pathways. However, the exact substrates that are degraded by autophagy and the metabolic pathways supported remain to be identified [57–59]. In cancer, autophagy has both tumor-suppressing and tumor-promoting functions. Healthy cells are thought to be protected from malignant transformation by autophagic responses, and carcinogenesis may involve a temporary loss in autophagy competence. Conversely, autophagy promotes tumor progression and therapy resistance in a variety of models [57, 59].

The concept that glucose deprivation induces autophagy which promotes survival was challenged by a report showing that glucose deprivation did not induce autophagy in four different cancer cell lines, and autophagy inhibition did not alter apoptosis and necrosis induction by low glucose [60]. In some cancer models, the p53 status switches the role of autophagy during tumor development. Mutant p53 ameliorated the inhibition of tumor growth by autophagy inhibition in some models, including a genetically engineered lung cancer model with lung-specific expression of mutant Kirsten rat sarcoma (KRAS), making tumors less autophagy dependent, but not in other models [57]. Thus, the protective role of autophagy under glucose starvation in cancers may depend on their genetic background or on other unknown factors like availability of alternative carbon and energy sources.

18.2.4 Lactate

Lactate, the glycolysis end product, is known to accumulate in cancers [61, 62]. It is produced primarily via glycolysis and exported into the extracellular space by monocarboxylate transporters, most importantly MCT4 [63]. Lactate has been shown to exert multiple effects on cancer cells, mostly by reducing the pH, including enhancement of invasion and metastasis, induction of apoptosis resistance, and others (for review see [63]). Elevated lactate levels were shown to correlate with poor prognosis and poor disease-free survival in several epithelial cancers, such as cervical, head and neck, NSCLC, and breast cancers [63].

Lactate is not only a metabolic waste product but may be consumed by cancer cells, especially under low glucose concentrations. This phenomenon was described in SiHa cervix squamous cell carcinoma cancer cells [64] and p53−/− HCT116 colon carcinoma cells [65]. We have shown that this is also an important mechanism in lung cancer cells [54]. Lactate is transported into the cell via a bidirectional transporter and monocarboxylate transporter 1 (MCT1) [66] and oxidized to pyruvate via lactate dehydrogenase [64], the same enzyme that catalyzes the reverse reaction during glycolysis. Thereafter, lactate may be converted to acetyl-CoA and serve as an energy fuel [64, 65] or feed into biosynthetic pathways. Due to the important role of MCTs in regulating local lactate accumulation and use, inhibitors of MCTs have been considered as anticancer therapeutic drugs. AZD3965, an MCT1 inhibitor, reduced SCLC tumor growth in a mouse model in vivo [67]. Clinical trials with AZD3965 in different cancers are on the way [67] (www.clinicaltrials.gov).

18.2.5 Therapeutic Strategies Targeting Hypoxic and Metabolic Adaptation in Lung Cancer

Therapeutic approaches to target the cancer stroma or cancer metabolism or to affect the metabolic microenvironment are intensively studied. A detailed discussion is beyond the scope of this chapter. Briefly, targeting metabolic tumor cell vulnerabilities present in rapidly growing tumors [48] or tumor cell vulnerabilities induced by anti-angiogenic therapies (e.g., activation of HIF) is under preclinical evaluation [20]. Furthermore, the use of hypoxia-activated prodrugs and HIF1α inhibitors in hypoxic cancers, like lung cancer, is a promising approach [20]. Cancer hypoxia, assessed by novel hypoxic tracers, is increasingly taken into account in the planning of radiotherapy [68]. On the other hand, tumor hypoxia may be potentially predictive for the efficacy of anti-angiogenic therapies, which is currently investigated in clinical studies [20]. Thus, analysis of the metabolic cancer microenvironment might not only help to uncover metabolic vulnerabilities in cancer cells but may also help in the clinical routine to predict response to therapy.

In summary, there is a remarkable heterogeneity among different tumors and within single tumors with respect to the access to O_2, glucose, and maybe other essential substances. Cancer cells use specific enzymes like PEPCK allowing them to make use of fuels like lactate under glucose-deprived conditions. However, this makes them also vulnerable to specific approaches using such enzymes as therapeutic target. Cancer cells may use hypoxia-induced metabolic changes to attain resistance to chemotherapy. The investigation of the basic mechanisms in vivo, but also in well-defined in vitro models, mimicking the in vivo situation, will be essential for future research.

References

1. Ramalingam SS, Owonikoko TK, Khuri FR. Lung cancer: new biological insights and recent therapeutic advances. CA Cancer J Clin. 2011;61(2):91–112.
2. Socinski MA. Cytotoxic chemotherapy in advanced non-small cell lung cancer: a review of standard treatment paradigms. Clin Cancer Res. 2004;10:4210s–4s.
3. Reck M, Heigener DF, Mok T, Soria JC, Rabe KF. Management of non-small-cell lung cancer: recent developments. Lancet. 2013;382:709–19.
4. Vaupel P. Hypoxia and aggressive tumor phenotype: implications for therapy and prognosis. Oncologist. 2008;13(Suppl 3):21–6.
5. Cantor JR, Sabatini DM. Cancer cell metabolism: one hallmark, many faces. Cancer Discov. 2012;2(10):881–98.
6. Harris AL. Hypoxia—a key regulatory factor in tumour growth. Nat Rev Cancer. 2002;2(1):38–47.
7. Ackerman D, Simon MC. Hypoxia, lipids, and cancer: surviving the harsh tumor microenvironment. Trends Cell Biol. 2014;24(8):472–8.
8. Vaupel P, Kallinowski F, Okunieff P. Blood flow, oxygen and nutrient supply, and metabolic microenvironment of human tumors: a review. Cancer Res. 1989;49(23):6449–65.
9. Cagle PT, Allen TC. Advances in surgical pathology: lung cancer. Philadelphia: Wolters Kluwer Health; 2010.
10. Vaupel P, Höckel M, Mayer A. Detection and characterization of tumor hypoxia using pO2 histography. Antioxid Redox Signal. 2007;9(8):1221–35.

11. Hardee ME, Dewhirst MW, Agarwal N, Sorg BS. Novel imaging provides new insights into mechanisms of oxygen transport in tumors. Curr Mol Med. 2009;9(4):435–41.
12. Torres Filho IP, Leunig M, Yuan F, Intaglietta M, Jain RK. Noninvasive measurement of microvascular and interstitial oxygen profiles in a human tumor in SCID mice. Proc Natl Acad Sci U S A. 1994;91(6):2081–5.
13. Karoor V, Le M, Merrick D, Fagan KA, Dempsey EC, Miller YE. Alveolar hypoxia promotes murine lung tumor growth through a VEGFR-2/EGFR-dependent mechanism. Cancer Prev Res (Phila). 2012;5(8):1061–71.
14. Le QT, Chen E, Salim A, Cao H, Kong CS, Whyte R, et al. An evaluation of tumor oxygenation and gene expression in patients with early stage non-small cell lung cancers. Clin Cancer Res. 2006;12(5):1507–14.
15. Graves EE, Vilalta M, Cecic IK, Erler JT, Tran PT, Felsher D, et al. Hypoxia in models of lung cancer: implications for targeted therapeutics. Clin Cancer Res. 2010;16(19):4843–52.
16. Yip C, Blower PJ, Goh V, Landau DB, Cook GJ. Molecular imaging of hypoxia in non-small-cell lung cancer. Eur J Nucl Med Mol Imaging. 2015;42(6):956–76.
17. Wilson WR, Hay MP. Targeting hypoxia in cancer therapy. Nat Rev Cancer. 2011;11(6):393–410.
18. Wohlkoenig C, Leithner K, Deutsch A, Hrzenjak A, Olschewski A, Olschewski H. Hypoxia-induced cisplatin resistance is reversible and growth rate independent in lung cancer cells. Cancer Lett. 2011;308(2):134–43.
19. Calabro E, Randi G, La Vecchia C, Sverzellati N, Marchiano A, Villani M, et al. Lung function predicts lung cancer risk in smokers: a tool for targeting screening programmes. Eur Respir J. 2010;35(1):146–51.
20. McIntyre A, Harris AL. Metabolic and hypoxic adaptation to anti-angiogenic therapy: a target for induced essentiality. EMBO Mol Med. 2015;7(4):368–79.
21. Shimoda LA, Semenza GL. HIF and the lung: role of hypoxia-inducible factors in pulmonary development and disease. Am J Respir Crit Care Med. 2011;183(2):152–6.
22. Liao D, Corle C, Seagroves TN, Johnson RS. Hypoxia-inducible factor-1alpha is a key regulator of metastasis in a transgenic model of cancer initiation and progression. Cancer Res. 2007;67(2):563–72.
23. Bolli GB. How to ameliorate the problem of hypoglycemia in intensive as well as nonintensive treatment of type 1 diabetes. Diabetes Care. 1999;22(Suppl 2):B43–52.
24. Rocha CM, Barros AS, Goodfellow BJ, Carreira IM, Gomes A, Sousa V, et al. NMR metabolomics of human lung tumours reveals distinct metabolic signatures for adenocarcinoma and squamous cell carcinoma. Carcinogenesis. 2015;36(1):68–75.
25. Rocha CM, Barros AS, Gil AM, Goodfellow BJ, Humpfer E, Spraul M, et al. Metabolic profiling of human lung cancer tissue by 1H high resolution magic angle spinning (HRMAS) NMR spectroscopy. J Proteome Res. 2010;9(1):319–32.
26. Duarte IF, Rocha CM, Barros AS, Gil AM, Goodfellow BJ, Carreira IM, et al. Can nuclear magnetic resonance (NMR) spectroscopy reveal different metabolic signatures for lung tumours? Virchows Arch. 2010;457(6):715–25.
27. Hirayama A, Kami K, Sugimoto M, Sugawara M, Toki N, Onozuka H, et al. Quantitative metabolome profiling of colon and stomach cancer microenvironment by capillary electrophoresis time-of-flight mass spectrometry. Cancer Res. 2009;69(11):4918–25.
28. Wang L, Chen J, Chen L, Deng P, Bu Q, Xiang P, et al. 1H-NMR based metabonomic profiling of human esophageal cancer tissue. Mol Cancer. 2013;12:12–25.
29. Ziebart T, Walenta S, Kunkel M, Reichert TE, Wagner W, Mueller-Klieser W. Metabolic and proteomic differentials in head and neck squamous cell carcinomas and normal gingival tissue. J Cancer Res Clin Oncol. 2011;137(2):193–9.
30. Birsoy K, Possemato R, Lorbeer FK, Bayraktar EC, Thiru P, Yucel B, et al. Metabolic determinants of cancer cell sensitivity to glucose limitation and biguanides. Nature. 2014;508(7494):108–12.
31. Walenta S, Snyder S, Haroon ZA, Braun RD, Amin K, Brizel D, et al. Tissue gradients of energy metabolites mirror oxygen tension gradients in a rat mammary carcinoma model. Int J Radiat Oncol Biol Phys. 2001;51(3):840–8.

32. Schroeder T, Yuan H, Viglianti BL, Peltz C, Asopa S, Vujaskovic Z, et al. Spatial heterogeneity and oxygen dependence of glucose consumption in R3230Ac and fibrosarcomas of the Fischer 344 rat. Cancer Res. 2005;65(12):5163–71.
33. Bensinger SJ, Christofk HR. New aspects of the Warburg effect in cancer cell biology. Semin Cell Dev Biol. 2012;23(4):352–61.
34. Nakamura W, Hosoda S. The absence of glucose in Ehrlich ascites tumor cells and fluid. Biochim Biophys Acta. 1968;158(2):212–8.
35. Fabian C, Koetz L, Favaro E, Indraccolo S, Mueller-Klieser W, Sattler UG. Protein profiles in human ovarian cancer cell lines correspond to their metabolic activity and to metabolic profiles of respective tumor xenografts. FEBS J. 2012;279(5):882–91.
36. Brown RS, Leung JY, Kison PV, Zasadny KR, Flint A, Wahl RL. Glucose transporters and FDG uptake in untreated primary human non-small cell lung cancer. J Nucl Med. 1999;40(4):556–65.
37. Vander Heiden MG, Cantley LC, Thompson CB. Understanding the Warburg effect: the metabolic requirements of cell proliferation. Science. 2009;324(5930):1029–33.
38. Schulze A, Harris AL. How cancer metabolism is tuned for proliferation and vulnerable to disruption. Nature. 2012;491(7424):364–73.
39. Cairns RA, Harris IS, Mak TW. Regulation of cancer cell metabolism. Nat Rev Cancer. 2011;11(2):85–95.
40. DeBerardinis RJ, Thompson CB. Cellular metabolism and disease: what do metabolic outliers teach us? Cell. 2012;148(6):1132–44.
41. Kalhan SC, Hanson RW. Resurgence of serine: an often neglected but indispensable amino acid. J Biol Chem. 2012;287(24):19786–91.
42. Wellen KE, Lu C, Mancuso A, Lemons JM, Ryczko M, Dennis JW, et al. The hexosamine biosynthetic pathway couples growth factor-induced glutamine uptake to glucose metabolism. Genes Dev. 2010;24(24):2784–99.
43. Warburg O, Wind F, Negelein E. The metabolism of tumors in the body. J Gen Physiol. 1927;8(6):519–30.
44. Yun J, Rago C, Cheong I, Pagliarini R, Angenendt P, Rajagopalan H, et al. Glucose deprivation contributes to the development of KRAS pathway mutations in tumor cells. Science. 2009;325(5947):1555–9.
45. Schlappack OK, Zimmermann A, Hill RP. Glucose starvation and acidosis: effect on experimental metastatic potential, DNA content and MTX resistance of murine tumour cells. Br J Cancer. 1991;64(4):663–70.
46. Beckonert O, Monnerjahn J, Bonk U, Leibfritz D. Visualizing metabolic changes in breast-cancer tissue using 1H-NMR spectroscopy and self-organizing maps. NMR Biomed. 2003;16(1):1–11.
47. Ellingsen C, Walenta S, Hompland T, Mueller-Klieser W, Rofstad EK. The microenvironment of cervical carcinoma xenografts: associations with lymph node metastasis and its assessment by DCE-MRI. Transl Oncol. 2013;6(5):607–17.
48. Boroughs LK, DeBerardinis RJ. Metabolic pathways promoting cancer cell survival and growth. Nat Cell Biol. 2015;17(4):351–9.
49. Pelletier J, Bellot G, Gounon P, Lacas-Gervais S, Pouyssegur J, Mazure NM. Glycogen synthesis is induced in hypoxia by the hypoxia-inducible factor and promotes cancer cell survival. Front Oncol. 2012;2:18.
50. Romero IL, Mukherjee A, Kenny HA, Litchfield LM, Lengyel E. Molecular pathways: trafficking of metabolic resources in the tumor microenvironment. Clin Cancer Res. 2015;21(4):680–6.
51. Icard P, Kafara P, Steyaert JM, Schwartz L, Lincet H. The metabolic cooperation between cells in solid cancer tumors. Biochim Biophys Acta. 2014;1846(1):216–25.
52. Berg JM, Tymoczko JL, Stryer L. Biochemistry. 7th ed. New York: W.H. Freeman; 2011.
53. Yang J, Kalhan SC, Hanson RW. What is the metabolic role of phosphoenolpyruvate carboxykinase? J Biol Chem. 2009;284(40):27025–9.

54. Leithner K, Hrzenjak A, Trotzmuller M, Moustafa T, Kofeler HC, Wohlkoenig C, et al. PCK2 activation mediates an adaptive response to glucose depletion in lung cancer. Oncogene. 2015;34(8):1044–50.
55. Mendez-Lucas A, Hyrossova P, Novellasdemunt L, Vinals F, Perales JC. Mitochondrial phosphoenolpyruvate carboxykinase (PEPCK-M) is a pro-survival, endoplasmic reticulum (ER) stress response gene involved in tumor cell adaptation to nutrient availability. J Biol Chem. 2014;289(32):22090–102.
56. Vincent EE, Sergushichev A, Griss T, Gingras MC, Samborska B, Ntimbane T, et al. Mitochondrial phosphoenolpyruvate carboxykinase regulates metabolic adaptation and enables glucose-independent tumor growth. Mol Cell. 2015;60(2):195–207.
57. White E. The role for autophagy in cancer. J Clin Invest. 2015;125(1):42–6.
58. Galluzzi L, Pietrocola F, Bravo-San Pedro JM, Amaravadi RK, Baehrecke EH, Cecconi F, et al. Autophagy in malignant transformation and cancer progression. EMBO J. 2015;34(7):856–80.
59. Kroemer G, Marino G, Levine B. Autophagy and the integrated stress response. Mol Cell. 2010;40(2):280–93.
60. Ramirez-Peinado S, Leon-Annicchiarico CL, Galindo-Moreno J, Iurlaro R, Caro-Maldonado A, Prehn JH, et al. Glucose-starved cells do not engage in prosurvival autophagy. J Biol Chem. 2013;288(42):30387–98.
61. Fan TW, Lane AN, Higashi RM, Farag MA, Gao H, Bousamra M, et al. Altered regulation of metabolic pathways in human lung cancer discerned by (13)C stable isotope-resolved metabolomics (SIRM). Mol Cancer. 2009;8:41.
62. Polet F, Feron O. Endothelial cell metabolism and tumour angiogenesis: glucose and glutamine as essential fuels and lactate as the driving force. J Intern Med. 2013;273(2):156–65.
63. Peppicelli S, Bianchini F, Calorini L. Extracellular acidity, a "reappreciated" trait of tumor environment driving malignancy: perspectives in diagnosis and therapy. Cancer Metastasis Rev. 2014;33(2–3):823–32.
64. Sonveaux P, Vegran F, Schroeder T, Wergin MC, Verrax J, Rabbani ZN, et al. Targeting lactate-fueled respiration selectively kills hypoxic tumor cells in mice. J Clin Invest. 2008;118(12):3930–42.
65. Boidot R, Vegran F, Meulle A, Le Breton A, Dessy C, Sonveaux P, et al. Regulation of monocarboxylate transporter MCT1 expression by p53 mediates inward and outward lactate fluxes in tumors. Cancer Res. 2012;72(4):939–48.
66. Halestrap AP. The monocarboxylate transporter family—structure and functional characterization. IUBMB Life. 2012;64(1):1–9.
67. Polanski R, Hodgkinson CL, Fusi A, Nonaka D, Priest L, Kelly P, et al. Activity of the monocarboxylate transporter 1 inhibitor AZD3965 in small cell lung cancer. Clin Cancer Res. 2014;20(4):926–37.
68. Toma-Dasu I, Dasu A. Quantitative hypoxia imaging for treatment planning of radiotherapy. Adv Exp Med Biol. 2014;812:143–8.

The manufacturer's authorised representative in the EU is Springer
Nature Customer Service Centre GmbH, Europaplatz 3, 69115 Heidelberg,
Germany. If you have any concerns regarding our products, please
contact ProductSafety@springernature.com

Printed and bound by CPI Group (UK) Ltd, Croydon, CR0 4YY
01/10/2026
02225869-0002